AP®
2015-2016
BIOLOGY
Theodore L. Johnson

streamline
San Francisco

This book is intended to be used as a review companion by students enrolled in an AP® Biology course, or who are preparing to take the AP® Biology exam.

For additional information, comments, or corrections contact apbiology@bassalearning.com

ABOUT THE AUTHOR

Theodore Johnson has worked as a science educator for more than 17 years. During this time, he has taught students from a variety of socioeconomic and cultural backgrounds as well as different levels of academic achievement. For the past 10 years, he has taught at Lowell High School in San Francisco, a high-achieving public school. Prior to moving to San Francisco, he taught in the Providence Public Schools (Rhode Island) where he helped plan and implement the district's first science magnet high school, and contributed to a district-wide standards-based curriculum redesign.

Mr. Johnson holds a B.Sc. in Chemistry (Tougaloo College), a M.Sc. in Biology (Brown University), and MA in Teaching (Brown University). Additionally, he completed a professional training in secondary school administration (Providence College) and has been certified as a school administrator as well as a biology, chemistry, and general science teacher in Rhode Island, California, and Connecticut. He has also worked on a variety of research projects in the fields of physical and organic chemistry, plant genetics, and immunology.

Aside from drawing on his substantive scientific knowledge, the author has put his expertise in curriculum design and instruction his to work in developing this book. He applied a narrative approach based on central themes in biological systems and guided by the AP Biology curriculum framework. This book is the outcome of an ongoing effort to develop an AP Biology course that is fully aligned to the new AP Biology curriculum, and that is student-centered. As you flip through the pages, you will appreciate its rich, colorful, and carefully composed illustrations that are accompanied by simple analogies and relevant real-life examples to bring the subject matter to life.

TABLE OF CONTENT

BIG IDEA 2

15 DYNAMIC HOMEOSTASIS 189

BIG IDEA 3

BIG IDEA 4

Introduction

ABOUT THE AP BIOLOGY EXAM

The most complete information about the AP Biology exam can be found on the College Board website (www.collegeboard.com). However, the information below is based on the most up-to-date information at the time of the writing and publication of this book.

Exam format

The AP Biology exam is given on the same day in two 90-minute sessions, each focusing on a different section of the exam. The first section is divided into a 63 question multiple-choice section, and a 6 question grid-in section. The second section is divided into 2 long free-response and 6 short free-response questions.

Question focus

The exam focuses on assessing the student's ability to apply quantitative skills and mathematics methods to identify trends, and answer questions related to experimental design, data analysis, and the application of statistics to justify results and conclusions. As such, students are expected to be able to explain, reason, and justify their responses based on the key concepts identified in the AP Biology Course and Exam Description document (link to document: https://goo.gl/dT7JSg).

- 63 multiple-choice questions are a mix of stand-alone questions and question sets based on a described scenario or data set.
- 6 Grid-ins are all stand-alone questions that integrate mathematical skills and require a quantitative response.
- 2 Long FRQs are stand-alone questions, one of which is based on a lab or a set of data.
- 6 Short FRQs are all stand alone, often with sub-questions that require a paragraph-length discussion.

Curriculum overview

The AP biology curriculum is divided into the following four Big Ideas (referred to in this companion book as evolution, Biological systems, information, and interactions, respectively):

- **Big Idea 1**: The process of evolution drives the diversity and unity of life.
- **Big Idea 2**: Biological systems utilize free energy and molecular building blocks to grow, reproduce, and maintain dynamic homeostasis.
- **Big Idea 3**: Living systems store, retrieve, transmit, and respond to information that is essential to life processes.
- **Big Idea 4**: Biological systems interact, and these systems and their interactions lead to complex property.

In the AP Biology Course and Exam Description document, which provides the curriculum framework, each Big Idea is segmented into larger themes called **Enduring Understanding** that are defined by a set of **Essential Knowledge** statements with clearly identified learning objective. This book begins with the topic of abiogenesis, or *the origin of life* (Big Idea I, Essential Knowledge D1 and D2), and continues on through all of the content of BI1. This content is not linearly aligned to the curriculum framework, but instead it follows a logical sequence focused on explaining how life emerged and how multicellular organisms derived their complexity through evolution by natural selection. This book is thorough and hits on all parts of the AP Biology Big Ideas. In addition, the Investigation companion book (not included with this purchase) provides a thorough description of each of the AP Biology investigations with sample data, analysis and conclusion based on the data, including descriptive statistics.

How is the exam scored?

When taking the exam it is best to attempt every question because only the number of correct answers counts toward your score. You are better off guessing on a difficult question and having a chance at a lucky point, than skipping it completely and surely loosing out on that point.

Session I questions (Multiple-choice and Grid-in) are automatically graded, while the FRQs are graded by teachers and professors at a grading convention in mid-June. The scores from the two sessions are combined and scaled from 1 to 5, with 5 being the highest score and 3 the cutoff passing point. The general score distribution is:

Performance	Scaled score	% of students
Excellent	5	19%
Good	4	16%
Passed	3	16%
Below	2	15%
Failed	1	34%

By earning a passing score, you may qualify for 8 college and tuition credits when you enroll in college. This will allow you to bypass an introductory college biology course and begin with a more advanced second year course. Be aware that colleges and universities have different policies regarding AP credits. Some do not accepts them at all.

HOW TO USE THIS BOOK

The book is divided into 23 subject area content chapters. Big Idea I is covered in chapter 1 – 8; Big Idea 2 is covered in chapter 9-15; Big Idea 3 is covered in chapter 16-22; and Big Idea 4 is primarily covered in chapter 24, but a significant portion of its content is addressed in various parts of the book.

The title page of each chapter includes a list of questions and performance tasks that all students are expected to be able to complete. Also, on the bottom left of the title page is a Big Idea key that identifies the specific part of the AP Biology curriculum that the chapter addresses – this is the curriculum alignment key. In order to fully appreciate how the book aligns with the AP biology curriculum, refer to page 8 – 96 of the Course and Exam Description (https://goo.gl/dT7JSg) and compare the alignment key with the relevant curriculum descriptor statements. A label of the Big Idea is placed at the top right corner of the right-side page, and the chapter title is position on the first page and along side the page number on each subsequent page.

Each chapter is broken up into sections with major heading. These sections may also be segmented into subsections with minor heading. The outside margin of each page contains an area that can be use for note taking. This note section may also include additional information related to the key concepts presented on the same page. Do not ignore the side information because they are important and critical parts of the curriculum.

Throughout the book, you will find elegant illustrations that are thoughtfully designed with the visual learner in mind. All illustrations are accompanied by a description or explanation. Study each and understand the underlying principle behind each, including how it relates to the specific Big Idea.

At the end of each chapter is a summary section that uses bullet points to summarize the main points from the chapter. The summary section is followed by a short end-of-chapter quiz that assesses the student's understanding. The answers and explanations to the quiz questions are provided.

DIAGNOSTIC TEST

Before beginning to use this book, you are encouraged to take the diagnostic test. It will evaluate what you already know and point out areas that require special attention. After taking the test, use the *answer and explanation* section to score your responses, and then refer Question Alignment Chart to determine where to focus your review.

How to score the diagnostic test?

Score the two sections separately to get the raw score. For Section 1, divide the number of correct answers by 69 and multiply by 50. For example, if you received 35 correct answers on the multiple choice and grid-in questions of Section 1, the calculation is as follows:

$$35/69 \times 50.4 = 25.4$$

Scoring of Section 2 is more difficult since each question is graded using its own rubrics. However, our method should provide a reasonable assessment of your performance on this section. The long FRQ questions are weighted equal to the short FRQ questions, at 25 raw points for each part. Use you own judgement, guided by the *Answer and Explanation* section, as a guide to determine whether your response is excellent (A), good (B), pass (C) or fail (F). Once you have assigned these letter values to each question refer to the chart below to convert the letter grade to points and calculate your total raw score for section 2:

Letter grade	Long FRQ	Short FRQ
A	4.2	12.5
B	3.4	10
C	2.5	7.5
F	1.7	5
No response	0	0

For example, if on the long FRQ questions you received a letter score of B and C, and the short FRQ you received B, B, F, A, C, and no response for question 8, your total score for section 2 is calculated as follows:

3.4 + 2.5 + 10 + 10 + 5 + 12.5 + 7.5 + 0 = 43.4.

You will then add your scores from the two section to get your combined raw score and use the following table to determine your scaled score:

Raw score	AP score	Translation
65 - 100	5	Excellent
55 - 64	4	Good
45 - 54	3	Passed
40 - 43	2	Possibly passed
0 - 39	1	Failed

For example, if your raw score for Section 1 was 25.4 and Section 2 was 43.4, your combined raw score is 25.4 + 43.4 = 68.9, and your AP score is 5.

Section I

Time: 90 minutes
Part A: 63 multiple-choice questions
Part B: 6 grid-in questions

Directions (Part A - multiple choice): For each of the following 63 questions or statements, chose the best of the possible responses. When you complete these questions, continue on to Part B.

1. According to the organic soup hypothesis, natural selection began:
   (A) shortly after the Earth and moon formed when the Earth's crust was still molten
   (B) during the late heavy bombardment
   (C) during the period of prebiotic chemistry
   (D) during the RNA world

2. Evidence that RNA-based life forms appeared before DNA-based life comes from the fact that
   (A) DNA is less stable than RNA
   (B) DNA is double stranded while RNA is single stranded and more susceptible to mutations
   (C) proteins are better function molecules than both DNA and RNA
   (D) DNA required the emergence of a cell nucleus that only evolved much later with eukaryotes

3. The first "true" cells to appear were:
   (A) RNA protocells
   (B) Prokaryotes
   (C) RNA protoeukaryotes
   (D) eukaryotes

4. During the RNA world, natural selection favors which of the following:
   i. molecules with strong molecular bonds
   ii. molecules with weak molecular bonds
   iii. molecules that behaved like catalysts by speeding-up reactions
   iv. self-replicators
   (A) ii, iii, and iv only
   (B) i, iii, and iv only
   (C) iii and iv only
   (D) none of the options since natural selection began after the emergence of the first "real" cell

5. The Miller-Urey experiment provided support for the RNA world hypothesis by demonstrating that:
   (A) water may have condensed in the early atmosphere to form the oceans and seas.
   (B) inorganic molecules can undergo chemical reactions using available free energy to produce organic molecules.
   (C) free energy available in the early earth system was available for use in chemical reactions.
   (D) the primordial soup was a collection of dissolved organic and inorganic molecules.

GO ON TO THE NEXT PAGE

6. Which of the following statements about the organic soup hypothesis is correct?
   i. the temporal window for life opened after the earth cooled
   ii. amino acids, CO_2, water, and nucleic acids were present prior to the .formation of the primordial soup.
   iii. the first cell appeared immediately after the earth cooled.
   (A) i only
   (B) ii only
   (C) iii only
   (D) all three statements are correct

7. The table below shows the percent homology of 9 different species to human p53 protein sequence. Which species shares the most recent common ancestor with humans?

Species	% homology
European flounder	46
Rhesus monkey	95
Channel catfish	48
Congo puffer fish	41

   (A) European flounder
   (B) Rhesus monkey
   (C) Channel catfish
   (D) Congo puffer fish

8. A gene pool is
   (A) a localized group of individuals capable of interbreeding.
   (B) all of the genes and their alleles in a defined population.
   (C) the 46 chromosomes that make up the human genome.
   (D) contains 2 copies of every gene within the genome.

9. In a population of 27 individuals, 18 express the dominant phenotype and 1/3 of the total population are carriers of the recessive allele. How many recessive alleles are there in the gene pool?
   (A) 6
   (B) 12
   (C) 27
   (D) none of the above

10. Which of the following is not a condition for Hardy-Weinberg equilibrium
   (A) Equal fitness among individuals of the population
   (B) Equal rate of migration among individuals
   (C) Chance (luck) mating only
   (D) Large population

11. If the population size is N, which of the following can be used to calculate the number of heterozygous individuals?
   i. $N \times 2pq$
   ii. $N \times (1 - p^2 - q^2)$
   iii. $2pq$
   iv. $1 - q - p$
   (A) i only
   (B) i and ii
   (C) iii and iv
   (D) none of the options

12. If the population size is N, which of the following is used to calculate the number of recessive alleles in the gene pool?
   (A) $N \times (2pq + 2q^2)$
   (B) $N \times q$
   (C) Both A and B are correct
   (D) None of the above

GO ON TO THE NEXT PAGE

13. A cladogram is based on the following 4 species: Tasmanian wolf, the gray wolf, human, and dog. Which species is the outgroup?
    (A) Tasmanian wolf
    (B) Gray wolf
    (C) Human
    (D) Dog

14. How many clades are there in the cladogram of #6
    (A) 1
    (B) 2
    (C) 3
    (D) 4

15. Which taxon is the sister group of the gray wolf?
    (A) Tasmanian wolf
    (B) Gray wolf
    (C) Human
    (D) Dog

16. Sea urchin has a high % of A/T relative to % G/C, while sarcina lutea has a high % of G/C relative to % A/T. Which of the following is true?
    (A) The DNA in sea urchin will denature more easily
    (B) The mRNA of sarcina luteal will form tight bonds with its complement
    (C) DNA replication in sarcina lutea occurs faster
    (D) All of the above

17. Analogous structure occurs through __, while homologous structures occur through __.
    (A) Convergence, divergence
    (B) Divergence, convergence
    (C) Adaptive radiation, punctuated equilibrium
    (D) None of the above

18. During the pre-RNA world, natural selection on molecules were based on their
    i.   structural stability.
    ii. ability to self-replicate.
    iii. functionality.
    iv. ability to store information.
    (A) i, iii, and iv
    (B) ii and iv
    (C) i and iii
    (D) all of the above

19. Which of the following animals has the highest mass specific metabolic rate?
    (A) elephant
    (B) wolf
    (C) house cat
    (D) mouse

20. Which of the following of Darwin's observation does not directly explain why competition increases in a population?
    i.   population growth rate = 0
    ii. limited resources
    iii. superfecundity
    iv. inheritance
    v.   steady population
    (A) i and iv
    (B) ii, iii, and v
    (C) iv only
    (D) all of the above statements can be used to explain increased competition in a population.

21. Which of the following statements about endotherms is true?
    (A) They are "cold blooded."
    (B) They will typically have a higher pulse rate than ectotherms.
    (C) They adjust their behavior to exchange heat with their surroundings.
    (D) There body temperature correlates directly with the surrounding temperature.

GO ON TO THE NEXT PAGE

22. Which of the following organisms has the highest whole-animal metabolic rate?
    (A) elephant
    (B) wolf
    (C) house cat
    (D) mouse

23. If the population size is N, which of the following can be used to calculate the number of individuals with the dominant phenotype?
    i. $N \times 2p$
    ii. $N \times (1 - q^2)$
    iii. $N \times (p^2 + 2pq)$
    iv. $N \times p^2$
    (A) i only
    (B) ii and iii
    (C) iii only
    (D) iv only

24. At the moment of death due to old age, which of the following would you expect to have the greatest entropy?
    (A) elephant
    (B) wolf
    (C) house cat
    (D) mouse

25. Recall that Work = Force x distance. In order to perform work on it's surrounding, the system must push (apply force on) surrounding matter over a distance. When you are thinking through a problem set, you are doing work. What are the system and the surrounding in this work scenario? (system; surrounding)
    (A) brain power; thoughts
    (B) finger muscle; pencil
    (C) nerve cells; neurotransmitters
    (D) none of the above

26. The earth is most similar to which of the following types of systems?
    (A) Isolated system
    (B) Closed system
    (C) Open system
    (D) Both A and C

27. What happens if there is an increased enthalpy change while temperature and entropy change remains constant?
    (A) Change in chemical energy increases
    (B) Change in Gibbs free energy decreases
    (C) $\Delta G > 0$
    (D) $\Delta G < 0$

28. The hypothesis that life may have originated in hydrothermal vents is supported by the fact that these vents:
    (A) have local barriers to diffusion of substances and can concentrate minerals needed by reactions in the pre-RNA world.
    (B) have a local source of energy
    (C) are open systems that can receive energy from the sun
    (D) A and B only

GO ON TO THE NEXT PAGE

29. Which of the following statements about living systems is true?
    i. There is a natural and universal tendency toward order
    ii. Maximum disorder is achieved at thermodynamic equilibrium between the system and surrounding
    iii. Energy is needed by biological systems to maintain a high entropy level
    iv. Death is a direct result of energy loss or of achieving maximum sustainable disorder
    v. Living systems must take in more energy than they use up for physiological processes
    (A) i, iii, and iv
    (B) ii, iv, and v
    (C) i and iii
    (D) All are correct

30. In biological system, chemical reaction are naturally selected for their ability to
    (A) Gather information from the surroundings
    (B) Store information
    (C) Transmit information into the future
    (D) All of the above

31. Which of the following statements are true about gene regulation?
    i. It is the cell's control of which genes are favored during natural selection?
    ii. It enhances the organism's energy efficiency
    iii. It allows the organisms to respond to its environment
    iv. It increases organization in biological systems
    v. It does not influence the organism's phenotype.
    (A) ii, iii, iv, and v only
    (B) i, ii, and iv only
    (C) ii, iii, and iv only
    (D) ii and iii only

32. Which of the follow is not a level of gene regulation in eukaryotic cells?
    (A) Chromatin packing
    (B) Assembly of nuclear membrane
    (C) mRNA transcription
    (D) polypeptide modification

33. Which of the following statements about the operon is incorrect?
    (A) They never include the regulator gene
    (B) They always contain a promoter region
    (C) They can include multiple structural genes
    (D) The operator is usually upstream from the promoter

34. Where in the operon does the repressor bind?
    (A) operator
    (B) Enhancer
    (C) Inducer
    (D) Promoter

35. Which of the following can cause the destabilization of the chromatin structure?
    (A) DNA methylation
    (B) Histone acetylation
    (C) Histone dimethylation
    (D) RNAi

36. Which of the following statements about a positive inducible operon is correct?
    i. The regulator is a repressor
    (A) The regulator is normally inactive
    (B) The structural genes are normally produced
    (C) A product of the biochemical pathway that is associated with the operon typically turns on the operon.
    (D) All of the above statements are correct.

GO ON TO THE NEXT PAGE

37. Which of the following is true about the lac operon?
    (A) The lac operon is a negative inducible operon.
    (B) Permease acts via a positive feedback.
    (C) Both are true.
    (D) Both are false.

38. The inducer of the lac operon is
    (A) Lactose
    (B) beta-galactosidase
    (C) glucose
    (D) none of the above

39. Which of the following statements about genomic imprinting is correct
    (A) All great ape species including human and chimpanzee share a similar imprinting pattern
    (B) All male somatic cells will show both the maternal and paternal imprint pattern
    (C) All females somatic cell will show only the maternal imprint pattern
    (D) Imprint patterns are reprogrammed after the gamete cells fuse.

40. In 1966, microbiologist Kwang Jeon noticed an unexpected infection of his amoeba colony by a strain of x-bacteria. While most of the amoeba became sick and died, some survived and thrived with live x-bacteria living within. When treated with antibiotics, both the x-bacteria and host amoeba died. These observations suggested that:
    (A) antibiotics usually kills both amoeba and bacteria.
    (B) the host amoeba had become symbiotic with the x-bacteria.
    (C) x-bacteria do not usually infect amoeba cells.
    (D) none of the above.

41. Which of the following statements are true?
    i.   Evolution occurs at the genetic level.
    ii. Heritable information is exclusively responsible for determining the behavior within and between cells, organisms and populations.
    iii. Both the organism's genes and the external conditions can influence behaviors such as biological rhythms, mating rituals, flowering, and animal social structures.
    (A) i only
    (B) i, ii, only
    (C) i, and iii, only
    (D) All of the statements are true.

42. Which of the following statements is true?
    i.   Organ systems that can sense and process external information are used to enhance survival, growth and reproduction in multicellular organisms.
    ii. The nervous system is critical for coordinating responses and behavior in mammals.
    iii. Loss of function and coordination within the nervous system can be fatal.
    (A) i only
    (B) i, ii, only
    (C) iii, only
    (D) All statements are true

43. Which of the following involves changes in behavior as a result of information exchange between individuals?
    i.   Fight or flight response
    ii. Predatory warnings
    iii. Protection of young
    iv. Plant response to herbivory
    v.   Avoidance responses
    (A) All statements are true
    (B) i, ii, iv, v only
    (C) i, ii, iv only
    (D) ii, iii only

GO ON TO THE NEXT PAGE

44. Which of the following can initiate the strongest avoidance response?
   (A) A toddler touches a hot stove and cries out.
   (B) A driver sees the yellow light and speeds up to avoid being caught by the oncoming red light.
   (C) A honeybee sees what appears to be a field of red flowers. She travels 2 miles to reach it only to discover that it was red confetti litter from the previous independence day parade.
   (D) All of the above are equal.

45. Which of the following statement is incorrect
   (A) Spraying by male lions is used to affect the behavior of others.
   (B) Coloration in flowers evolved to attract humans as cultivators.
   (C) Bees use their stingers as a predatory warning device.
   (D) All of the statements are correct.

46. Which of the following statement is correct
   (A) For increased fitness, the artic fox changes color in the winter months, becoming white.
   (B) Innate behavior is only influenced by the individual's genes.
   (C) All learned behaviors are independent of genetics.
   (D) All of the statements are correct

47. Which of the following statement is correct
   (A) A dog's avoidance of an electric fence is a good example of innate behavior, rather than learned behavior.
   (B) The cloning of Dolly the sheep must have involved the artificial production of totipotent cells
   (C) In temperature-dependent sex determination, more males are produced at higher temperatures.
   (D) All of the above statements are correct.

48. Cooperative behavior tends to increase the fitness of individuals and the survival of the whole population. Which of the following behavior illustrates this concept?
   i.   Pack behavior in animals
   ii. Herd, flock and schooling behavior in animals
   iii. Predatory warning
   iv. Colony and swarming behavior in insects
   (A) i, ii, ii only
   (B) i, ii, iv only
   (C) ii, iii only
   (D) All statements are true

49. When the nerve impulse arrives at the presynaptic dendritic membrane
   (A) Voltage-gated calcium channels open
   (B) The influx of calcium causes the synaptic vesicles to fuse with the presynaptic membrane.
   (C) Potassium ions rapidly diffuse out of the presynaptic cell and bind their ligand receptors on the post-synaptic membrane to propagate the nerve impulse to the next neuron.
   (D) Neurotransmitters are packaged into synaptic vesicles.

50. Which of the following molecules is involved in cell-to-cell recognition.
   (A) Major histocompatibility complexes (MHC)
   (B) Hormones
   (C) Membrane bound proton pumps
   (D) Antibodies

GO ON TO THE NEXT PAGE

51. Which of the following statements about lichen is true
    (A) It is made of algae and cyanobacteria living in a symbiotic relationship
    (B) It is an example of mutualistic symbiosis.
    (C) Although, like plants, the lichen is stationary and can perform photosynthesis, it is not considered a plant.
    (D) All of the above statements are true.

52. The function of cholesterol include
    (A) Cushioning organs
    (B) Body insulation
    (C) Energy storage
    (D) Stabilizing the phospholipid bilayer.

53. Density-dependent factors that effect population growth include
    (A) Natural disasters
    (B) Human activity
    (C) Parasitism and disease
    (D) Forest fires

54. The function of carbohydrates include
    (A) Cushioning organs
    (B) Body insulation
    (C) Energy storage
    (D) Stabilizing the phospholipid bilayer.

55. Which of the following occurs during RNA processing?
    (A) Alternative splicing
    (B) Addition of the 3' poly A tail
    (C) Additional of the 5' methyl cap
    (D) All of the above

56. Which of the following is involved in stabilizing the tertiary structure of the protein
    (A) Peptide bonds
    (B) Hydrogen bonds
    (C) Disulfide bonds
    (D) Phosphate bonds

57. Which of the following RNAs are involved in cytoplasmic degradation of nuclear derived mRNA
    (A) miRNA
    (B) siRNA
    (C) snoRNa
    (D) rRNA

58. Okazaki fragments
    (A) Grow in the 3' to 5' direction.
    (B) Are fused to form the lagging strand.
    (C) Are collectively produced by a single DNA polymerase.
    (D) All of the above.

59. Which of the following statements makes for a good null hypothesis?
    (A) The Observed ratio is sufficiently close to the expected ratio so as to validate the data
    (B) The expected data and the observed data are within statistical range of each other
    (C) Any deviation of the observed data from the expected data is due to chance
    (D) All of the above statements are good null hypothesis statements

60. Your teacher believes that the student population at school is 2/3 girls. You perform a survey of 100 students and discovered that 38% are boys. Is your teacher statistically correct (use chi square analysis)?
    (A) Yes
    (B) No
    (C) Not enough information to solve the problem
    (D) The sample size is too small to provide a reliable assessment.

GO ON TO THE NEXT PAGE

61. The gene for the testosterone receptor protein is X-linked. What can be expected if a frameshit mutation occurred in this gene?
    (A) XX individuals will express male characteristics.
    (B) XY individuals will express female characteristics.
    (C) XY individuals will show no effect since the mutation will be recessive, and would require two copies of the mutant allele to be expressed.
    (D) XØ individuals will express male characteristics.

62. A scientist investigating eagle predation on white-gray- and black-furred mice introduces 50 white, 50 gray, and 50 black mice into a habitat containing black volcanic rocks. She returns a week later and discovers that there are 7 black, 11 grey, and 9 white mice remaining. What is her likely response to this data?
    (A) She is surprised because the white mice should have greater exposure to UV radiation and should have died from cancer at a higher rate, significantly reducing their numbers relative to the other two coats of mice.
    (B) She expected to see this result because eagles are not familiar with white mice and probably did not think they were edible prey.
    (C) She is surprised because she had expected that the black mice would have the highest and white mice the lowest survivorship due to camouflage.
    (D) Not surprised. The results most likely supported her initial hypothesis.

63. During photosynthesis,
    (A) water and oxygen are used to produce glucose.
    (B) the Calvin cycle require light.
    (C) light absorption occurs in the thylakoid membranes
    (D) $NADP^+$ and $FADH_2$ are used to transport electrons to the dark reactions for glucose synthesis.

GO ON TO THE NEXT PAGE

Directions (Part B - Grid-in): For each of the following 6 question, make the necessary calculations and record your numeric answer on the provided grid. You may use only a four-function calculator and the provided formula sheet.

64. The following data was generated from the mitosis lab where the affect of auxin and lectin on onion root growth was evaluated. What is the expected interphase frequency? Provide answer to the nearest hundredth?

	Interphase	Mitosis
Control	194	32
Lectin treated	150	118
Auxin treated	200	106

65. In 2012, the city of Monrovia, Liberia had a population with blood types distribution as follows (assume HWE): A = 254,841; B = 371,340; AB = 305,810; O = 38,833. (Alleles are IA, IB, io). What is the frequency of AB individuals? Give your answer to the hundredth decimal.

66. Three lab groups collect data on males and females students at school. One group reports 30% boys, another reports 40% boys, and the last group report 35%. Calculate the mean to the tenth decimal

67. For the following two DNA sense strands, calculate the percent homology between their corresponding polypeptides. Report your answer as a whole number.
Sense DNA strand A:
5'-ATCGCCCTGCACAGUGGGTGA-3'

Sense DNA strand B:
5'-ATCGCCCTGCAGAGUGGGTGA-3'

68. If at death, a fossil contains 6 grams of C-14, after 3 half lives, how much C-14 would be detected? Report your answer to the hundredth decimal.

69. PKU is a recessive disorder that occurs in 49 of every 10,000 births. What is the frequency of the normal allele? Report your answer to the tenth decimal.

GO ON TO THE NEXT PAGE

Time: 90 minutes
Part A: 2 Long free-response questions
Part B: Short free-response questions

Directions (Part A - Long FRQs): Use the first 10 minutes to read over the 8 questions and writing notes in the margin. However, you may not begin answering the question until after the 10 minutes.

Use complete sentences to answer each of the following questions. You may use diagrams and illustrations as supporting evidence of your mastery of the relevant concepts, but these alone are not sufficient. Be a detailed as possible without including information that is irrelevant to the question.

1. The chloroplast is a critical component of plant and many photosynthetic organisms. It structure and function has been extensively studied and a clear link between the two has been defined.

   (A) Describe the structure of the chloroplasts, including its various membrane and any internal compartments that it may contain.

   (B) Explain how these structures support the chloroplast in its ability to efficiently harness light energy and use that energy to synthesize glucose. Your explanation must reflect on the light reaction and Calvin cycle, and the relevant substructures that make these reactions possible.

   (C) Describe how oxidative phosphorylation is coupled to energy absorption by the light absorbing pigment molecules.

2. Evolution is one of the central themes in biology. The Hardy-Weinberg equilibrium provides a model for modeling the mechanism by which evolution occurs.

   (A) Use the Hardy-Weinberg model to explain evolution.
   (B) Describe the conditions required for the Hardy-Weinberg equilibrium.
   (C) Use specific examples to provide the rationale for each of the conditions.

GO ON TO THE NEXT PAGE

Directions (Part B - Short FRQs): Use complete sentences to answer each of the following questions. You may use diagrams and illustrations as supporting evidence of your mastery of the relevant concepts, but these alone are not sufficient. Be a detailed as possible without including information that is irrelevant to the question.

3. In 2012, the city of Monrovia had a population with blood types distribution as follows (assume Hardy-Weinberg equilibrium): A = 254,841; B = 371,340; AB = 305,810; O = 38,833. (Alleles are I^A, I^B, i^o)

   (A) What is the recessive io allele frequency?

   (B) What is the frequency of the AB phenotype?

   (C) How many individuals are heterozygous for the recessive allele?

4. After analyzing the data from the artificial selection investigation (Investigation 1), you noticed a pattern that suggests the expression of a single gene trait. Two sets of data (height and trichome density per plant) from 4 lab groups are shown below.

Trichome density (trichomes/cm)

Plant	group 1	group2	group 3	group 4
1	5	5	6	18
2	6	4	7	19
3	5	4	6	4
4	7	5	17	18
5	6	6	7	19
6	6	6	6	18
7	5	5	6	17
8	7	6	5	3

Plant height (cm)

Plant	group 1	group2	group 3	group 4
1	8.2	8.1	8.1	8.1
2	9	7.4	7.9	8.1
3	7.3	7.3	7.9	7.5
4	8.1	7.9	8.1	8.0
5	8.4	8.4	8.2	8.3
6	9	9	8.4	8.8
7	7.6	8.3	9.4	8.4
8	8.5	7.9	10.2	5.3

   (A) Which data set shows a distribution pattern of a single gene trait? Justify your response.

   (B) Which group of the first data set has the greatest standard deviation? Justify your response.

GO ON TO THE NEXT PAGE ⟩

5. Based on the DNA sequence 5' -T-T-A-G-C-T-G-G-G-C-A-T- 3' answer the questions below:
   (A) Is the sequence for the sense or antisense DNA strand? Justfy your response.
   (B) What is the mRNA sequence?
   (C) What is the polypeptide sequence?
   (D) What is the last codon in the sequence?
   (E) A substitution mutation occurs in the third position of the second codon. The new sequence
       has an adenine in that position. What impact will this mutation have on the protein sequence?
       Justify your response.

6. The flying phalanger, a marsupial, and the placental flying squirrel are not closely related, yet they
   share many traits. Explain how the two species could have converged.

7. Compare and contrast the processes of transcription and translation in eukaryotes and prokaryotes. Be sure to reflect on relevant structural and functional similarities and differences.

8. An experiment was conducted to measure the reaction rate of the human salivary enzyme α-
   amylase. 10 mL of a concentrated starch solution and 1.0 mL of α-amylase solution were placed
   in a test tube. The test tube was inverted several times to mix the solution and then incubated at
   25°C. The amount of product (maltose) present was measured every 10 minutes for an hour. The
   results are shown in the table below.

Time (minutes)	Maltose concentration (μM)
0	0
10	5.1
20	8.6
30	10.4
40	11.1
50	11.2
60	11.5

   (A) Graph the data.
   (B) Calculate the rate of the reaction from 0 to 30 minutes and explain how and why the reaction
       rate changed after 30 minutes.
   (C) Draw, label, and explain another curve on the graph for a reaction with double the α–amylase
       concentration.
   (D) Explain how can a competitive inhibitor (Inhibitor X) of this α–amylase be used to negatively
       regulate the enzyme.

30

DIAGNOSTIC TEST ANSWER FORM

NAME:_____

Section 1: 63 Multiple choice & 6 Grid-in questions

1 Ⓐ Ⓑ Ⓒ Ⓓ   31 Ⓐ Ⓑ Ⓒ Ⓓ   61 Ⓐ Ⓑ Ⓒ Ⓓ

2 Ⓐ Ⓑ Ⓒ Ⓓ   32 Ⓐ Ⓑ Ⓒ Ⓓ   62 Ⓐ Ⓑ Ⓒ Ⓓ

3 Ⓐ Ⓑ Ⓒ Ⓓ   33 Ⓐ Ⓑ Ⓒ Ⓓ   63 Ⓐ Ⓑ Ⓒ Ⓓ

4 Ⓐ Ⓑ Ⓒ Ⓓ   34 Ⓐ Ⓑ Ⓒ Ⓓ

5 Ⓐ Ⓑ Ⓒ Ⓓ   35 Ⓐ Ⓑ Ⓒ Ⓓ   64

6 Ⓐ Ⓑ Ⓒ Ⓓ   36 Ⓐ Ⓑ Ⓒ Ⓓ

7 Ⓐ Ⓑ Ⓒ Ⓓ   37 Ⓐ Ⓑ Ⓒ Ⓓ

8 Ⓐ Ⓑ Ⓒ Ⓓ   38 Ⓐ Ⓑ Ⓒ Ⓓ

9 Ⓐ Ⓑ Ⓒ Ⓓ   39 Ⓐ Ⓑ Ⓒ Ⓓ

10 Ⓐ Ⓑ Ⓒ Ⓓ   40 Ⓐ Ⓑ Ⓒ Ⓓ   67

11 Ⓐ Ⓑ Ⓒ Ⓓ   41 Ⓐ Ⓑ Ⓒ Ⓓ

12 Ⓐ Ⓑ Ⓒ Ⓓ   42 Ⓐ Ⓑ Ⓒ Ⓓ

13 Ⓐ Ⓑ Ⓒ Ⓓ   43 Ⓐ Ⓑ Ⓒ Ⓓ

14 Ⓐ Ⓑ Ⓒ Ⓓ   44 Ⓐ Ⓑ Ⓒ Ⓓ   65

15 Ⓐ Ⓑ Ⓒ Ⓓ   45 Ⓐ Ⓑ Ⓒ Ⓓ

16 Ⓐ Ⓑ Ⓒ Ⓓ   46 Ⓐ Ⓑ Ⓒ Ⓓ

17 Ⓐ Ⓑ Ⓒ Ⓓ   47 Ⓐ Ⓑ Ⓒ Ⓓ

18 Ⓐ Ⓑ Ⓒ Ⓓ   48 Ⓐ Ⓑ Ⓒ Ⓓ   68

19 Ⓐ Ⓑ Ⓒ Ⓓ   49 Ⓐ Ⓑ Ⓒ Ⓓ

20 Ⓐ Ⓑ Ⓒ Ⓓ   50 Ⓐ Ⓑ Ⓒ Ⓓ

21 Ⓐ Ⓑ Ⓒ Ⓓ   51 Ⓐ Ⓑ Ⓒ Ⓓ

22 Ⓐ Ⓑ Ⓒ Ⓓ   52 Ⓐ Ⓑ Ⓒ Ⓓ

23 Ⓐ Ⓑ Ⓒ Ⓓ   53 Ⓐ Ⓑ Ⓒ Ⓓ

24 Ⓐ Ⓑ Ⓒ Ⓓ   54 Ⓐ Ⓑ Ⓒ Ⓓ   66

25 Ⓐ Ⓑ Ⓒ Ⓓ   55 Ⓐ Ⓑ Ⓒ Ⓓ

26 Ⓐ Ⓑ Ⓒ Ⓓ   56 Ⓐ Ⓑ Ⓒ Ⓓ

27 Ⓐ Ⓑ Ⓒ Ⓓ   57 Ⓐ Ⓑ Ⓒ Ⓓ   69

28 Ⓐ Ⓑ Ⓒ Ⓓ   58 Ⓐ Ⓑ Ⓒ Ⓓ

29 Ⓐ Ⓑ Ⓒ Ⓓ   59 Ⓐ Ⓑ Ⓒ Ⓓ

30 Ⓐ Ⓑ Ⓒ Ⓓ   60 Ⓐ Ⓑ Ⓒ Ⓓ

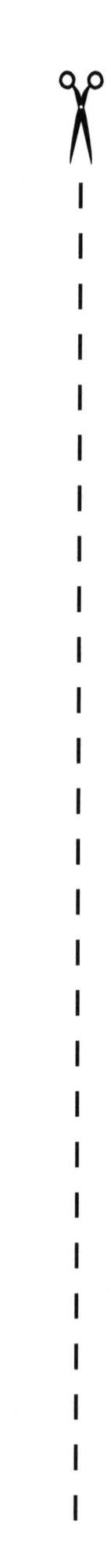

ANSWERS AND EXPLANATIONS

Section I

1. C

2. A

3. B

4. B

5. D

6. A; statement ii is wrong because the amino acids and nucleic acids are products of the organic soup, or at least formed while the organic soup was taking shape. Statement iii is wrong because the first cells only emerged after the the primoardial soup and RNA world came into existence.

7. B, because rhesus monkey has the highest % homology with human, meaning that the two share the highest similarity in their p53 gene relative to human and the other species.

8. C

9. D because there are 18 - 12 = 6 recessive individuals and and 1/3(18) = 6 heterozygous individuals. Number of recessive allele is equal to the heterozygous plus twice the homozygous = 6 + 2(6) = 18.

10. B

11. C because number of heterozygous individuals equal to the total population (N) multiplied by the heterozygous frequency (2pq). $p^2 + 2pq + q^2 =$, therefore $2pq = 1 - p^2 - q^2$, and $N(2pq) = N(1 - p^2 - q^2)$

12. C because the number of recessive alleles is equal to the population size (N) multiplied by the recessive allele frequency (q), or the number of heterozygous, $N(2pq)$, plus twice number of recessive individuals, $2Nq^2$.
    # recessive alleles = $Nq = N (2pq+2q^2)$

13. A because the Tasmanian wolf is a marsupial and the only none placental mammal.

14. D

15. D

16. A because the A/T rich DNA regions have 2 H-bonds between complemental nucleotides, as compared to 3 H-bonds in G/C rich regions. Fewer H-bonds in A/T rich regions allows the complementary strands to separate more easily.

17. A

18. D

19. D because the smaller the body weight the less efficient the organism in its metabolism.

20. D

21. C

22. A

23. B because the number of individuals who express the dominant phenotype includes all of the homozygous dominant (Nq^2) and heterozygous ($N2pq$) individuals, or $N(q^2 + 2pq)$. Since $p^2 + 2pq + q^2 = 1$, $p^2 + 2pq = 1 - q^2$. Therefore, $N(q^2 + 2pq) = N(1 - q^2)$.

24. D because the mouse has the lowest concentration of matter relative to its surrounding.

25. C because the propagation of the nerve impulse facilitates thought processes. The nerve impulse is propagated when the presynaptic neuron releases neurotransmitters into the synaptic cleft. Once released, the neurotransmitters become a part of the nerve cell's environment.

26. B because, although the Earth should be classified as an open system, it is more similar to a closed system because there is a negligible exchange of matter, but extensive exchange of energy with the surrounding solar system.

27. C

28. D

29. B

30. D

31. C

32. B

33. D

34. A

35. B

36. B

37. A

38. D because allolactase, an isomer of lactose, and not lactose is the actual inducer.

39. C

40. A

41. C

42. D

43. A

44. A

45. B because flower pigmentation evolved to attract pollinators, not humans.

46. A

47. B

48. D

49. B

50. A

51. D

52. C

53. C

54. D

55. D

56. C

57. A

58. B

59. C

60. A because the Chi square value is 0.71, which is lower than the critical value of 3.84.

61. B is correct because, although testosterone is produced, the mutation in testosterone receptor results in the production of a non-functional receptor that cannot propagate the signal from the hormone. Consequently, male sex-determination is reduced, resulting in the expression of female characteristics.

62. C

63. C

64. $194/226 = 0.86$

65. N = population size
$= 254,841 + 371,340 + 305,810 + 38,833$
$= 970,824$
AB freq = AB ÷ N
$= 254,841 ÷ 970,824 = 0.26$

66. $(30+40+35)/3 = 35$

67. $20/21 = 95\%$ homology. There are 20 out of 21 identical nucleotides between the two sequences.

68. Final mass = Initial mass $\times 0.5^{\text{half lives}}$
Final mass $= 6 (0.5)^3 = 0.75$ grams

69. $q^2 = 49/10000 = 0.0049$
$q = (0.0049)^{0.5} = 0.07$

$p = 1 - 0.07 = 0.93$

Section II

1. Chloroplasts are the photosynthetic organelles of plants. It contains an outer and inner membrane, and internal structures with called thylakoids that are specialized compartments much like the cytoplasmic organelles. The thylakoids are separated from the rest of the chloroplast by a thylakoid membrane.

   The outer membrane, which separates the chloroplast from the rest of the cell, is porous and permeable to most small molecules and ions. The inner membrane is separated from the outer membrane by an intermembrane space and is less permeable to substances. To facilitate selective permeability, the inner membrane has embedded transport proteins that move certain cytoplasmic proteins and glucose into and out of the chloroplast. The space within the inner membrane is called the stroma and is analogous of the cytoplasm of the cell.

   The thylakoid membrane segments that chloroplast into the photosynthetic thylakoid compartments. Think of the thylakoids as the chloroplast's own mini-organelles that are specialized for capturing light energy. The thylakoid space contains the lumen fluid that fills stacked subcompartmental discs called granum. Grana connect to each other via tunnel-like bridge structures called stromal lamellae that transport nutrients and other material needed to sustain the chloroplast in it light capturing role.

   Photosynthesis occurs through 2 reactions. The first is the thylakoid membrane's light-dependent reaction that absorbs light energy, and stoma's Calvin cycle that synthesizes glucose with energy from ATP. ATP is produced by a separate coupled reaction (oxidative phosphorylation) of the thylakoid membrane. Light is absorbed by the thylakoid through special functional units called photosystem I and II that use light absorbing pigments including chlorophyll a and b, and carotene, each capable of absorbing a unique wavelength of light.

2. Evolution is the change in the genetic make-up of a population over time. It can present qualitatively as a change in the phenotype distribution within the population, or quantitatively as the change in the allele frequencies in the gene pool. When the allele frequency remains unchanged between generations, the population is said to be in Hardy-Weinberg Equilibrium and not evolving. HWE is a hypothetical situation that rarely occurs in nature; however, it is useful for studying evolution.

   In order for HWE to occur, there are 5 conditions that must be met by the population: (1) there can be no mutations appearing in the population, because new mutations can introduce new allele and altering the allele frequencies of the population; (2) there must be random mating so that there is no sex selection due to preferential mating. Sex selection can cause some individuals to reproduce at higher rates than other, allowing them to contribute their alleles to the next generation's gene pool at a higher frequency; (3) there can be no natural selection that favors certain traits over others, because like sex selection, certain allele will be favored resulting in a disproportionate contribution of those favored allele to the next generation; (4) there must be a large population so that the impact of genetic drifts is reduced. Genetic drift can cause random changes in the allele frequency of the population; and (5) there must be a closed population to migration so that there is no gene flow that can introduce or remove alleles.

If a gene has only 2 alleles, the letters p and q are used to represent the frequency of the dominant and recessive alleles respectively. Since the sum of all the allele frequencies for a given gene must add to 1, the following equations apply:

$p + q = 1 \Rightarrow (p + q)^2 = p^2 + 2pq + q^2 = 1$
where, p is the dominant allele frequency, q the recessive allele frequency, p^2 the homozygous dominant genotype frequency, 2pq is the heterozygous genotype frequency, and q^2 the homozygous recessive genotype frequency. The equations are used to calculate the allele frequencies from one generation to the next to determine whether there are any changes.

3. (A) r = 0.2
    $p^2 + q^2 + r^2 + 2pq + 2pr + 2qr = 1$
    p = frequency of I^A; q = frequency of I^B
    r = frequency of i^o; r^2 = freq. of O blood
    n = population size
    = 254,841 + 371,340 + 305,810 + 38,833
    = 970,824
    $r^2 = 38,833/970,824 = 0.04$
    $r = (0.04)^{1/2} = 0.20$
    (B) 2pq = 305,810/970,824 = 0.32

    (C) N(2pr + 2qr) = 310,664
    # heterozygous i^o = AO + BO blood
    = N (2pr + 2qr)
    # homozygous i^o = 2 × O blood
    = $2Nr^2$ = 77,666
    Total alleles = 2N = 2 × 970,824
    = 1,941,648
    Total i^o alleles = r x total # alleles
    = 2Nr = 1,941,648 x 0.02 = 388,330
    N (2pr + 2qr) = $2Nr - 2Nr^2$
    = 388,330 - 77,666 = 310,664
    <u>Alternative solution</u>
    From (A), 2pq = 0.32; r = 0.20
    p + q + r = 1; p + q = 1 – r = 0.80
    $(p + q)^2 = (0.80)^2 = 0.64$
    $p^2 + 2pq + q^2 = 0.64$
    $p^2 + q^2 = 0.64 - 2pq$
    $p^2 + q^2 = 0.64 – 0.32 = 0.32$
    $p^2 + q^2 + r^2 + 2pq + 2pr + 2qr = 1$

0.32 + 0.04 + 0.32 + 2pr + 2qr = 1
2pr + 2qr = 0.32; 0.32 × 970,824
= 310,664

4. (A) The first data set on trichome density because there seems to be a dual phenotype pattern of high trichome density or low trichome density that may be caused by a two-allele system. It lacks the bell curve distribution often seen with polygenic traits. (B) Even without calculating the standard deviation, you can predict that Group 4 has the largest value since it has a significantly wider data spread from trichome density of 3 to 19 trichomes/cm. This can be verified by calculating its standard deviation, which is 6.4, compared to 0.78, 0.78, and 3.6 for the other three data sets.

5. (A) Assuming that it is the start of the gene, the sense strand should have an ATG triplet at one of its end to correspond with an AUG start codon. This strand instead has a TAC triplet, which complements ATG. This suggests that it is the antisense codon.

    (B) 5'A-U-G-C-C-C-A-G-C-U-A-A 3'
    Use U instead of T, and label 3' and 5' ends.
    (C) Met-Pro-Ser (Stop)

    (D) U-A-A

    (E) The mutation will have no impact because it is a silent mutation due to codon redundancy caused by base pair wobble that results in a violation of Chargaff's rules when the third nucleotide in the codon is matched to more than one nucleotide from the anticodon. It can also be due to some tRNAs containing the nucleotide called inosine that can complement U, C, and A.

6. Marsupials underwent adaptive radiation in Australia, evolving into the largest diversity of extant marsupials. Although marsupial and

placental mammals are of divergent genetic lineages, similar ecological niches between Australasia and other geographic locations inhabited by placentals led to the morphological convergence of species from the two lineages. The flying phalanger and the flying squirrel are adapted to forest ecosystems and are both specialized for gliding from between trees. Although they appear to share the same superficial features, the ability to glide are independently derived characters that may be mistakenly assumed to have originated in a common ancestor of the two groups.

7. The central dogma governs how information flows in both prokaryotes and eukaryotes. In prokaryotes, the process is simpler, with both occurring simultaneously in the cytoplasm.

    In eukaryotes, transcription occurs in the nucleus to produce a pre-mRNA that undergoes further processing (exon splicing, 3' cleavage, addition of 5' methyl cap and 3' poly A tail) before leaving the nucleus for the cytoplasm.

    Translation of the mRNA occurs on ribosomes that are either free-floating (both prokaryotes and eukaryotes) or bound to the rough ER (eukaryotes only).

    Since in eukaryotes, transcription and translation are compartmentalized in the nucleus and cytoplasm, they do not occur simultaneously as they do in prokaryotes.

8. (A)
    The curve represent a reaction that goes to completion in half the time because there is twice the enzyme concentration. Consequently, its reaction rate is doubled.
    (B) Rate = $10.4 \div 30 = 0.35\mu M/min$

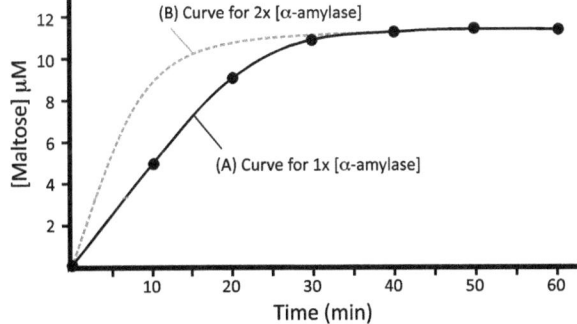

(C) As the starch is metabolized, its concentration falls below that of the competitive inhibitor. The inhibitor than inhibits the enzyme (negative regulation) by binding the active site to block the substrate.

QUESTION ALIGNMENT CHART

Chapter	Topic	Questions
1	Abiogenesis	1 - 6,18,30
2	Origin of life	21, 41
3	Evidence of evolution	7, 17, 41, 68
4	Phylogenetics	13-15
5	Hardy-Weinberg equilibrium	8 - 12, 23, 41, 65, FRQ 2, 3
6	Selection processes	FRQ 4
7	Speciation	41, FRQ 6
8	Endosynbiosis	40
9	Thermodynamics	24 - 29
10	Energy in living systems	19, 22
11	Matter in living systems	
12	Membranes	52
13	Cell compartmentalization	63, FRQ 1
14	Energy capture	63, FRQ 1
15	Dynamic homeostasis	64
16	Information flow	67, FRQ 5, 7
17	Cell reproduction	64
18	Patterns of inheritance	61, 62, 65, 69
19	DNA replication	16, 58
20	Transcription and translation	55 - 57, 61
21	Gene regulation	31 - 39
22	Cell communication	49, 50
23	Biological interactions	41 - 48, 51 - 54
Labs	Chi square, statistics, reaction rate	59, 60, 66

BIG IDEA 1

Evolution

- **ABIOGENESIS**
- **ORIGIN OF SPECIES**
- **EVIDENCE OF EVOLUTION**
- **PHYLOGENETIC TREE**
- **HARDY-WEINBERG**
- **SELECTION PROCESSES**
- **SPECIATION**
- **ENDOSYMBIOTIC THEORY**

01 Abiogenesis

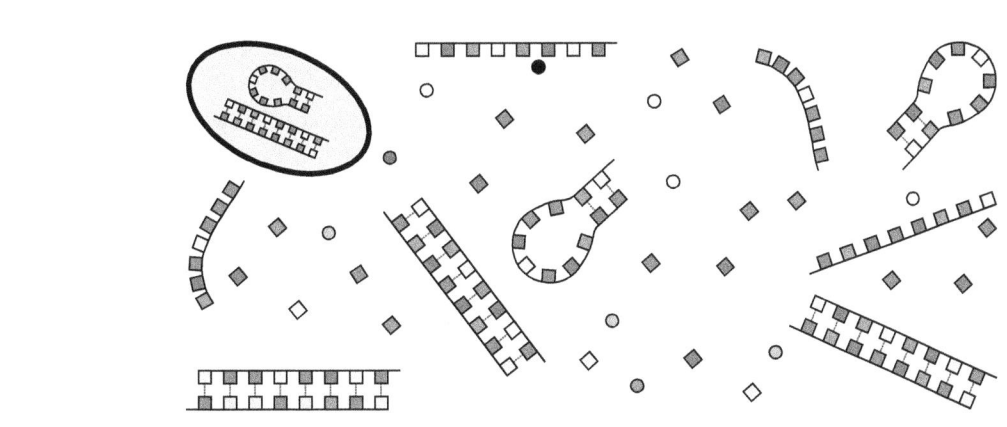

ALL STUDENTS MUST BE ABLE TO ANSWER THESE QUESTIONS

1. Why was the idea of spontaneous generation rejected?
2. How is our current understanding about the origin of life shaped by the Pasteur and Miller-Urey experiments?
3. What is the time-line for geologic and geochemical events leading up to the evolution of life?
4. Why did abiogenesis require a global cooling period and a reducing (oxygen-free) atmosphere?
5. How did natural selection on inorganic molecules lead to the RNA protocell?
6. What are 3 essential features of the first heritable molecule?
7. What is the RNA world hypothesis?
8. What makes DNA more structurally stable than RNA?
9. Based on our understanding of molecular biology, why is RNA (rather than DNA or protein) a better candidate for the first heritable material?
10. How does the Miller-Urey experiment support the RNA world hypothesis and the organic soup model?
11. What are the supporting geologic evidence for the models and hypotheses about the origin of life and the common ancestry of all know life forms?

ALL STUDENTS MUST BE ABLE TO COMPLETE THE FOLLOWING TASKS

1. Use diagrams and supporting evidences to explain the RNA world hypothesis.
2. Analyze a provided set of data to determine whether it is consistent with the RNA world hypothesis and the known history of Earth.

 Big Idea 1D1, 1D2

SPONTANEOUS GENERATION

The origin of life on Earth is still an unsolved mystery. Until recently, the idea of spontaneous generation - that life can spontaneously and abruptly arise from inorganic sources – was widely accepted. Support came from observations of the sudden appearance of maggots on spoiled meat, or microbes in broth. With the development of the scientific method, scientists used controlled experiments to test the claims of spontaneous generation. Among them was the French microbiologist Louis Pasteur, whose classic pasteurization experiment disproved the hypothesis.

Pasteur's experiment

Pasteur's experiment tested whether microbes can spontaneously appear in broth that was pasteurized and kept free of contaminants. Following pasteurization by heating, the broth was incubated in a gooseneck flask that allowed air in, while trapping microbes and dust in the bend of the neck. After more than a year, the broth remained unspoiled. When the neck was broken to remove the microbe trap, it quickly became contaminated and spoiled. The experiment demonstrated that preexisting microbes were needed to spoil the broth and provided no evidence of the existence of spontaneous generation.

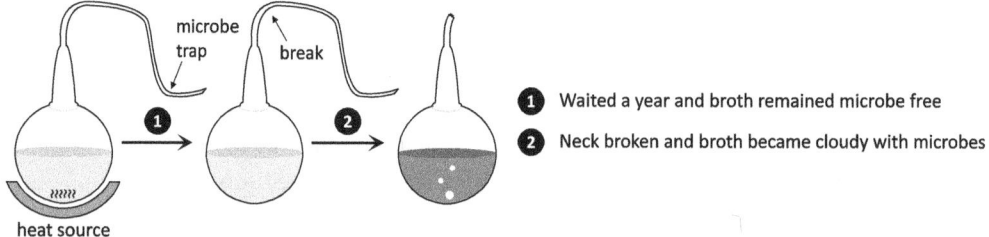

① Waited a year and broth remained microbe free

② Neck broken and broth became cloudy with microbes

EARLY EARTH HISTORY

Information gathered through scientific investigation is used to build a narrative about how life may have originated through the process called **abiogenesis**. Unlike spontaneous generation that is based on the abrupt emergence of advanced living organism, abiogenesis is a natural and gradual process with a sequence of geophysical, chemical, and biological changes that lead to the evolution of the first photosynthetic cyanobacteria from inorganic matter. According to this narrative, Earth formed roughly 4.5 billion years ago and was followed by a cooling event and abiogenesis that produced the first cell a billion years later.

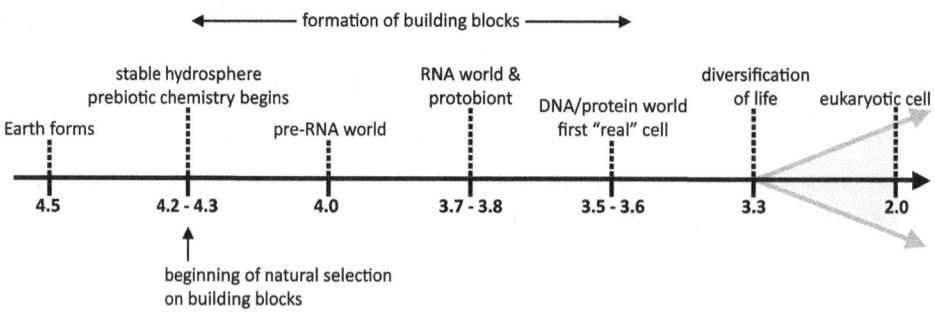

Global cooling event

Why did abiogenesis take a billion years to produce the first cell? The answer to this question is that the early Earth was extremely hot and inhospitable to life. The excess free energy was due the lack of a hydrosphere that protected the Earth from intense solar radiation and heavy asteroid bombardment. Once things settled down and there was a cooling period that allowed prebiotic chemistry to begin forming the inorganic compounds that would eventually combine to produce the organic compounds of life.

NOTES

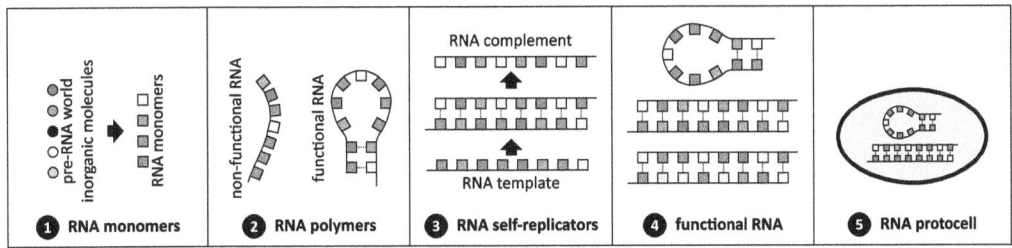

RNA world hypothesis

What were the series of chemical and biological changes that lead to the evolution of the first cell? The RNA world hypothesis is an attempt to answer this question. According to this hypothesis, natural selection began during prebiotic chemistry after global cooling, and progressively lead increasingly complex organization of systems.

Natural selection

Heavy rains and storms dissolved atmospheric gases and returned them to the ocean to create the **organic soup** where prebiotic chemistry occurred. The result was the formation of a wide variety of new molecules, some more structurally stable than others. The stable molecules persisted for longer periods, while less structurally stable molecules quickly degraded.

Evolution of self-replicators

Some molecules were able to form polymers by linking together monomer subunits. The earliest of these was the RNA that was able to store information. It was also able to copy this information by self-replicating, making RNA the first genetic material. In some cases, the information stored in the RNA molecules instructed it fold into shapes that give it special functional properties. The most important functions was an improved ability to respond to the environment (gather information) and to self-replicate. Evidence to support this functional versatility of RNA exists today in a form of enzymatic RNA called **ribozymes** that can catalyze reaction and self-replicate. Continued natural selection led to the emergence and evolution of the first RNA based life form called the protobiont.

The first genetic material

The RNA world hypothesis relies on 3 essential features that must have been present in the first heritable molecule:
- The ability to form **polymers** for information storage purposes
- **Structural stability** to limit the degradation and alteration (mutation) of the

polymer so that any stored information is sufficiently conserved over time.

- **Biochemical functionality**, including the ability to self-replicate so that the information store information is amplified over time.

Why was DNA not the original genetic molecule?

As a double-stranded molecule, DNA is more structurally stable than RNA, and has a greater tolerance for a harsh chemical environments. This makes DNA a better for storing heritable information. However, DNA lacks biochemical functionality, and cannot self-replicate – DNA replication requires the protein called DNA polymerase.

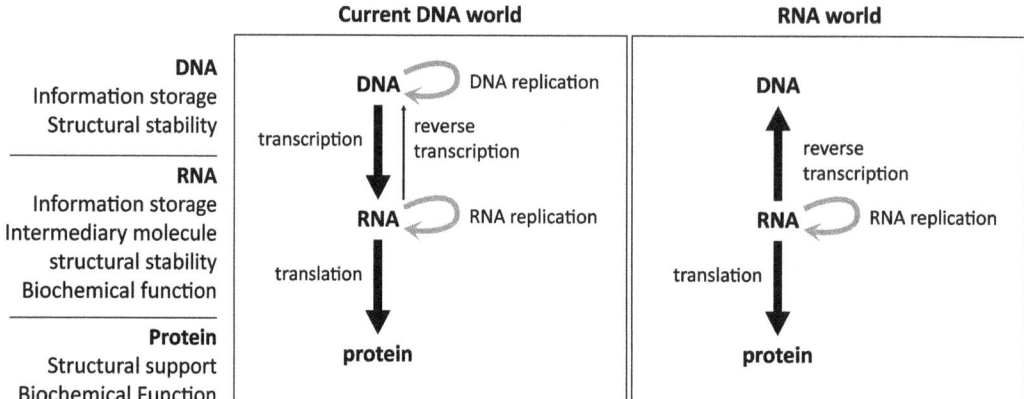

Why was protein not the original genetic molecule?

Proteins are the principal functional molecules in living systems, acting in various life-supporting capacities, including:

- Transport (hemoglobin proteins transports oxygen)
- Structural support (collagen proteins strengthens ligaments)
- Cellular communication (insulin protein regulates blood glucose levels)
- Enzymes (lactase enzyme catabolizes lactose)
- Movement (actin and myosin proteins facilitate muscle contraction)
- Storage (ferritin protein stores iron).

Although they are critical components of all known life forms, protein lack the ability to self-replicate, and are functionally dependent on a 3-D structure that is highly sensitive to changes in temperature and pH.

Why was RNA the original genetic molecule?

RNA possesses all 3 essential features of the original genetic material:

- RNA is functionally active as evident by tRNA (transfer amino acids to ribosomes), rRNA (critical functional component of ribosomes), and RNAi (inhibit gene expression); and RNA can self-replicate as evident by self-replicating RNA called ribozymes.
- RNA is a polymer that can store genetic information (HIV stores its genetic information as RNA, not DNA);
- RNA display structural stability lower than DNA, but higher than proteins. It's intermediate stability allowed to the rapidly evolve in the primordial soup.

During the RNA world, because of it ability to fulfill all 3 roles as a genetic material, RNA was favored by natural selection. When DNA and proteins emerged, these early roles of RNA diminished because DNA is more resistant to mutation and protein is a more functionally versatile molecule.

The evolution of DNA and proteins

DNA and protein likely evolved from reverse transcription (RNA to DNA) and translation (RNA to protein) mechanism (refer to the previous figure). Although displaced from its RNA world roles, RNA has been evolutionarily conserved across all biological domains because neither DNA nor proteins can independently perform all life-sustaining functions. Specifically, the functional roles of mRNA, tRNA, and rRNA remain critical to the flow of information from DNA, the genetic molecule, to proteins, the functional molecule. There is no mechanism for the direct exchange of information between the DNA and protein, thus RNA is evolutionarily conserved to function as an intermediary.

SUPPORT FOR THE PRIMORDIAL SOUP HYPOTHESIS

The idea that organic molecules can form without cellular life is a critical component of the organic soup model and RNA world hypotheses. The Miller-Urey experiment provided empirical evidence to support these two interconnected ideas. By mimicking the conditions of early Earth, it addressed the question of whether prebiotic chemistry could have produced the precursor organic molecules of life in the absence of cells.

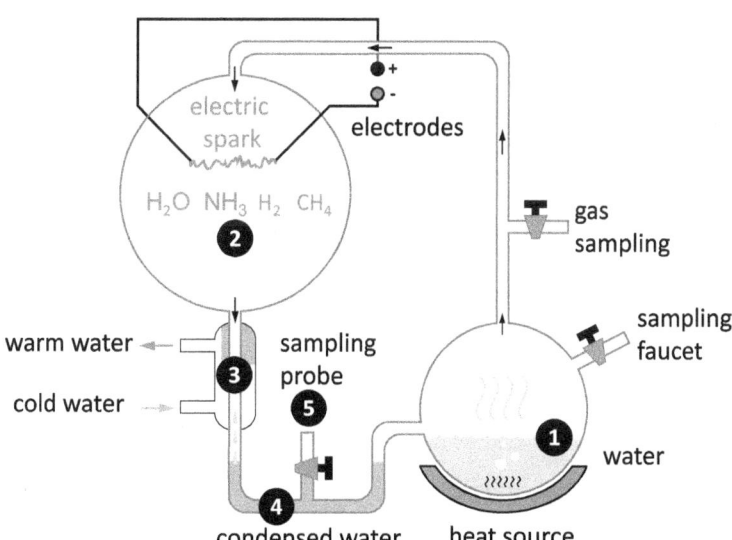

In the Miller-Urey experiment, free energy as heat was used to evaporate water into a chamber containing water vapor, ammonia, hydrogen, and methane – mimicking the early hydrosphere. In this chamber, electrical spark – mimicking lightning - was continuously applied to power endergonic reactions. As water condensed and precipitated, it dissolved molecules from the gas to form the "organic soup." After a week, samples of the condensed liquid of the "organic soup," was analyzed and discovered to contained organic compounds including amino acids. Further investigations have detected nucleic acids.

REDUCING ATMOSPHERE
The early Earth had a reducing atmosphere (comprised of H_2, NH_4, H_2O, and CH_4) that was free of oxygen. Reducing atmospheres are made of substances called reducing agents. Unlike oxidizing agents (ie., oxygen), reducing agents are less reactive and do not typically initiate reactions. If the early atmosphere was oxidizing, many of the newly formed compounds would have been quickly destroyed when oxygen forcibly initiated a reaction with them. In a reducing atmosphere, newly formed compounds were safe and could remain in the primordial soup for longer periods of time before decaying.

The sequence of events leading to the evolution of life

NOTES

ORGANIC SOUP = PRIMORDIAL SOUP
The terms organic soup and primordial soup are used interchangeably.

- Formation of Earth and moon; geologic period of heavy asteroid bombardment and volcanism produce excess free energy.
- Global cooling due to decreased bombardment and volcanism; formation of the early atmosphere with the gases (CH_4, NH_3, H_2O, and H_2) released by Volcanoes; water condensation and precipitation to form seas.
- Prebiotic chemistry initiates the formation of a diversity of inorganic compounds to form the "organic soup."
- Free energy powers the formation of organic compounds using the inorganic precursors of the organic soup.
- Natural selection on organic compounds leads to the evolution of the basic building blocks of life (amino acids, base purines, sugars, fatty acids), and the the self-replication polymers.
- Evolution of first RNA based life form, the RNA protobiont
- Evolution of first prokaryotes (photosynthetic cyanobacteria), followed by prokaryotic heterotrophs.
- Photosynthetic release of oxygen by cyanobacteria transforms the Earth's atmosphere from reducing to oxidizing atmosphere, causing a mass extinction of many prokaryotes. Surviving species evolved adaptations such as cellular respiration that takes advantage of oxygen to harness energy from organic compounds.
- Evolution of the first eukaryotic cell by endosymbiosis.

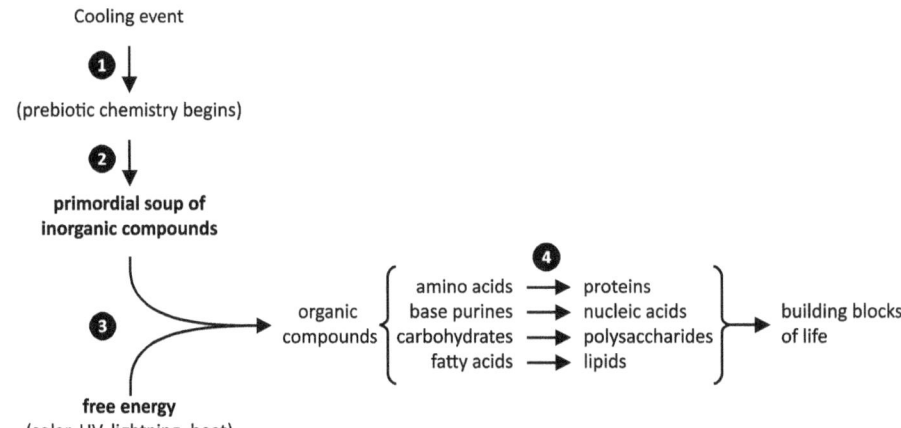

DEEP SEA GEOTHERMAL VENTS

Following prebiotic chemistry, the first organic compound form by an energy absorbing endergonic reaction that occurred in the primordial soup. The reaction that produced this organic compound can be represented as $A + B \rightarrow C$, where A and B are inorganic precursor molecules, and C is the first organic compound.

Land, air, or sea?

This first biochemical reaction would have occurred in a liquid medium because a liquid solvent provides increased opportunity for A and B to diffuse toward each other and come into close contact. Although particles in gases can also diffuse, they tend to be further apart, limiting the chance that the precursor molecules would have interacted. Since water is the most abundant liquid on Earth, and because it is a universal solvent that readily dissolves many other substances, it is likely to have been the solvent of A and B.

At the time when the first organic compound was forming, the Earth had an abundance of liquid water. Because of this large volume of water, the global concentration of A and B would have been negligible. This is because the various bodies of water on Earth are **open systems**, allowing for the free flow of energy and matter. At such low concentrations, even with diffusion, it was improbable that A and B would find each other. Neither **closed** nor **isolated systems** would have been ideal because they would block substances from reaching each other, and isolated systems quickly exhaust any available free energy needed by endergonic reaction.

NOTES

CLOSED & ISOLATED SYSTEMS
Neither closed nor isolated systems would have been ideal for abiogenesis because they block substances from reaching each other, and isolated systems quickly exhaust any available free energy needed by endergonic reaction.

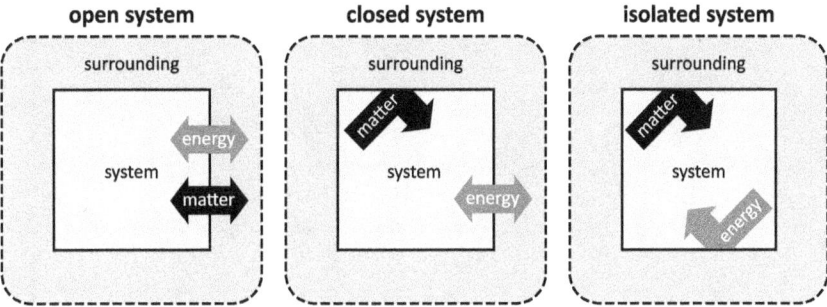

What was needed was a local body of water that had direct access to a free energy source and precursor inorganic compounds (A and B), and that was semi-closed to limits the outward diffusion of energy and matter. Under these conditions, the substances A and B would have been maintained at a high local concentration that favored the formation of the first organic compound.

The surface of the oceans has access to lots of free solar energy, but it lacks geologic barriers to diffusion and direct access to sources of inorganic precursors. **Deep-sea hydrothermal vents** have direct access to geothermal free energy from the Earth, they produce and release lots of precursor inorganic compounds that can form organic molecules, and they have geological vents that limit diffusion and allow local concentrations of precursor compounds to build-up.

SUMMARY OF KEY CONCEPTS

- Although Earth formed 4.6 billion years ago, the first cell (photosynthetic cyanobacteria) did not appear until a billion years later (3.6 billion years ago) after the environment cooled and the Earth became more hospitable for life.
- Early Earth had a reducing oxygen-free atmosphere with excess free energy.
- The organic soup contained the inorganic precursor molecules that were needed to form the building blocks (amino acids, nucleotides) for organic molecules (DNA, RNA, proteins, carbohydrates). This hypothesis is supported by the Miller-Urey experiment.
- RNA may have been the earliest genetic material formed by nucleotide monomers linking up to form polymers that could store information, were biochem-

ically functional, and could self-replicate. As a single stranded molecule RNA has relatively mutation rates (as compared to double stranded DNA), resulting in the RNA self-replicators undergoing rapid evolution.

■ These early reactions may have occurred in a solution (organic soup model) where some self-replicators became enclosed in a lipid capsules to form the RNA protocell, and then evolving to the first prokaryotic cell.

■ Deep-sea hydrothermal vents are ideal locations for the formation of the first organic compounds and abiogenesis.

CHECK YOUR UNDERSTANDING

1. Based on the current body of information, a key component of the early Earth that led to the emergence of life was

   (A) an excess of free energy.
   (B) an oxygen-rich atmosphere.
   (C) the natural selection on inorganic molecules.
   (D) the presence of deoxyribonucleotides.

2. Which of the following statements correctly reflects components of the RNA world hypothesis?

   (A) DNA appeared before RNA.
   (B) Natural selection favored self-replicating polymers.
   (C) Proteins are the only organic molecules that can catalyze reactions.
   (D) Reverse transcription of DNA, led to the emergence of RNA.

3. Which of the following provides support for why DNA is not the first heritable molecule?

   (A) DNA can undergo replication.
   (B) DNA is more structurally stable than RNA.
   (C) DNA is unable to catalyze biochemical reactions.
   (D) DNA has a high rate of mutation, relative to RNA.

4. Which of the following provides support for why RNA is likely the first heritable molecule?

   (A) Ribozymes can self-replicate.
   (B) Different forms of RNA (tRNA, rRNA, mRNA) each serve a unique function in protein translation.
   (C) RNA is less reliable than DNA at storing genetic information.
   (D) All of the above.

5. Which of the following can store and pass on genetic information?

   (A) DNA
   (B) DNA and RNA
   (C) Proteins
   (D) All of the above

6. Which of the following statements best explains why RNA has been evolutionarily conserved across biological domains?

   (A) Ribozymes are critical components of all cell types and is necessary for replication of regulatory RNA molecules.
   (B) The genetic information stored in DNA must be transcribed to RNA before translation into a biochemically functional protein.
   (C) RNAi can inhibit gene expression in eukaryotic cells
   (D) As a polymer, RNA can also store genetic information.

7. Which of the following statements about the organic soup model is incorrect?

   (A) It is based on the assumption that global cooling and the onset of prebiotic chemistry occurred before the formation of inorganic compounds.
   (B) It assumes that the formation of the organic soup required free energy from the environment.
   (C) The molecules that made up the organic soup were formed when available free energy caused naturally occurring molecules to break-down into their basic component parts, such as base purines, amino acids, sugars, and fatty acids.
   (D) All of the statements are correct.

8. Deep sea geothermal vents are ideal locations for the emergence of the first life forms because

   (A) They are extremely hot environments.
   (B) They are fully opened systems where energy and matter are easily and freely exchanged with other systems.
   (C) They have direct access to free energy released by the Earth's internal geologic activity, and lots of inorganic molecules that can be used to form precursor organic compounds.
   (D) All of the above.

ANSWERS AND EXPLANATIONS

1. (C) Natural selection initially favored stable inorganic molecules that could better withstand the harsh conditions of the early Earth environment. Of these, those that formed polymers and had self-replicating capabilities, and could therefore store and pass on information, were most favored. These self-replicating polymer became the first heritable molecules - RNA a surviving member of this class of molecules. (A) is incorrect because an excess amount of free energy was a barrier to the emergence of life. Even the most stable inorganic compounds cannot maintain structural stability once a threshold tolerance level for free energy is exceeded. (B) is incorrect because an oxygen-rich atmosphere would have strong oxidizing properties. This means that newly formed molecules will be rapidly destroyed by the highly reactive oxygen molecules, making it nearly impossible for life-sustaining organic compounds to form. In today's oxygen-rich atmosphere, living organisms can survive because they have had billions of years to evolve special adaptations that take advantage of the high reactivity of oxygen. (D) is incorrect because, although DNA is critical to life as we know it, the first protocell relied on RNA, not DNA.

2. (B) See previous question for explanation. (A) is wrong because RNA evolved first, not DNA. (C) is wrong because ribozymes, which are catalytic RNA, have been proven to catalyze reactions, including self-replication. mRNA, tRNA, rRNA, and RNAi are also all different types of functional molecules, perform specific life-supporting tasks within cells. (D) is wrong for two specific reasons: DNA did not lead to the emergence of RNA and, reverse transcription produces DNA from RNA, not RNA from DNA - direct transcription produces RNA from DNA.

3. (C) The first heritable molecule needed to be able to catalyze biochemical reactions, specifically self-replication. DNA is unable to self-replicate. (A) is incorrect because, although DNA can replicate, the process requires special enzymes called DNA polymerase. Although (B) is an accurate statement, it is not relevant to answering the question. In fact, the lower structural stability of RNA relative to DNA, may have allowed RNA to evolved and diversify more quickly to give rise to the more stable DNA molecule. (D) is an inaccurate statement since RNA mutates faster.

4. (D) Both (A) and (B) highlight the biochemical functionality of RNA. Although proteins are the primary functional molecules, cells still rely on RNA to perform critical life-supporting functions, including protein translation and gene regulation. The first heritable molecule would have needed similar functional capabilities. (C) is also correct because, according to the RNA world hypothesis, DNA has replaced RNA as the primary heritable molecule. In order to natural selection to favor DNA over RNA as a heritable molecule, it must have improved information storing capabilities, which comes from enhanced structural stability due to DNA's double-strandedness.

5. (B) is correct. Although, as polymers, proteins, RNA, and DNA can all store information, only DNA and RNA can serve as templates for replication. To be considered heritable, a molecule must not only store information, it must also undergo replication so that its information can be passed into the future. (C) is incorrect because proteins cannot directly pass on information via replication, and are therefore not heritable molecules.

6. (B) is correct. It is the main premise of the central dogma of molecular biology, which describes the flow of information from DNA to RNA to protein. Although (A) may also be correct since there is some evidence that ribozymes are important components of ribosomes and may be involved in linking amino acids together, its exact role has not been fully described. Furthermore, the second part of statement (A) is wrong. (C) is an accurate statement, but it is irrelevant to the question. (D) is also an accurate statement, but the fact that RNA can store genetic information is not relevant to its evolutionary conservation since DNA is better at storing information.

7. (C) is an incorrect statement about the organic soup hypothesis. According to the organic soup hypothesis, the available free energy was used to synthesize precursor molecules that underwent further endergonic reactions to form organic compounds.

8. (C) is correct. Remember that the emergence of life required free energy and inorganic molecules. (A) is incorrect because an extremely hot environment create unstable conditions that are unfavorable to life. (B) is incorrect because the most ideal environment for the emergence of life would have been one that had access to free energy, and a source of precursor inorganic molecules, and that was a semi-closed so that the precursor molecules could concentrate.

02 Origin of Species

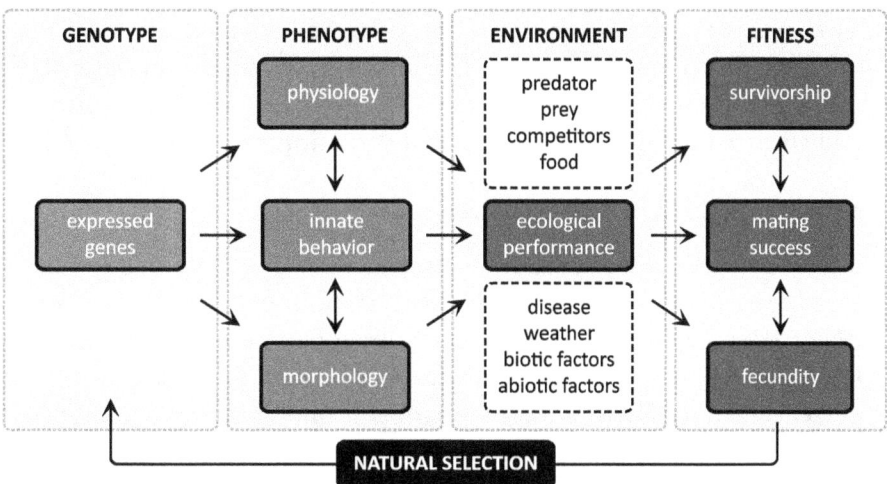

ALL STUDENTS MUST BE ABLE TO ANSWER THESE QUESTIONS

1. What are different ways of defining evolution?
2. How has empirical evidence from geology and the fossil record been used to challenge the traditional views about the origin of species?
3. Describe Lamarck's hypothesis on evolution, including its two principles? What was the primary flaw(s) with this hypothesis?
4. What were Darwin's 5 observations and 3 inferences that formed for basic framework of his theory on evolution by natural selection?
5. What does the term "fitness" mean?
6. How does genetic variation and mutation influence natural selection?
7. How can changes in the environment influence the rate and direction of evolution in a population?
8. How has Darwin's theory been updated to accommodate our understandings about genes and its influence on individual fitness?
9. What is an adaptation, and how does it influence natural selection?
10. What are similarities and differences between natural and artificial selection?
11. How do humans use artificial selection to impact variation in other species?

ALL STUDENTS MUST BE ABLE TO COMPLETE THE FOLLOWING TASKS

1. Explain how the evolution of a population or species is connected to changes in the environment.
2. Analyze a provided set of data to identify the role of natural selection in the evolution of a population.

 Big Idea 1A1a-e, 1A2d

EVOLUTION IS
A process that results in heritable changes in a population over many generations.

A change in the allele frequencies of a gene pool from one generation to the next.

A change in the inherited characteristics of populations over multiple generations.

A descent with modification.

The descent of different species from a common ancestor over many generations.

EVOLUTION IS NOT
The gradual process that produced the present diversity of plant and animal life from the earliest and most primitive organisms.

The change in an individual over time.

A way to explain how higher forms of life have gradually arisen from lower forms

A change in the physical characteristics of an individual over its lifetime.

The direction of evolution refers to how natural selection favors phenotypes over time. For example, if the population evolves over multiple generation to become progressively taller, the direction of evolution is toward the taller phenotype.

WHAT IS EVOLUTION?

Evolution is a process of change in the observable characteristics of a population over time. Evolution always refers to the collective population or species, not to individuals. This means that evolution never occurs at the individual level. Instead, it requires an interbreeding population of multiple individuals that pass on the traits through genetic inheritnce to descendant populations. By comparing the characteristics of descendants and ancestral populations, the rate and **direction of evolution** can be determined.

CHALLENGES TO TRADITIONAL VIEWS

The question of the origin of life remains unresolved, although every society and culture have tried to address it. Some use creation myths and folk tales, while other use philosophical thought and the scientific method to explore ideas about life's origin. In line with creation myths of his time, Aristotle described species as being fixed in their current state, unable to change. In modern times, many religions still share this belief in a perfect creation that occurred about 4000 years ago. Carolus Linnaeus who is considered the father of modern taxonomy, described the unique adaptations of species as evidence of a divine creation, with each species having been created to the specific purpose for which they are so well adapted.

The fossil record supports evolution

There are vast amounts of scientific evidence to support the theory of evolution. The most valuable evidence is the distribution of fossilized remains of dead organisms in various layers of sedimentary rocks (or **strata**). The fossil record is analogous to snapshot photographs of organisms that lived during various time periods of Earth's history. Recovered fossils can be dated and compared morphologically with both **extinct** and **extant** species to determine their evolutionary relationships. For example, with age estimates of fossilized remains of extinct primates, coupled with morphological comparison, scientists are able to map out the evolutionary history of our species (*Homo sapiens*), and show possible relationship with extinct species such as *Australopithecus, Homo Habilis, Homo erectus,* and *Neanderthal.*

Catastrophism and mass extinction events

During his study of strata layers, the geologist George Cuvier observed that each boundary between geologic strata contained different types of fossils. Based on this observation, he wondered what would have caused the abrupt changes in the types of living organism from one time period (represented by a particular strata) to another time period (represented by an adjacent strata). He suggested that catastrophic events were responsible for mass extinction between the periods of the strata deposition. This became the foundation of the idea know as Catastrophism, which highlights how the environment can pressure species to change (evolve) or go extinct. Catastrophism suggests that with abrupt and intense environment changes, species tend to go extinct as the lose that adaptation to the local environment.

Gradualism, uniformitarianism and the ever-changing Earth

The geologists James Hutton and Charles Lyell, challenged the view that the Earth and universe was only thousands of years old by explaining how the geologic changes such as erosion, mountain formation, and sediment deposition happen through slow continuous actions (**gradualism**) that are still operating today (**uniformitarianism**). Since gradualism and uniformitarianism require time on the scale or millions or billions of years, Earth too is likely millions or billions of years old.

LAMARCK'S HYPOTHESIS OF EVOLUTION

One of the earliest attempts to explain how species might evolve was proposed by the French naturalist, Jean Lamarck. Lamarck's hypothesis of evolution was based on two principles: use and disuse, and transmission of acquired traits.

The **principle of use and disuse** was built on the idea that organism changed in respond to their environment. The change was based on how much an organism used a specific organ or body part. Increased use caused the organ to change, becoming more finely adapted to the specific task that the organism needed to be successful in its environment. Less use (disuse) would result in a loss of functionality, and eventual loss of the organ.

According to the **principle of transmission of acquired traits,** the changes in the organism could be acquired in its lifetime and transmitted directly to any offspring through inheritance. Accordingly, when the ancestral giraffe's environment changed those giraffes that stretched their necks to reach the leaves of the trees had access to more food resulting in their neck changing to become longer. According to Larmarck's hypothesis, these acquired characteristic could be transmitted to their offspring.

DARWIN'S THEORY OF EVOLUTION

Charles Darwin developed the theory of evolution by natural selection based on his observations of biodiversity and biogeography as he traveled around the world on the, and his reflections on the work of other naturalists.

Darwin's observations

During his travels, Darwin made 5 key observations about the organisms that he observed in nature:

1. Species tend to produce more offspring (**over-reproduction** or **super fecundity**) than the environment can support.
2. In various natural communities, there tends to be a **steady population size**.
3. Each environment has **limited resources** available to the population.
4. There are individual variations in the traits observed in each population.
5. The observable traits are passed down by **inheritance** from parent to offspring.

Darwin's 3 inferences

Darwin used his 5 observations to make 3 inferences about how evolution by natural selection might occur.

1. Super fecundity and a population kept in check (steady population) by limited resources, causes **intense competition** among individuals.
2. Individuals whose inherited traits make them better adapted to the environment (individual variation), will tend to out compete others for the limited resources, and will have increased survival and reproduction (**survival of the fittest**).
3. The unequal ability of individuals to survive and reproduce will lead to the accumulation of favorable traits in the population over successive generations (**evolution of time**).

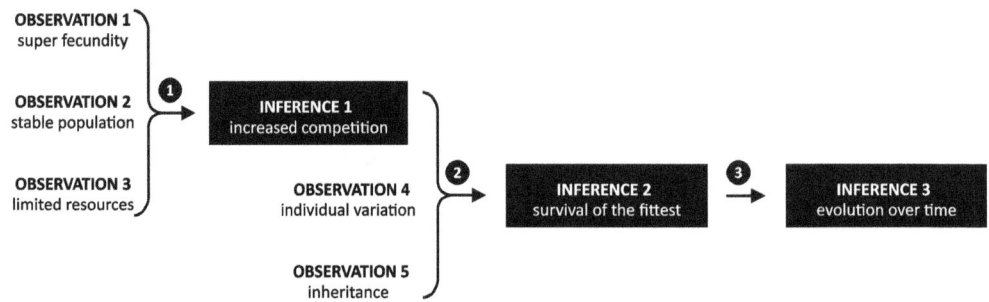

INTRODUCTION TO NATURAL SELECTION

Darwin's observations and inferences form the basis of his grand idea of evolution by natural selection. He imagined a world in which organisms were in a constant struggle for their existence. How successful an individual was at surviving depended on inherited characteristics. He therefore inferred that there were something passed from parent to offspring, although he knew nothing about DNA and genes.

Today the theory of evolution by natural selection has been updated to include new information about the nature of biological systems. As illustrated in to following figure, the modern view of evolution incorporates our understanding that **genes** carry genetic information from parent to offspring. The expression of these genes determines physical characteristics (**phenotype**), including the organism's **physiology, morphol-**

ogy, and **innate behavior**. These 3 factors then influence **ecological performance**, and ultimately, **fitness** – the ability to survive, attract a mate, and reproduce. Natural selection favors the genes of the most reproductively successful individuals because it is through reproduction that genetic inheritance is passed from parent to offspring.

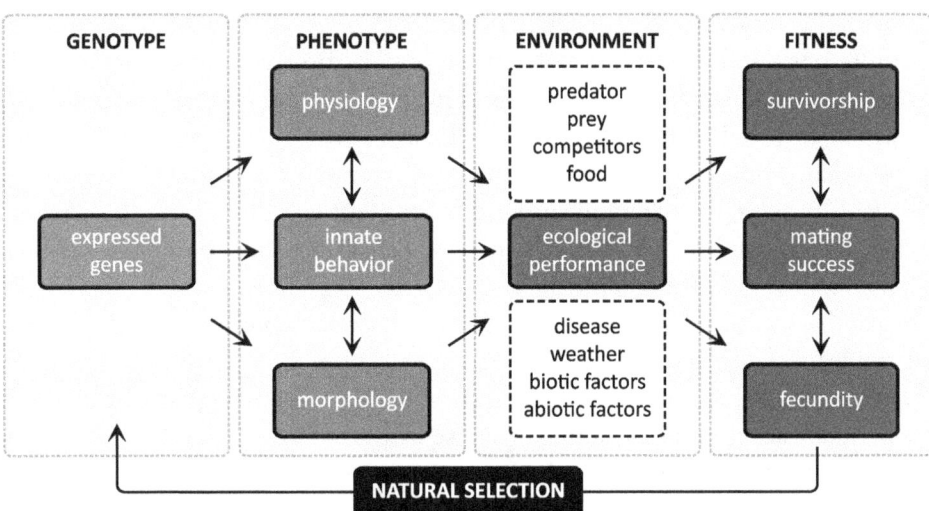

Natural selection versus artificial selection

Natural and **artificial selection** use similar mechanisms to drive evolution. Both require a diverse population with individuals expressing different combination of traits. This variation in population comes from an accumulation of mutations over time.

In natural selection, the environment provides the selective pressure that favors those individuals with inherited adaptations that enhance their fitness. But, in artificial selection the human breeder provides the selective pressure, favoring those individuals with the breeder-preferred traits such as taste, oil content, texture, and color.

The outcome in both cases is a less diverse population with reduced genetic variation in the gene pool of successive generation. Natural selection causes extinction, or **adaptive radiation** and **punctuated equilibrium** to produce extant wild species, while artificial selection produces domesticated species with characteristics that benefits the human population.

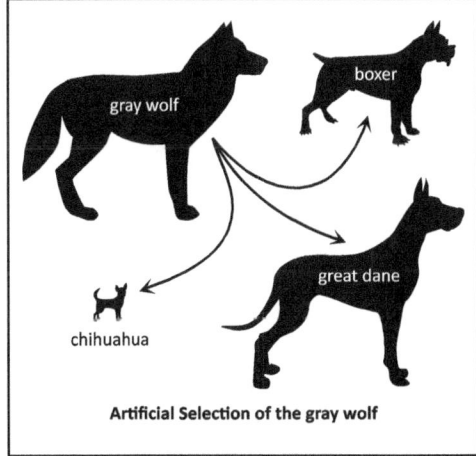

Artificial Selection of the gray wolf

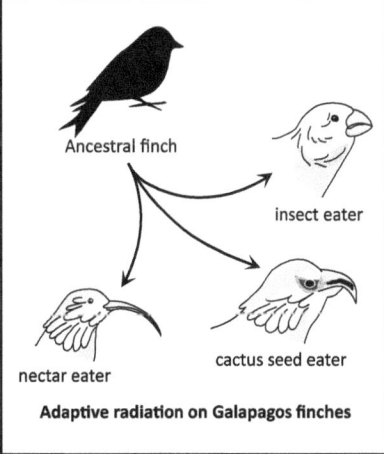

Adaptive radiation on Galapagos finches

SUMMARY OF KEY CONCEPTS

- Evolution is the change in the genetic makeup of a population over time.
- Limited resources, competition, super fecundity, individual variation and inheritance are all factors that influence an individuals fitness, and evolution.
- Fitness is a measure of survival and reproductive success.
- Natural selection requires individual variations in the population and a diverse gene pool.
- The variation of a gene pool comes from the accumulation of mutations over time.
- Over time, a particular population will become well adapted to an environment that remains stable. As the population adapts to the environment, it's rate of evolution will decrease.
- In an unstable environment, any changes in the environment is a disruption the equilibrium and can cause the population to lose its adaptive advantage and either readapt to the new environment by evolving, or risk going extinct.
- A more genetically diverse population is less vulnerable to catastrophic events (changing environment) that can lead to extinction.
- Humans can impact variation in the gene pools of other species through artificial selection, which reduces genetic variation. For example, the use or pesticides and antibiotics introduces a selective pressure on targeted pests or pathogens that reduces their genetic variation.

CHECK YOUR UNDERSTANDING

1. Which of the following statements is not an example of evolution?

    (A) A population of birds has an average beak length that is 0.5 inches longer than an ancestral population.

    (B) A group of individuals that have gained an additional 20% muscle mass over a period of 6 months of weight lifting.

    (C) A species of monkeys that have acquired a relatively shorter tail length than all other closely related species of monkeys.

    (D) A population of African-American shows a significantly lower incidence of the sickle cell allele relative to all West African populations.

2. Which of the following can be used to challenge the view that each organism is a perfect creation and never changes?

    (A) Fossil distribution in strata.

    (B) The formation of mountains.

    (C) Catastrophic events such as a volcanic eruption.

    (D) Observations of gradualism and uniformitarianism.

3. Which of the following statements correctly reflects the assumptions of Lamarck's hypothesis?

    (A) A population of giraffes that is preyed on by lions will become progressively faster runners as the faster giraffes pass their genes to their offspring.

    (B) A weight-lifter's son will be born with increase muscle mass relative to other newborns.

    (C) Both are correct.

    (D) Neither are correct.

4. Which of the following is not part of Darwin's theory of evolution by natural selection.

    (A) Populations tend to have high infant mortality rates.

    (B) Population sizes tend to fluctuate.

    (C) The environment tends to have a steady supply of resources.

    (D) Individuals in a population tend to differ.

5. Competition among individuals of a population is driven by

    (A) resource limitation.

    (B) physiology and morphology.

    (C) the ability to attract a mate.

    (D) All of the above

6. Ecological performance is influenced by

   (A) local predators.
   (B) parasites and disease.
   (C) expressed genes
   (D) All of the above.

7. Which of the following statements is wrong?

   (A) Newborn suckling in mammals is an innate behavior.
   (B) Mating success refers to the ability of individuals to produce offspring.
   (C) Fitness is only counted for individuals who have offspring.
   (D) None of the statements are wrong.

8. Which of the following statements refers to artificial selection, but not natural selection.

   (A) It requires a diverse starting population.
   (B) It relies on mutations.
   (C) The selective pressure comes from the breeder.
   (D) All of the above.

ANSWERS & EXPLANATIONS

(B) is correct because traits that an individual acquire during their lifetime such as increased muscle mass from weight-lifting cannot be transmitted to offspring. Choice (A) an example of evolution because it is describing changes in the characteristics of a population relative to its ancestral population. Choices (C) and (D) are inferring that populations have changed from an ancestral population, assuming that the other descendant groups of long-tailed monkeys and West Africans, respectively, are displaying a characteristics present in the ancestors.

(A) is correct since different strata should has been shown to contain contrastively different types of fossils, suggesting that species have changed, or gone extinct. The other three choices are relevant for geologic events and, although they influence evolution, do not directly involve living organisms.

(B) is correct because is reflect Lamarck's idea that an acquired trait can be passed on to offspring. Choice (A) reflects Darwin's theory of natural selection since the faster giraffes will have a better chance of surviving. This will also increase the fecundity, allowing them to pass on their fast-running genes to their offspring.

(B) is correct because Darwin observed that the size of populations tended to remain steady. Choice (A) is the same as the observation that populations tend to display superfecundity. Choice (C) refers to limited resources, and choice (D) to individual variation in the population.

(D) is correct.

(D) is correct.

(B) is correct because mating success does not equate to reproductive success. Mating success only refers to the ability to attract a mate. However, not all copulation attempts will result in reproductive success. Choice (A) is an accurate statement because newborn mammals suckle without being taught. Choice (C) is an accurate statement because fitness requires survival, mating success, and fecundity (reproductive success). If either of these are missing, fitness = 0.

(C) is correct.

03 Evidence of evolution

 Big Idea 1A4,1B1a

Support for evolution comes from various scientific disciplines, including biogeography, geology, archeology (fossil record), comparative anatomy (homologous and analogous structures, and vestigial organs), genetics, embryology, and taxonomy.

BIOGEOGRAPHY

Biogeography is the study of the **geographic distribution** of **extant** and **extinct** species. It is used to evaluate the claims of the evolution by examining organisms and their habitats in order to discover links between the adaptable traits that organisms display and the environments to which they are adapted. While the distribution of extant species can be observed today, in order to study extinct species scientist must turn to the fossil record and geology to gather information about the past biodiversity, and climate conditions, respectively.

Why consider geology?

Geology is the science of the physical properties of the Earth. It focuses on how rocks form and undergo change over time. It explains the **continental drift** of **Pangaea** to form the current landmasses, **tectonic uplift** to form mountains, and asteroid bombardment that caused past climate change and mass extinctions. Simply put, geology is critical to our understanding of Earth's history, and our predictions of future patterns of evolutionary change. Through its influence on climate (temperature, water distribution, and atmospheric pressure), changes in the physical nature of Earth can cause the environment to shift in such a way as to pressure species to adapt, migrate, or face extinction.

Pangaea

Pangaea was a supercontinent that formed about 300 million years ago, but has since drifted apart to form the major landmasses of today. Species living on Pangaea could migrated across **continental plates** without the need to cross oceans. Evidence for this is found in the fossil record, which confirms that Pangaea had the 4 ancestral species of all living reptiles, mammals, and terrestrial plants. For example, fossils from geographic regions spanning the oceans proves that *mesosaurus* lived in both South America and Africa. As a freshwater reptile, it could not survive a swim across the salty Atlantic ocean. This suggests that the two land masses was once connected, and that after they separated and environmental conditions changed, sub-groups of *wmesosaurus* underwent evolutionary changes to adapt to their unique local environments.

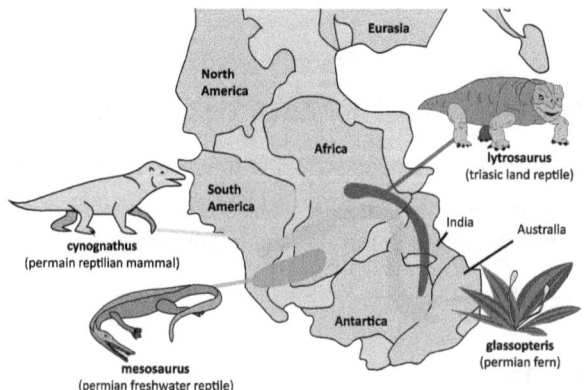

NOTES

BIOGEOGRAPHY QUESTIONS

How did certain ecosystems and species emerge at specific geologic times?

Where have various types of ecosystems and species existed, and what environmental conditions supported their emergence?

How did certain species become disperse in their specific habitats, ecosystems, and geographic locations?

What events may have caused specific ecosystems or species to go extinct?

What does the current and past patterns of geographic distribution of species tell us about evolution?

Are the adaptable physical characteristics of organisms evolutionarily predictable for their local geography and climate conditions?

Migration

As a population splits and subgroups migrate to different geographic regions, they experience different environmental conditions that provides the selective pressure for natural selection. For example, the fossil record shows that the ancestor of all extant species of the camel family including the South American llama, Asian Bactrian camel, and the common Arabian camel originated in South Dakota about 50 million years ago. The population split, with a subgroup migrating to South America, and another across the Bering land bridge to Asia. The Asian population split again, with a subgroup continuing on to Arabia and North Africa.

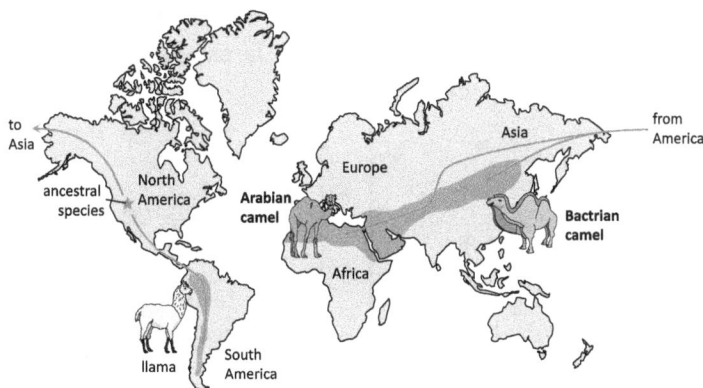

The new environments, being different from South Dakota and from each other, provided varying selective pressures for the evolutionary **divergence** of these closely related species. Existing phenotypic differences reflect each subspecies' unique adaptations to their local environments. For example, the Bactrian camel is adapted a colder regions, and therefore evolved with a thicker fur than the Arabian camel.

Marsupial distribution

The fossil record shows that **marsupial mammals** originated in North America and later extended their range to become the most successful globally distributed mammals. They reached Australia through the Antarctic land bridge prior to continental drift and Antarctic glaciation. When the more competitive **placental mammal** evolved and expanded its range, the continents had already separated. Except for the isolated (and therefore protected) Australasian marsupials, placental mammals drove all other species of marsupials extinct (except the North American *opossum* and the South American *monito del monte*).

COMPARATIVE ANATOMY

Comparative anatomy is the study of the similarities and differences in the physical structures of organisms. It is used to evaluate the claims of evolution by comparing distantly related species that **converged** with superficially similar **analogous structures** adapted to similar function, or closely related species that **diverged** with functionally different **homologous structures** that share anatomical similarities. As illustrated in the figure below, an ancestral lineage (A) can split when subgroups (B and C) migrate or begin occupying different niches. As the groups adapt to the unique niches, they diverge from each other. Likewise, when unrelated species (A1 and B1) begin occupying similar niches, the can converge with similar body structure that are adapted to serve similar functions. Although the descendants (A2 and B2) might look similar, they remain separate species.

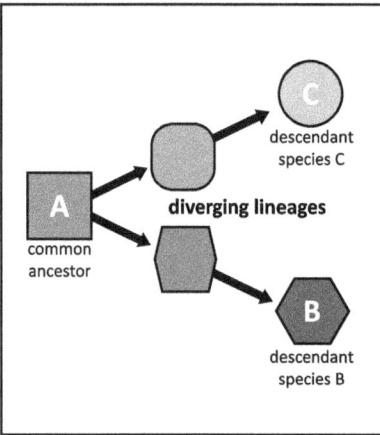 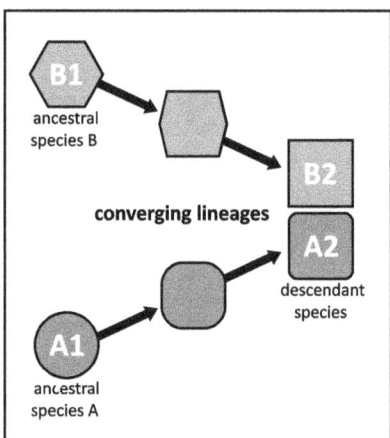

Homologous structures

When comparing closely related organisms that are adapted to different **habitats** (i.e., marine and terrestrial mammals), some structures that originally appeared in their common ancestor may diverge to serve different functions. For example, the human hand, whale flipper, and the cats front legs are derived from the forelimbs of a common mammalian ancestor. However, because each organism is adapted to a different niche, these body structures have change to serve appropriately serve each in its respective **niche**. Although the structures appear uniquely different from each other, this difference is only superficial as they have clear observable similarities in their skeletal anatomy. Body structures that serve different functions but have a common ancestral origin are called homologous structures.

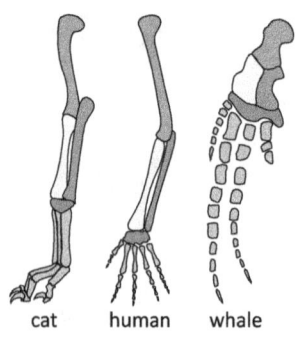

cat   human   whale

Vestigial organ

Some homologous structures lose their function to become **vestigial organs**, which non-functional remnants of structures that were functional in an ancestral species. Vestigial organs evolve when a selective pressure reduced its benefit. The human appendix is a nonfunctional homologue to the lumen of large herbivores, and the non-functional whale pelvic bone is homologous to the pelvis terrestrial vertebrates that use it for locomotion.

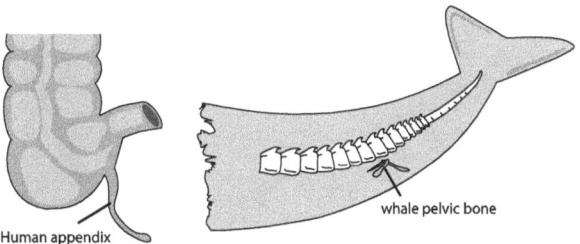

Human appendix   whale pelvic bone

Analogous structures

When comparing distantly related organisms that are adapted to similar habitats, it is common to observe superficial similarities in their physical structures. For example, the penguin (bird), dolphin (mammal), and shark (fish) are distantly related, but they have structure - wings, flippers, and fins - that look similar. Each of these structures have an independent evolutionary origin, as is evident by differences in their underlying skeletal anatomy. As modified forelimbs that are adapted for swimming, they are functionally similar analogous structures. The wing and flipper likely arose through convergent evolution when the ancestral species of the dolphin and penguin moved from land to water, occupying a habitat similar to the shark

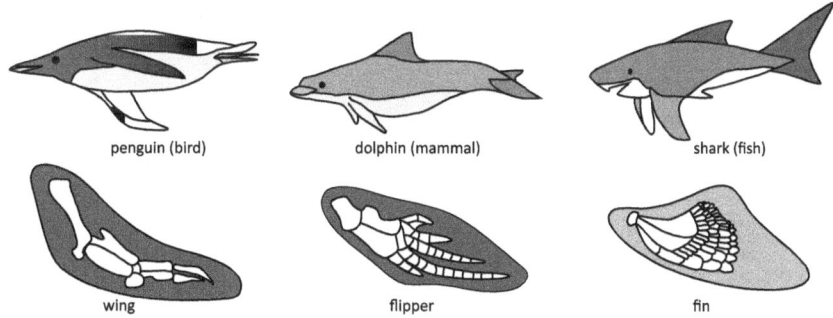

penguin (bird)   dolphin (mammal)   shark (fish)

wing   flipper   fin

GENETIC HOMOLOGY

Identical twins share 100% of their DNA sequence and are most closely related to each other than to any other. Less related organisms have lower degree of homology in their DNA or protein sequences. Comparison of organisms based on the similarity and difference in DNA or protein is referred to as **genetic homology**, and is more reliable than those based on comparative anatomy. This is because unrelated anatomical or morphological structures that are analogous, can be mistaken for structures derived from a common origin. This can lead to incorrect grouping of species and their placement on the **phylogenetic tree**.

The illustration below is called a cladogram is based on protein homology data of human hemoglobin protein relative to that of 5 other species. Of these, the macaque has the highest genetic homology with only 8 amino acids differences, while the lamprey has the lowest with 125 differences. It can therefore be concluded that the order of relatedness of these species to human is macaque, dog, bird, frog, and lamprey.

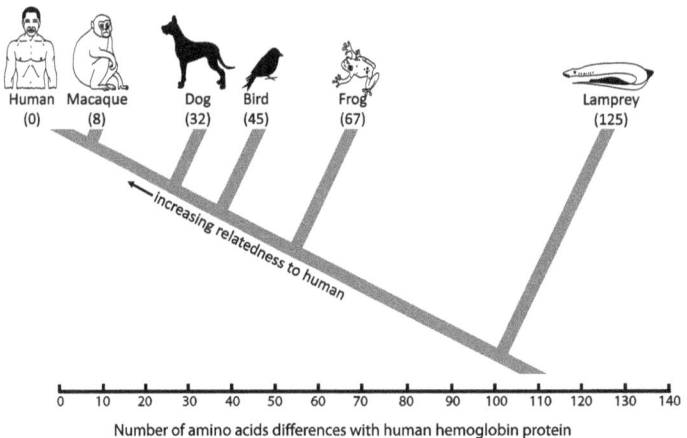

Number of amino acids differences with human hemoglobin protein

TAXONOMY

Taxonomy is the method of classifying organism based on their evolutionary relationship, with the related organism assigned to the same groups called **clades**. Traditionally, organisms were assigned to taxonomic clades based on morphological similarities their structures. However, as already explained, analogous structures due to convergence can complicate this method, leading to incorrect grouping. Likewise, the divergence between two closely related species can result in homologous structures that are so different in their morphology that they get mistakenly assigned to different clades.

In addition to traditional methods, modern taxonomy takes advantage of genetic and protein homology, and physiology to ensure proper grouping of organisms. Once comparisons are made, clades are used to construct, or reconstruct, branches of the phylogenetic tree (cladograms). The cladogram above is based on protein homology from a single hemoglobin gene. The one below shows the taxonomic relationship between the three extant branches of mammals and the distinguishing **derived characters** that separate the lineages. It is based on comparative anatomy that has been confirmed with genetic homology.

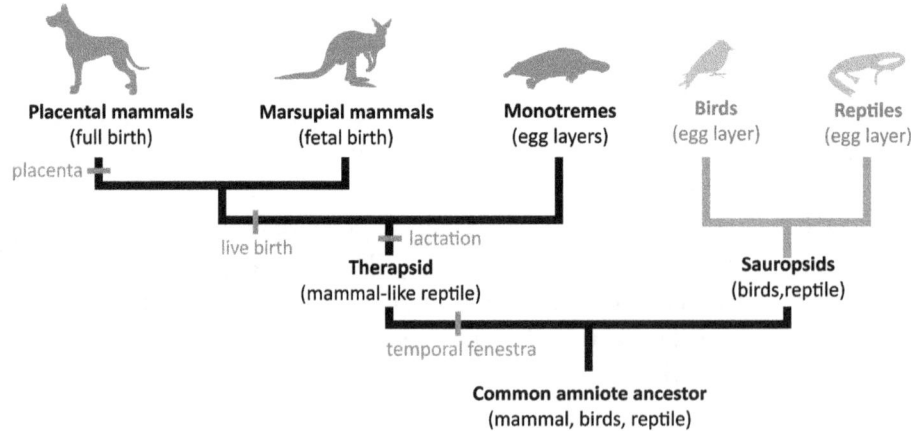

Challenges with classification based on morphology

Marsupials underwent adaptive radiation in Australia, evolving into the largest diversity of extant marsupial species. Although marsupial and placental mammals are of divergent genetic lineages, similar ecological niches between Australasia and other geographic locations inhabited by placental mammals led to the morphological convergence of species from the two lineages. The modified cladogram below shows examples of species that, although distantly related, have similar adaptive morphology. Based on morphologic classification alone, the flying phalanger and the squirrel could be incorrectly placed in the same clade, instead of the phalanger and Tasmanian wolf.

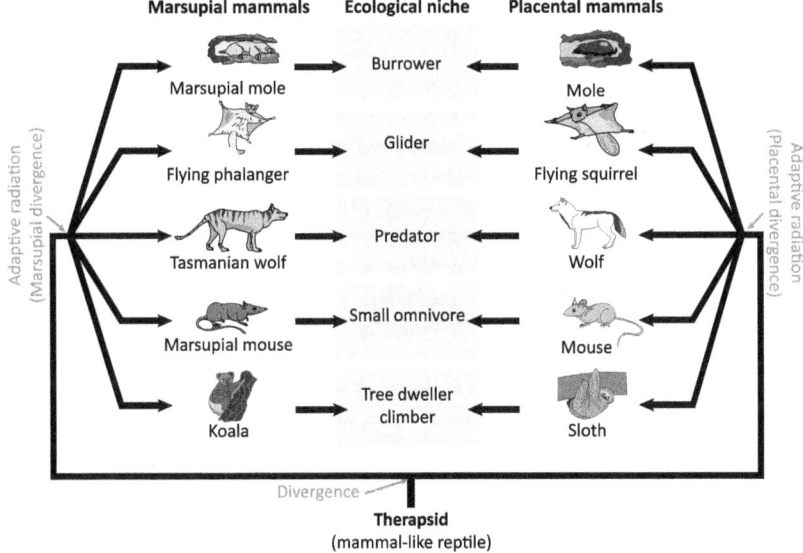

COMPARATIVE EMBRYOLOGY

In most species, there is a marked morphological difference between the **embryo** and the adult. When comparing distantly related species, the adults are usually clearly distinguishable from each other. For example, a human adult has very little physical similarity with an adult chicken; a chicken has feathers and a beak, while a human has hair, exposed skin, and teeth. However, when the embryos of chicken and human are compared we find that during the first several stages of development leading up to the limb buds stage when the extremities begin forming, they appear remarkably similar in morphology. Other more closely related species, such as macaque or chimpanzee, remain morphologically similar far past the limb bud stage and into the late fetus stage of embryonic development. Scientists can use variation in embryonic morphology as a comparative tool for grouping organisms according to their evolutionary relationship, with more closely related species experiencing morphologic divergence in later stages of embryonic development.

THE FOSSIL RECORD

The fossil record provides the most abundant evidence of evolution. It is a physical record of species that lived millions of years ago. It is used to infer other information, including when and how extinct species disappeared, and how extant species evolved. Even with the vast resource, there remains many missing links to explaining how species changed through time, including how our own species, Homo sapiens, evolved and why most of our closest relatives went extinct.

Transitional fossils

One of the challenges related to the evolution of Homo Sapiens is explaining our transition from an aquatic finned species to a terrestrial **tetrapod**. Where are the missing links? The answer can be found in fossils of **transitional species** that show a progressive and sequential transformation from an aquatic species to a four limb land animal (or tetrapod). These transitional fossils are important in explaining how contrasting morphological features of our extinct ancestor (finned species) and the descendant **extant** species (Homo Sapiens) may have occurred. The following figure illustrates this point with the evolution of modern tetrapods from the extinct pre-historic lobe-finned fish called *eusthenopteron*. With each subsequent transitional species, there appears to be a sequence of adaptations that give way to the limbs.

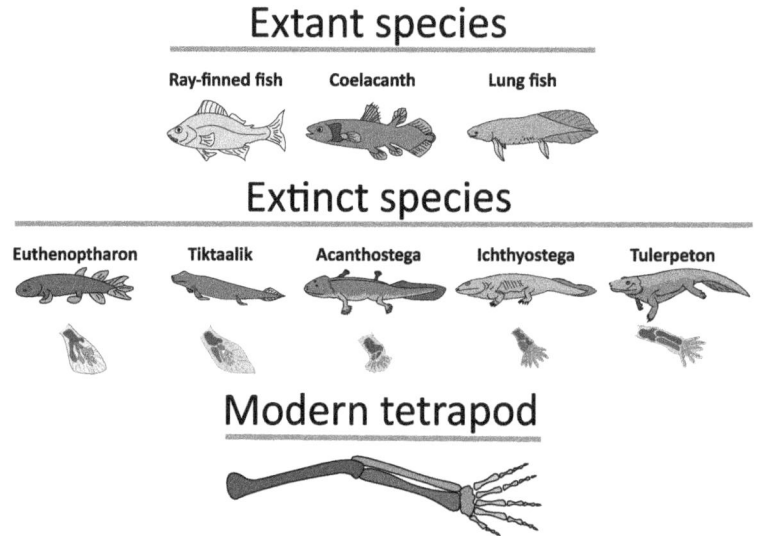

Fossil dating

A fossil is a record of when an organism lived, and the fossil's deposit location is a record of where it lived. Determining when a species lived depends on the accuracy and reliability of the methods used to date fossils. The various fossil dating methods apply two general strategies called **relative dating** and **absolute dating**.

Relative dating

Relative dating provides an approximate age of the fossil and is not used to determine its exact age. There are two methods for relative dating, biostratigraphy and fluorine dating.

Biostratigraphy relies on known **index fossils** to comparatively date unknown fossils. An index fossil must be from a known species that was widely distributed in different geographic areas, but that lived over a short geologic time period. The following figure is a schematic that shows relative dating with biostratigraphy. Notice that the older fossils are buried in the deeper stratum, and that the index fossil is confined to a single stratum. Based on the positions of the various fossils, they can be ordered by ages from the oldest fossil (deepest stratum) to the youngest (top stratum).

In **fluorine dating**, the age of the fossil is determined by the amount of fluorine

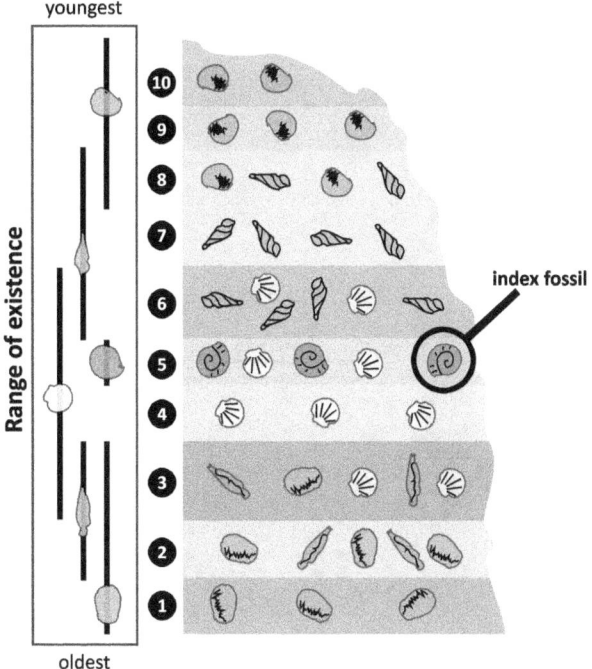

absorbed by the fossil from the surrounding soil, with the relative percent of fluorine increases over time, the percent nitrogen decrease. Fluorine dating is used where the rock strata have been disturbed as a result of geologic uplift or earthquake. Under these conditions, relative dating by biostratigraphy is not possible because the older fossils might have been relocated above younger fossils. However, the relative percent of fluorine should still correspond with relative age since fluorine absorption is not dependent on stratum depth. Fluorine dating is not used to determine absolute age because the amount of fluorine in the soil can vary from one location to another.

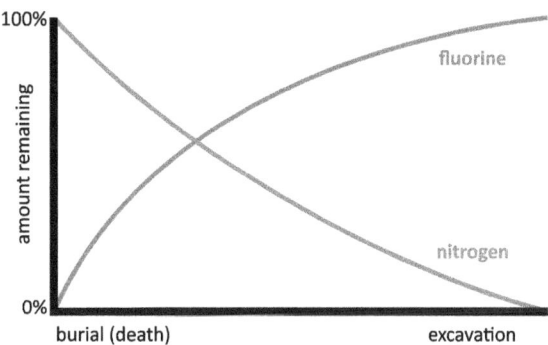

Absolute dating

Absolute dating is used to determine the exact age of a fossil. The three main types of absolute dating methods are:

- Radiocarbon dating (dates organic matter 100 – 50,000 years old)
- Potassium-argon dating (dates rocks between 100,000 to 4.5 billion years old)
- Uranium series (dates young rocks 10 million to 4.5 billion years)

Absolute dating relies on radioactive decay of naturally occurring isotopes of elements. The figure shows **radiocarbon dating** where carbon-14 decays to nitrogen-14 with a **half-life** of 5730 years. This means that after each 5730 years, half of the original C-14 that was stored in organism's tissue at death had decayed to N-14. The percent of C-14 wpresent at excavation can be used to calculate how many half-lives have passed, thereby determining the age of the fossil.

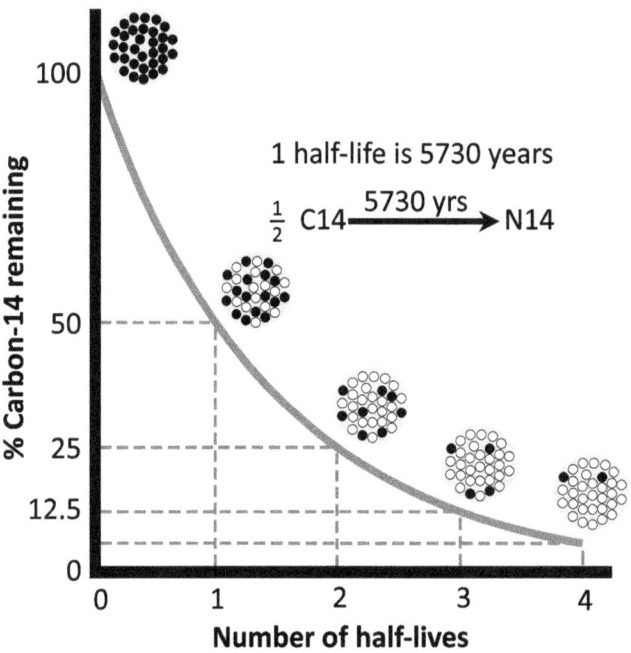

SUMMARY OF KEY CONCEPTS

- Evidence for evolution comes from multiple scientific disciplines including, biogeography, geology, fossils, comparative anatomy/morphology (analogous and homologous structure, and vestigial organs), molecular biology (DNA, RNA, proteins homology), embryology, and taxonomy.
- Morphological comparison alone can lead to the incorrect assignment of organisms to clades.
- Modern taxonomic classification incorporates protein and genetic homology, and other methods to ensure the correct grouping of organisms.
- Transitional fossils are the missing links between extant species and their distant relatives.
- The age of fossils can be determined using absolute dating techniques with radioactive isotope and relative dating techniques with biostratigraphy (undisturbed sites) or fluorine dating (disturbed sites).

CHECK YOUR UNDERSTANDING

1. Which of the following are sources of evidence for evolution?

   (A) Fossils of *lytrosaurus* have been discovered in Africa, Antarctica, and India.
   (B) Marsupials remains are common found in the Arabian desert.
   (C) Snakes have pelvis.
   (D) All of the above.

2. Which of the following statements is incorrect?

   (A) When a subgroup of a population splits and move to a colder, dryer region, that subgroup will tend to diverge from the original population.
   (B) The breakup of Pangaea resulted in an increase in biodiversity.
   (C) Marsupials likely appeared before placental mammals.
   (D) Analogous structures arise from divergent evolution.

3. Which of the following statements is correct?

   (A) Genetic homology based on DNA sequence is always the same as genetic homology based on amino acid sequence.
   (B) Lactation is a derived character of the monotremes, placental, and marsupial mammals.
   (C) Both are correct.
   (D) Neither are correct.

4. Which of the following is most closely related to the kangaroo?

   (A) Chicken
   (B) Wolf
   (C) Opossum
   (D) Flying phalanger

5. Which of the following statements about relative dating is incorrect?

   (A) At an archeological site where *tulerpeton* is located in a higher stratum than *euthenoptharon*, bio-stratigraphy would be the ideal dating method.
   (B) Fluorine content in a fossil correlates directly with its relative age.
   (C) The older fossils are located in deeper strata at an undisturbed site.
   (D) All of the above are correct.

ANSWERS & EXPLANATIONS

1. (D) is correct. Choice (A) is evidence that Pangaea existed since that three locations are now separated by ocean. Choices (B) is evidence that marsupials had a wider global distribution than they currently have. Choice (C) is highlighting a vestigial organ found in snakes, which suggests that snakes are descended from an ancestor that had legs.

2. (D) is correct because analogous structures arise through convergent evolution (not divergent evolution) when comparative species occupy similar niches.

3. (B) is correct because all mammals produce milk by lactation. Monotremes such as the platypus, is another class of mammals, and perhaps the earliest to appear. (A) is incorrect because, due to codon redundancy, some changes in the DNA sequence may not result in changes in the amino acid sequence. The protein sequence tends to have greater homology than the DNA sequence.

4. (D) is correct because the kangaroo and the flying phalanger are both Australasian marsupials. (A) is wrong because the chicken is a bird, (B) is wrong because the wolf is a placental mammal, and (D) is wrong because, although the opossum is a marsupial, its has been isolated from the kangaroo from a longer time than the flying phalanger. The order of relatedness would be D>C>B>A.

5. (D) is correct. All are correct statements. In (A), it appears that the site is undisturbed since the older fossil (*euthenoptharon*) is located deeper than the younger fossil (*tulerpeton*), therefore biostratigraphy is appropriate. If it was a disturbed site, fluorine dating would be preferred. (B) and (C) are both accurate statements.

04 Phylogenetic tree

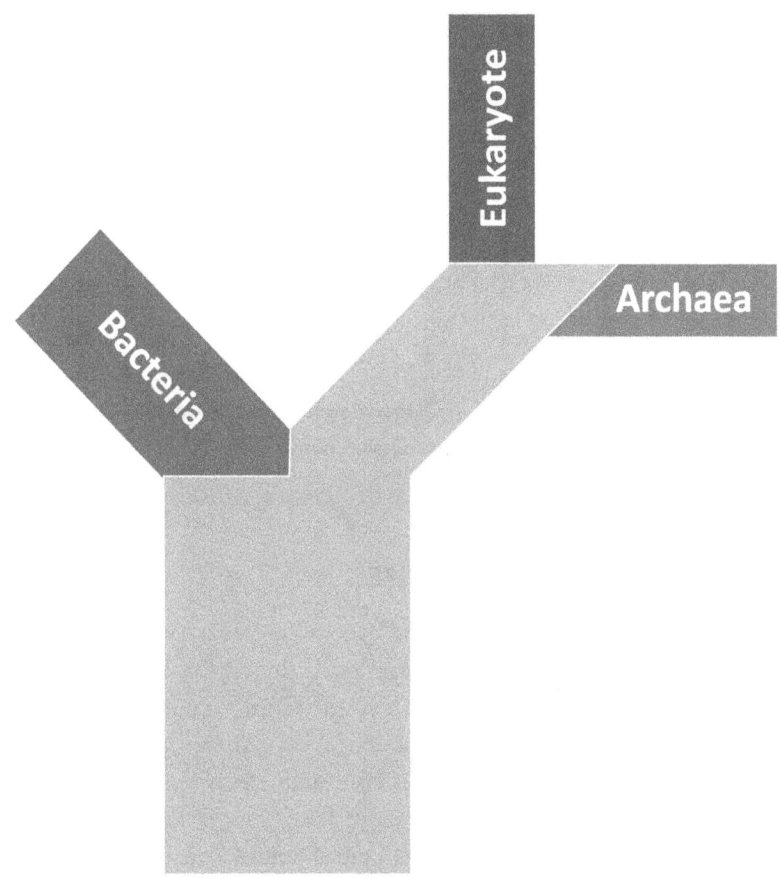

ALL STUDENTS MUST BE ABLE TO ANSWER THESE QUESTIONS

1. What sort of information does the phylogenetic tree contain?
2. How is an outgroup distinguished from sister groups?
3. What are the 3 domains of life and what are their distinguishing features?
4. What are the 8 biological levels of taxonomic classification?
5. How is a phylogentic tree organized?
6. What is the relationship between taxa, clades, cladograms, and the phylogenetic tree?

ALL STUDENTS MUST BE ABLE TO COMPLETE THE FOLLOWING TASK

1. Use a provided set of data to construct a cladogram.
2. Construct a cladogram based on heart morphology.

 Big Idea 1B2

A **phylogenetic tree** is a diagram that illustrates the evolutionary lineages of biological species and the relationships of these lineages and species to each other. For example, the primate families of hominid (human, chimpanzee, and gorilla, and orangutan) and lemurid (all lemur species of Madagascar), and the cat family can be compared and illustrated on a phylogenetic tree. The two primate families, being more closely related, diverged from a more recent common ancestral branch of the tree. The ancestral primate branch diverged from the cat branch further back in time. The cat family is the **outgroup** since it is least closely related to the others. If the four hominid species were compared, the human and chimpanzee branches should show the most recent divergence, followed by gorilla, and orangutan (outgroup).

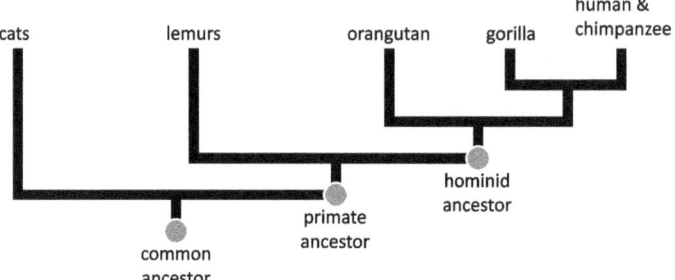

WHAT IS A PHYLOGENETIC TREE?

The phylogenetic tree is a diagram that looks like a branching tree. Each branch represents a separate evolutionary lineage, with the end tips representing the various extinct and extant species. It begins with a single trunk that extends back to the first prokaryote – the photosynthetic cyanobacteria. 3 primary branches diverge from the trunk to form the 3 **domains of life** - **Bacteria**, **Achaea**, and **Eukaryote**. The domains continue splitting into later lineages to form kingdoms, phyla, classes, orders, families, genera, and finally the various extinct and extant species. As a graphical representation of the evolutionary history of life, the phylogenetic tree shows the lineages or species that broke off and went extinct, and those that continued splitting to produce Earth's living biodiversity.

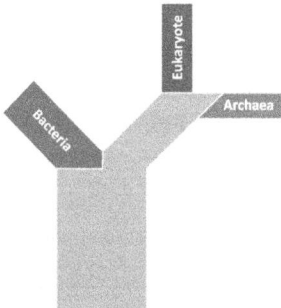

The structure of phylogenetic trees

Phylogenetic trees are used to show the evolutionary relationships between two or more taxonomically related groups of organisms. These groups are called taxa, and can include a single population, an entire species, a genus, or any other level of the taxonomic hierarchy of biological classification. Features of a phylogenetic diagram are **common ancestors**, **nodes**, **taxa**, **outgroups** and **sister groups**.

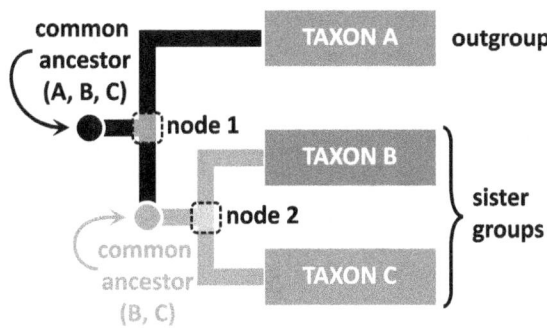

Clades

If a phylogenetic tree is analogous to a physical tree, a clade is a collection of terminal branches that make up the larger branches (cladograms) of the whole tree. General rules for constructing a phylogenetic tree includes:

- A clade contains 2 or more taxa that share a common ancestral node.
- Cladograms have a minimum of 1 clade and 1 outgroup; they contain at least 2 sister groups with derived characters that distinguish them from an outgroup.
- Cladograms vary in size depending on the number of taxa they contain.
- All taxa of a clade share a node.

The following cladogram shows the evolutionary relationship between 5 taxa. Notice that all taxa of the 4 different clades point back to a common node. For example, Node 1 is shared by all taxa of Clade 4, while Node 4 is shared in Clade 2.

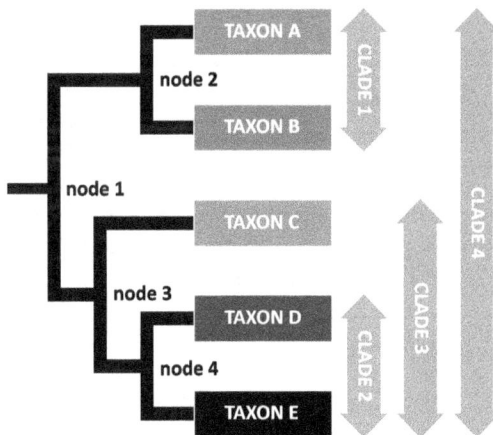

HOW TO CONSTRUCT A CLADOGRAM

Below is an outline of the general procedure for constructing a cladogram:

1. Decide on a group of organisms that are being compared
2. Identify a list of relevant derived characters.
3. Make a comparison chart of organisms versus derived characters
4. Place a check next to the organisms for each derived character.
5. Construct the Venn diagram based on the checklist.
6. Use the Venn diagram to construct a cladogram.

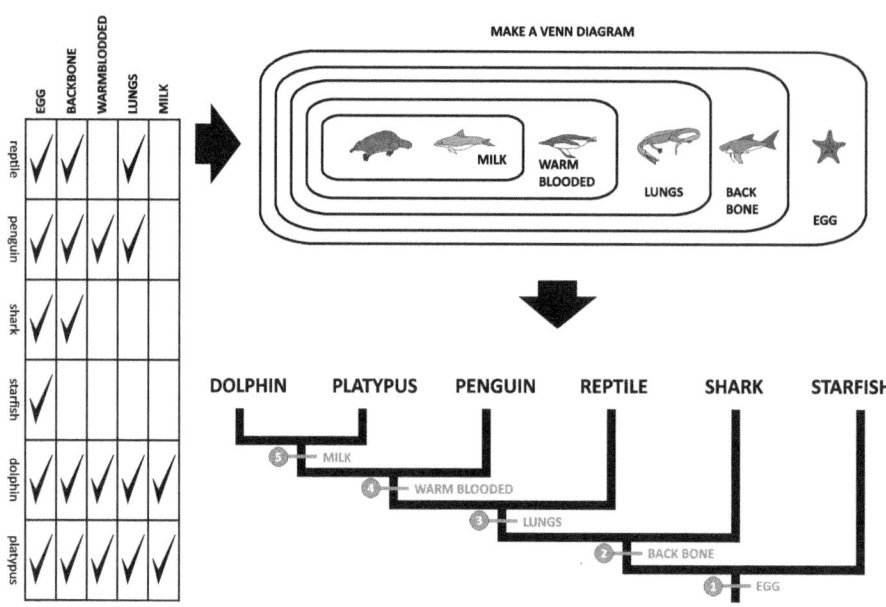

Evolution of the heart chambers

The following cladogram compares the evolutionary relationship between 5 classes of the phylum, **chordate**: mammals, birds, amphibians, reptiles, and fish. Notice that this cladogram compares taxonomic classes of a single phylum, and the relationships are inferred based only on heart morphology.

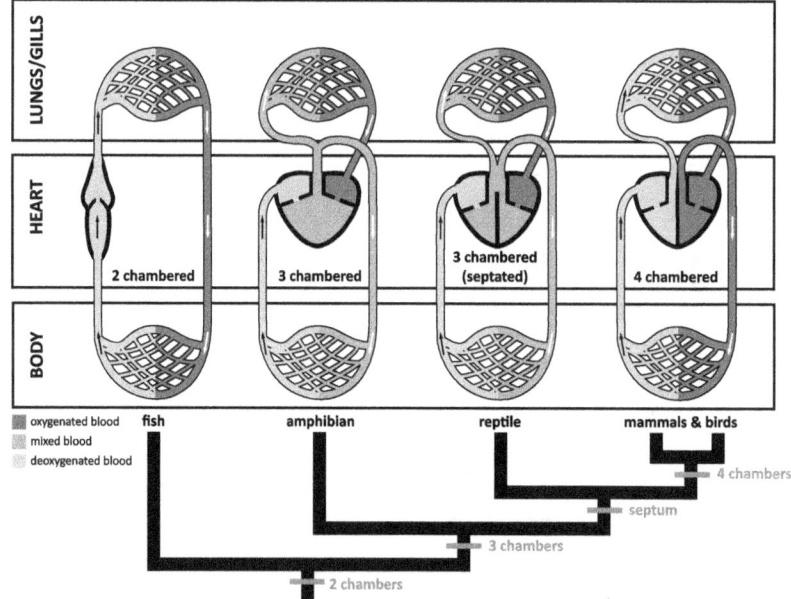

SUMMARY OF KEY CONCEPTS

- Phylogenetic trees and cladograms are graphical representations of evolutionary history of species, showing the traits that are either derived or lost during evolution.
- Phylogenetic trees and cladograms are used to show when speciation events occurred during evolutionary history.
- The accuracy of phylogenetic trees and cladograms relies on information from multiple sources including morphological homology between species (extinct and extant), genetic homology, and computer analysis (see Investigation 3).

CHECK YOUR UNDERSTANDING

1. The branch of science that deals with the naming and classification of organisms is

    (A) cladistics
    (B) phylogeny
    (C) taxonomy
    (D) none of the above.

2. The various levels of biological organization differ from each other in

    (A) the expanse of geographic distribution of the species they contain.
    (B) the shared characteristics of the organisms that they include.
    (C) the genetic complexity of the organisms that they include.
    (D) All of the above.

Use the following figure to answer the remaining questions.

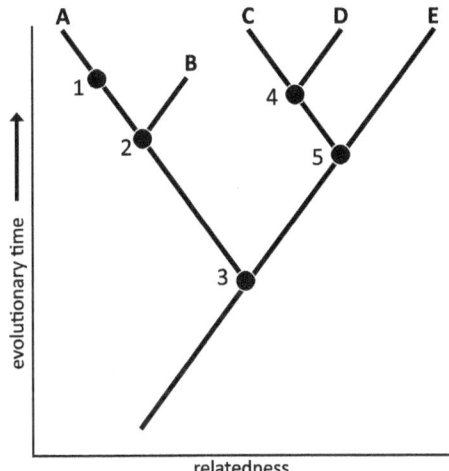

3. Which of the following nodes represents a common ancestor of taxon C, D, and E only?

    (A) 2
    (B) 3
    (C) 4
    (D) 5.

4. Which is the outgroup when comparing species B, C, D, and E?

    (A) A
    (B) B
    (C) C
    (D) D

5. The two most closely related extant species are?

   (A) A and B
   (B) C and D
   (C) D and E
   (D) A and E

6. Species A has a unique character that is not shared with any of the other species. At which point on the cladogram did this character emerge?

   (A) 1
   (B) 2
   (C) 3
   (D) 4

7. Which species has gone extinct?

   (A) A
   (B) B
   (C) C
   (D) D

ANSWERS & EXPLANATIONS

1. (C) is correct.

2. (B) is correct. There are distinct characteristics that all organisms in the eukaryotic domain share in common such as internal organelles and the endomembrane system. Likewise, all organisms within each of the hierarchical levels have some shared characteristics.

3. (D) is correct.

4. (B) is correct because species B is the only one that does not share the derived characters from common ancestor at node 5.

5. (B) is correct because their shared node at 4 split most recently.

6. (A) is correct.

7. (B) is correct because it is the only species not show at the present time, which is represented at the very top.

05 Hardy-Weinberg Equilibrium

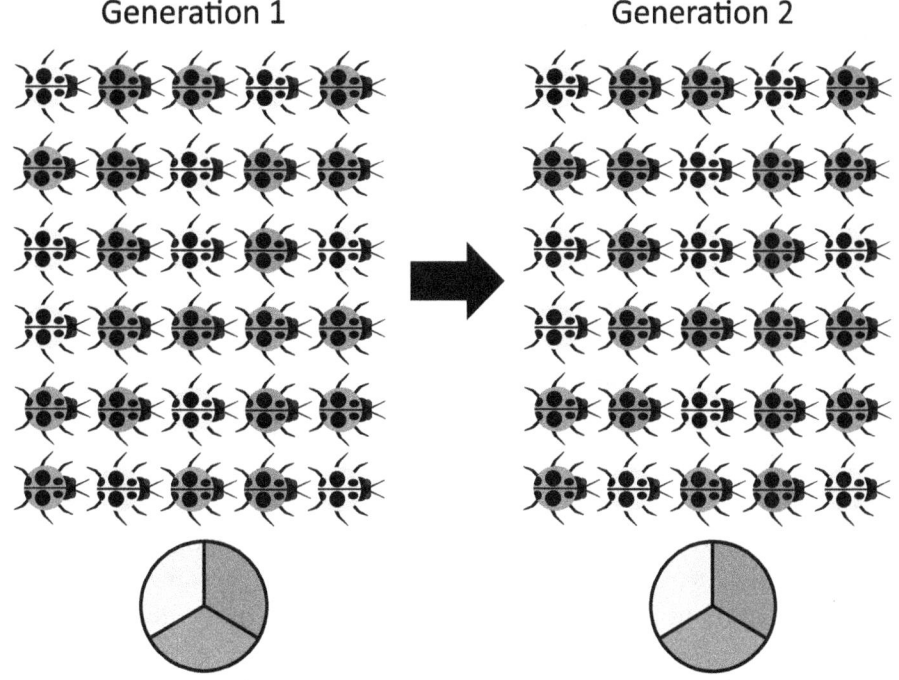

Generation 1   Generation 2

ALL STUDENTS MUST BE ABLE TO ANSWER THESE QUESTIONS

1. What is the relationship between evolution and phenotype frequency?
2. What is the relationship between evolution and genotype frequency?
3. What are alleles and how do they influence phenotypes?
4. Under what conditions is the recessive allele expressed?
5. How is the number of dominant and recessive alleles in the population calculated?
6. What are the five conditions of Hardy-Weinberg equilibrium?
7. What is the rationale for each condition of Hardy-Weinberg equilibrium?
8. What do the variables p and q represent, and why is the sum of these two variables equal to 1?
9. What do the expressions p^2, $2pq$, and q^2 represent, and why is the sum of these three expressions equal to 1?

ALL STUDENTS MUST BE ABLE TO COMPLETE THE FOLLOWING TASKS

1. Use a provided set of data to calculate the number of each allele in the population.
2. Apply the Hardy-Weinberg equation to solve a problem based on provided information.

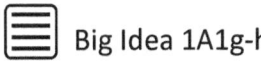 Big Idea 1A1g-h

EVOLUTION IS LINKED TO PHENOTYPE FREQUENCY

Evolution is a change in the observable physical characteristics of a population. For a given population, its characteristics are defined by the relative **phenotype frequency** among individuals. For example, if in a population of beetles there are 30 green and 10 yellow individuals, the phenotype frequency for the color traits is calculated as follows:

$$\text{Green} = 30 \div 40 = 0.75$$

$$\text{Yellow} = 10 \div 40 = 0.25$$

This shows that green is distributed at a higher frequency of 0.75, meaning that 75% of the individuals express the green allele. In a similar manner, the phenotype frequency of the next generation is determined and a comparison to a previous generation can be made. When comparing two different generations for the same **gene pool**, changes in the phenotype frequency from one generation to the next is usually (not always) an indication of evolution.

Genotype frequency

Although evolution is linked to phenotype frequency, it is the **genotype frequency** that really matters. Each individual receives two **alleles** of each gene. The **dominant allele** is always expressed when present, while the **recessive allele** remains hidden in the heterozygous individuals, and is only expressed in the homozygous state.

Genotype state	Allele 1	Allele 2	Phenotype
homozygous	dominant (A)	dominant (A)	dominant
heterozygous	dominant (A)	recessive (a)	dominant
homozygous	recessive (a)	recessive (a)	recessive

Change in phenotype frequency can occur without evolution

To appreciate how the phenotype distribution can change without changing the genotype frequency, consider the beetle population where the green allele (G) is dominant over the yellow allele (g). Also, assume that the genotype distribution is 10 **true breeding** green (GG), 20 heterozygous (Gg), and 10 recessive (gg).

Calculate the number of each allele in the population

> **Total alleles** = 2 × population size = 80

> **Dominant alleles** = 2 × homozygous dominant + heterozygous = 40

> **Recessive alleles** = 2 × recessive + heterozygous = 40

Calculate the allele frequencies (p and q)

> Dominant allele frequency or **p** = dominant alleles ÷ total alleles = 40 ÷ 80 = 0.5

> Recessive allele frequency or **q** = recessive alleles ÷ total alleles = 40 ÷ 80 = 0.5

Lets assume that in the next generation, the population size stays at 40 individuals, but all of the beetles are now green. It may appear that evolution has occurred because the phenotype frequency has clearly changed, with the yellow allele having an apparent lethal disadvantage. However, genetic testing shows that all of individuals are heterozygous (Gg). In this case, the phenotype frequency for green individuals changed from 0.75 to 1.0, but their allele frequency remained unchanged, with p = q = 0.5. Since the allele frequency remained unchanged, no evolution has occurred.

HARDY-WEINBERG EQUILIBRIUM

When the allele frequency remains unchanged between generations, the population is said to be in Hardy-Weinberg equilibrium. HWE is a hypothetical situation where there is no evolution occurring. Although rare in nature, it is a useful tool for studying evolution.

5 Conditions of Hardy-Weinberg Equilibrium & rationale

If HWE was to occur, there are 5 conditions that must be met by the population:
- There can be no mutations appearing in the population, because new mutations can introduce new allele and altering the allele frequencies of the population.
- There must be random mating so that there is no sex selection due to preferential mating. Sex selection can cause some individuals to reproduce at higher rates than other, allowing them to contribute their alleles to the next generation's gene pool at a higher frequency.
- There can be no natural selection that favors certain traits over others. The rationale for this is similar to sex selection since certain allele will be favored resulting in a disproportionate contribution of those favored allele to the next generation.
- There must be a large population so that the impact of genetic drifts is reduced. Genetic drift can cause random changes in the allele frequency of the population.
- There must be a closed population to migration so that there is no gene flow that can introduce or remove alleles.

Hardy-Weinberg equation

If a gene has only 2 alleles, the letters p and q are used to represent the frequency of the dominant and recessive alleles respectively. Since the sum of all the allele frequencies for a given gene must add to 1, the following equations apply:

$$p + q = 1$$
$$(p + q)^2 = 1^2$$
$$p^2 + 2pq + q^2 = 1$$

where, p is the dominant allele frequency, q the recessive allele frequency, p^2 the homozygous dominant genotype frequency, 2pq is the heterozygous genotype frequency, and q^2 the homozygous recessive genotype frequency.

HARDY-WEINBERG CONDITIONS

Generation 1   Generation 2

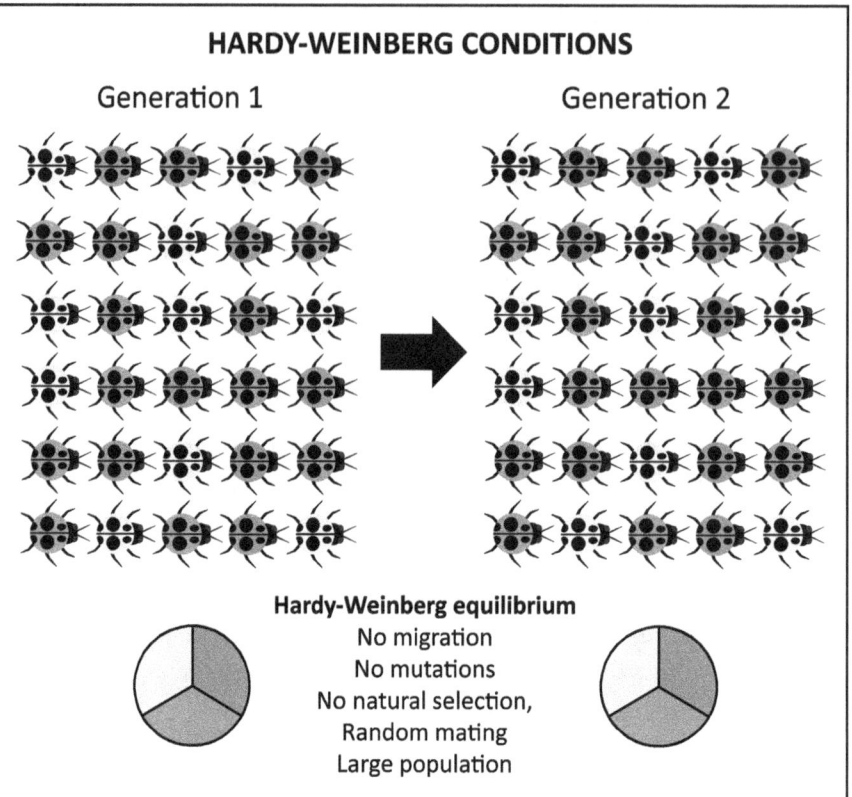

Hardy-Weinberg equilibrium
No migration
No mutations
No natural selection,
Random mating
Large population

EVOLUTION CONDITIONS

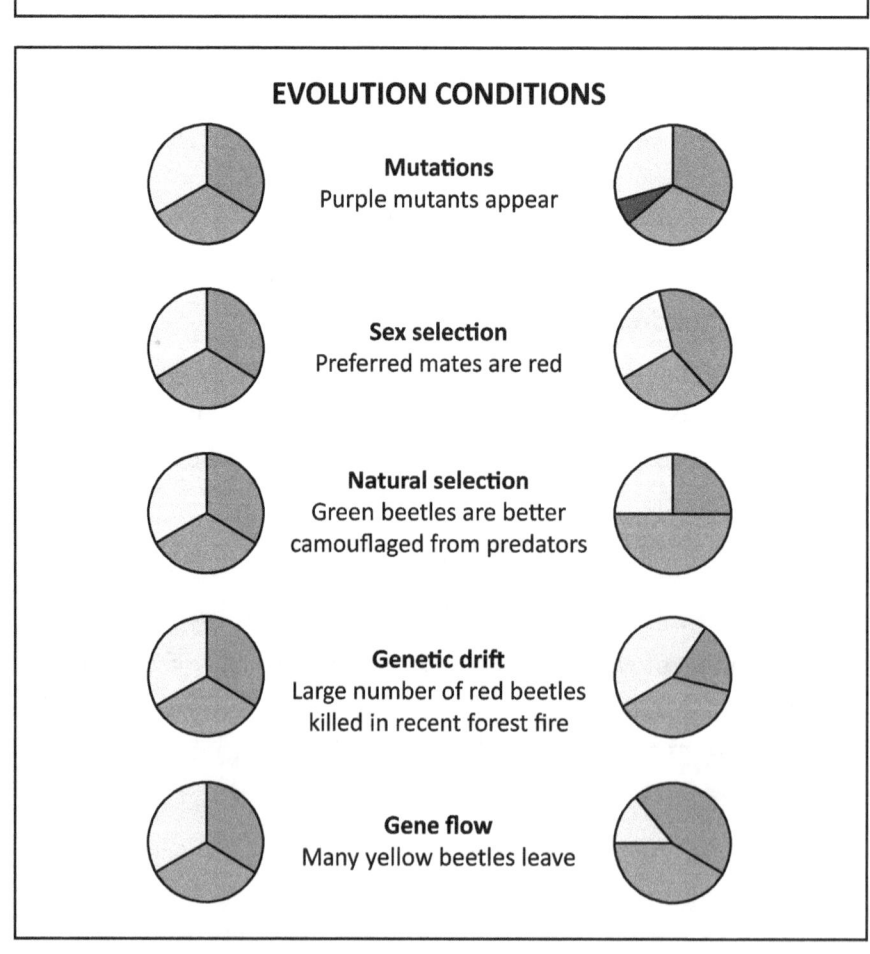

Mutations
Purple mutants appear

Sex selection
Preferred mates are red

Natural selection
Green beetles are better
camouflaged from predators

Genetic drift
Large number of red beetles
killed in recent forest fire

Gene flow
Many yellow beetles leave

APPLICATION OF HARDY-WEINBERG EQUILIBRIUM

Phenylketonuria (PKU) is a recessive disorders caused by the inheritance of a mutant allele of the enzyme phenylalanine hydroxylase. The normal functioning enzyme is used for metabolism of excess amounts of the amino acid called phenylalanine. In PKU patients, there are unusually high concentrations of the amino acid in the body, resulting in a series of neurological disorders. Assuming that the occurrence of PKU is 1 per 10,000 births, calculate (a) the frequencies each allele, and (b) the probability that an individual will be a carrier, normal, or affected by the condition.

Since the occurrence of PKU is 1 per 10,000 births,

$q^2 = 1 \div 10,000 = 0.0001 = 0.01\%$ (affected individuals)

$q = (0.0001)^{0.5} = 0.01$ (mutant allele frequency)

$p = 1 - q = 1 - 0.01 = 0.99$ (normal allele frequency)

$p^2 = (0.99)^2 = 0.98 = 98\%$ (normal individuals)

$2pq = 2 \times 0.99 \times 0.01 = 0.020 = 2\%$ (carrier)

SUMMARY OF KEY CONCEPTS

- The 5 conditions for Hardy-Weinberg equilibrium are:
    1. large population
    2. closed population
    3. no mutation
    4. no sex selection
    5. no natural selection.
- A change in allele frequency is evidence of evolution.
- In the Hardy-Weinberg equation:
    1. p is the dominant allele frequency
    2. q is the recessive alleles frequency
    3. p^2 is homozygous dominant genotype frequency
    4. 2pq is heterozygous genotype frequency
    5. q^2 is homozygous recessive genotype frequency

CHECK YOUR UNDERSTANDING

1. There is a population of 20 red and 30 white flowers of the same species. What is the frequency of white flowers? Assume that the white allele is recessive.

   (A) 1.5

   (B) 0.67

   (C) 0.6

   (D) 60%

2. If the white allele for flower color from question 1 is recessive, what is the dominant allele frequency?

   (A) 0.4

   (B) 0.8

   (C) 0.2

   (D) 5.0×10^{-2}

3. How many flowers are carriers (heterozygous) of the recessive allele?

   (A) 48

   (B) 30

   (C) 18

   (D) 9

4. How many recessive alleles are there in the population?

   (A) 77

   (B) 38

   (C) 23

   (D) 12

5. Which of the following can affect the gene pool by giving selective preference to certain individuals?

   (A) sex selection

   (B) migration

   (C) forest fire

   (D) none of the above

6. Which of the following can affect the gene pool by randomly deleting some alleles?

   (A) sex selection

   (B) mutation

   (C) forest fire

   (D) predation

7. Which of the following can affect the gene pool by introducing new alleles into the population?

   (A) sex selection
   (B) genetic drift
   (C) forest fire
   (D) none of the above

ANSWERS & EXPLANATION

1. (C) is correct. $q^2 = 30/50 = 0.6$

2. (C) is correct. $q = (0.6)^{1/2} = 0.77$; $p = 1 - 0.77$ $p = 0.23$ or 0.2

3. (C) is correct. $2pq = 2(0.77)(0.23) = 0.17$; # heterozygous = $n2pq = 50(0.17) = 17.7$ or 18

4. (A) is correct. $q = 0.77$; recessive alleles = $2nq = 2(50)(0.77) = 77$

5. (A) is correct. Only sex selection and natural selection can act on individual phenotype and affect fitness.

6. (C) is correct. Only gene flow due to migration and genetic drift can randomly affect allele frequency. Forest fire can cause genetic drift by random death of individuals who are caught in the fire.

7. (D) is correct. Only mutations and immigration can introduce new alleles into the gene pool.

06 Selection processes

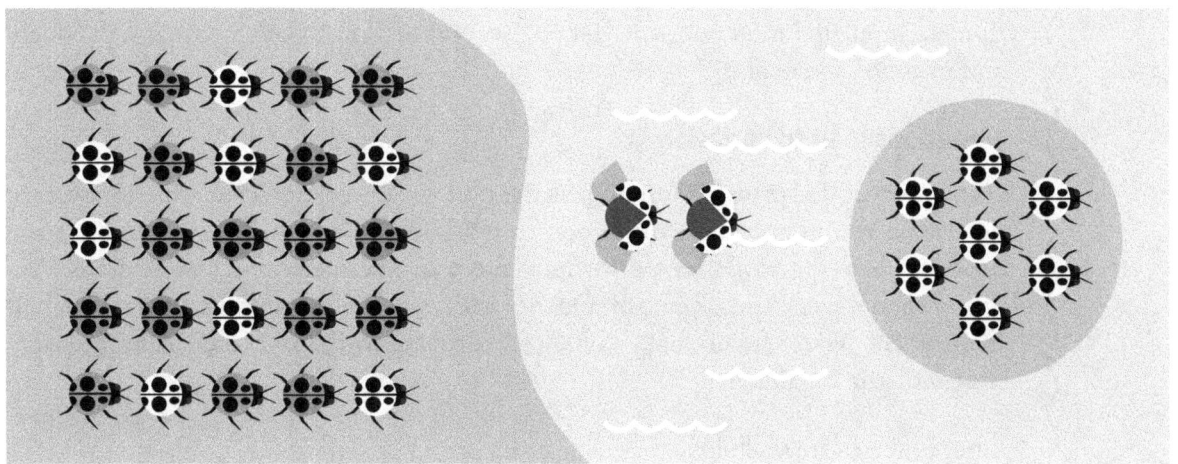

ALL STUDENTS MUST BE ABLE TO ANSWER THESE QUESTIONS

1. How do chance and random events have influence evolution?
2. Why are genetic drift, gene flow, and population size classified as chance events?
3. How does environmental change influence evolution by natural selection?
4. How do mutations and genetic recombination contribute to phenotypic variation?
5. Why are mutations considered the ultimate source of all variation in the population?
6. How does phenotypic variation effect the fitness of individuals and populations?
7. What are some examples of how humans impact variations in other species?
8. What are the 5 selections processes that can drive evolution?
9. Why is sex selection considered a type of natural selection?
10. How does genetic drift?
11. What are the 3 modes of natural selection?
12. What are the similarities and differences between gene flow and genetic drift?

ALL STUDENTS MUST BE ABLE TO COMPLETE THE FOLLOWING TASKS

1. Determine the mode of natural selection driving evolution for a provided set of data.
2. Apply Hardy-Weinberg equilibrium to determine the effects of genetic drift on a population
3. Make predictions about various selections processes on the evolution of a population.

   Big Idea 1A1f, 1A2, 1A3

VARIATION AND EVOLUTION

Evolution requires a population with preexisting phenotypic variation among individuals. Phenotypic variation leads to differential survival and reproduction, causing the gene pool to change in favor of those with traits that offer a competitive advantage. In populations that lack variation, all members have equal fitness, so there is no selective preference for one individual over another. There are 2 sources of phenotypic variation: accumulated mutation, and genetic recombination through sexual reproduction and meiotic crossover.

Mutations lead to variation

Genetically coded proteins function as the biological workhorses of the organism and determine the phenotype. Analogous to workhorses that pull plows and carriages, proteins are responsible for performing most of the metabolic and structural tasks that sustain the cell and organism. One form of a particular protein might work better than others. These functionally enhanced forms contribute to increasing the overall fitness of the organism.

Emergence of new alleles

DNA segments called **genes** code for proteins. Each gene originated as a single allele that coded for one form of the protein. Over evolutionary time, errors in DNA replication or DNA repair mechanisms introduce mutations that accumulate in the gene. Mutations can produce alternative alleles that code for new forms of the proteins by causing an amino acid of the native protein to be switched with another. Such a switch can affect the structure of the protein and influence its function. If the change improves ecological performance (**positive mutation**) the individual's fitness is enhanced, and if it reduces ecological performance (**negative mutation**), fitness is reduced. Overtime, the various derived forms become the different alleles.

Mutation is the ultimate source of all genetic variation

Without mutation, the population remains homogenous with only one form of each gene, and will lack biodiversity. Mutations were necessary for cyanobacteria to diversify into the variety of extinct and extant species. Although mutations are often viewed through a negative lens, a population that is void of mutations is at a competitive disadvantage. Such a population is less resilient to changing environmental conditions that can lead to extinction.

Genetic recombination leads to individual variation

Genetic recombination relies on existing mutations and does not directly influence the variation in the gene pool. Think of the gene pool as a fixed deck of 52 cards with all of the allele analogous to the individual types of cards. Similar to how the dealer distributes 10 randomly selected cards to each player, genetic recombination randomly distributes the alleles from the gene pool to individuals of the population. The dealer does not increase variation in the deck of cards by adding new suits or card types. He only shuffles the available cards so that distribution is random. Likewise, genetic recombination is a kind of **gene shuffling** that distributes a unique set of alleles to each individual. In most eukaryotes this occurs by sexual reproduction involving meiotic **crossover**, followed by **fertilization**.

SELECTION PROCESSES

Once there is variation in the population, selections processes tend to favor some individuals over others. This sort of favoritism reduces genetic variation by removal of the unfavorable alleles over time. There are 5 major selection processes that drive evolution:

- Natural selection
- Sexual selection
- Artificial selection
- Genetic drift
- Gene flow

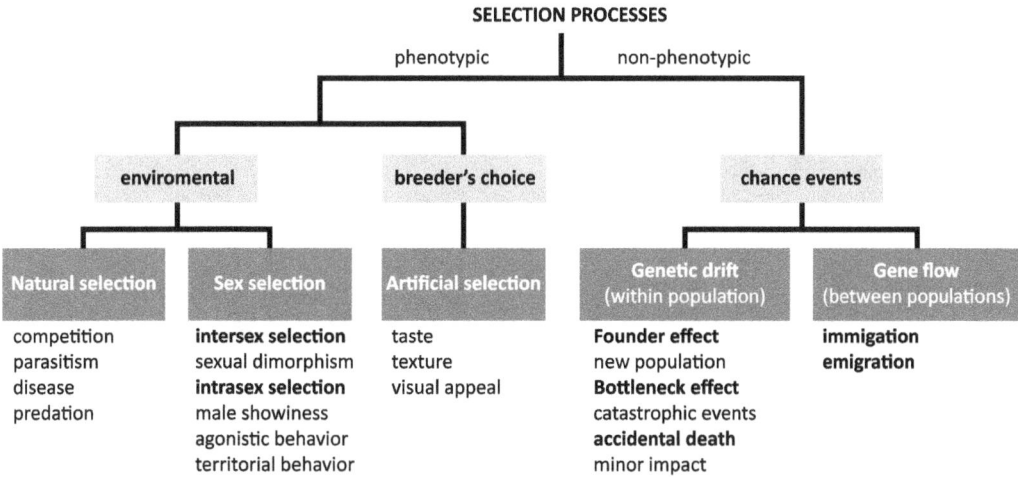

Selection based on genes and phenotypes

Natural, artificial, and **sex selection** rely on genetic and phenotype variation. In natural and sexual selection, the selective pressure comes from the environment, while in artificial selection it comes from the human breeder. Individuals that express the preferred phenotypes are selected at a higher frequency than others. For example, lionesses tend to copulate with male lions that display certain innate behavior (mating rituals, or agonistic territorial behavior) or phenotypes (mane size, muscle mass) that they find attractive. Those male lions that displayed the favored characteristics will have increased opportunities to mate and pass on their genes to the next generation.

Modes of natural selection

Natural selection is due to environmental pressure that drives the population to shift in one of three ways. In **directional selection** individuals at one extreme of the phenotypic spectrum are favored. In the figure, directional selection favors lighter pigmented birds, resulting in a shift on phenotype spectrum to the left. In another instance, the darker pigmented birds might be favored, causing a shift to the right. In **disruptive selection** individuals at both extremes of the phenotypic spectrum are favored. This leads to a disruption of the phenotype spectrum to favor the lighter and darker pigmented birds while the medium pigmented birds decline. **Stabilizing selection** is a caused by a strong selection pressure that favors the intermediate pigmented birds and works against the extreme phenotypes (darker and lighter pigmented birds). In each case, the environment changed, disrupting the preexisting equilibrium between the bird population and the environment.

Artificial selection is similar to natural selection, except the selection pressure come from the human breeder. Sex selection

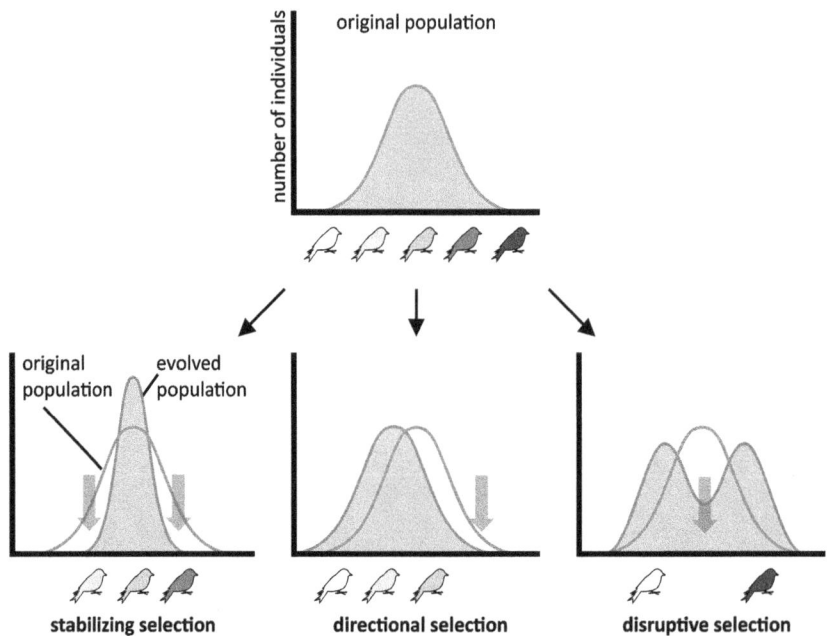

Sex selection

Sex selection is a type of natural selection that is often separated into its own category. In classic natural selection the selection pressure is external to the population, meaning that it arises from factors such as resource competition, predation, disease, and parasitism. In sexual selection it is internal, meaning that it is directly associated with the ability of individuals to interact with each other in securing mates or successfully copulating. Although sexual selection influences individual fecundity, it is not the same as reproductive success because every copulation attempt is not guaranteed to produce an offspring.

There are two types of sexual selection: **intersexual selection**, and **intrasexual selection**. Intersexual selection, also called mate choice, is where individuals of one sex (usually females) are choosy about their mates. For example, female lions chose to mate with a minority of male lions based on physical and behavioral characteristics observed in males. The choosier sex has a greater impact on the evolution of other sex and can cause the evolution of observable physical and behavioral differences between the sexes. These observable differences are collectively referred to as **sexual dimorphism**. For example, compared to human females, males tend to have deeper voices, greater muscle mass, and increased body and facial hair.

In intrasexual selection, individuals of one sex (usually males) compete among themselves for mates of the opposite sex. Competition can occur through aggressive ago-nistic and territorial behaviors.

Selection based on chance events

Gene flow and **genetic drift** cause changes in the gene pool of a population for rea-sons that are independent of individual fitness and genetics.

Gene flow involves two or more preexisting populations that exchange alleles through movement of individuals between the populations. In figure below, the **emigration** of a yellow beetle out and **immigration** of blue beetle in will cause the phenotype distribution to change with a reduced distribution of yellow and an addition of blue individuals.

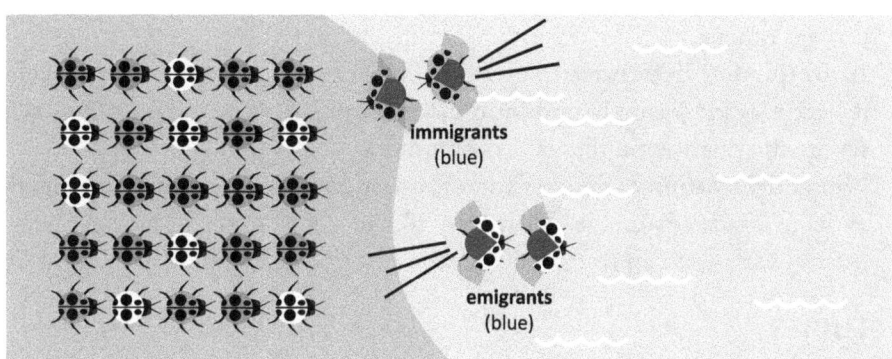

Genetic drift involves only one preexisting population. It is caused by the chance re-moval of individual from the population as a result of accidental death, catastrophic events that kill off a large portion of the population (**bottleneck effect**), or migration out of the original population to found a new population (**founder effect**).

The figure below illustrates the founder effect where two yellow beetles become the founders of a new population. The new population has a strikingly different pheno-type distribution from the original population. It reflects the genetic makeup of the founders. Where the founder effect focuses on the formation of a new population, the bottleneck effect describes changes in an existing population when a large number of beetles were killed by a wildfire. Although the fire does not discriminate on the basis of pigmentation, many of the yellow beetles survived by chance.

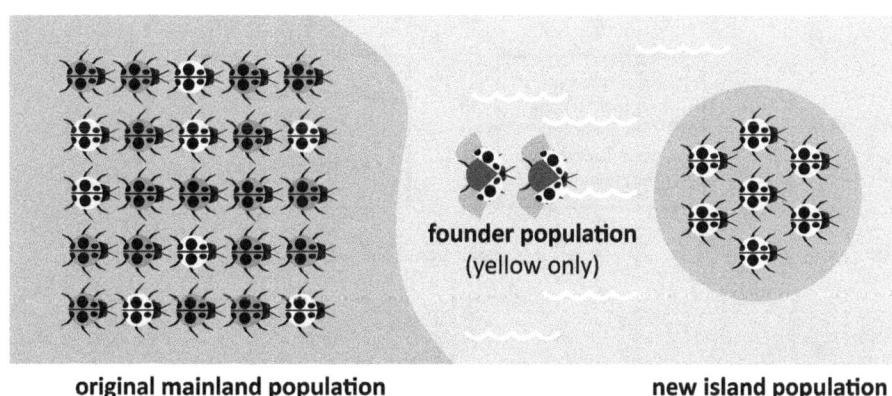

original mainland population
(red, yellow, green)

new island population
(yellow only)

PRESERVATION OF GENETIC VARIATION

Without genetic variation, there can be no evolution. The lower the genetic variation, the more vulnerable the population is to catastrophic events, resulting in an increased risk of extinction. Natural selection relies on existing phenotypic variation in the population that is preserved by the following natural processes.

- **Diploidy** exists in most eukaryotes. It is where each individual of the species inherits 2 copies of each gene. It maintains genetic variation in the form of hidden recessive alleles.

- **Heterozygote advantage** occurs when heterozygous individuals have a higher fitness than do either homozygous dominant or homozygous recessive. The two alleles involved show incomplete dominance patter of expression, with heterozygous individuals having an intermediate phenotype that with improved fitness.

- In **frequency-dependent selection**, the fitness of a phenotype declines as it becomes increasingly common in the population. In this process, selection favors the phenotype that is less common.

- **Neutral variation** is genetic variation that appears to confer no selective advantage or disadvantage. However, if the environment changes, previously neutral variation can potentially have a positive or negative impact on fitness.

CASE STUDIES

Global warming and genetic drift

Flowering plants and pollinators have coevolved with a **mutualistic symbiotic** relationship. The plants provide pollinators such as bees with nutritious sugar that they can use to nourish the colony, and the pollinators transport flower pollen between plants, allowing flower populations to benefit from genetic recombination.

Global warming negatively affects this relationship by altering the coordinated temporal appearance of pollinators and spring flower buds. In the previously established equilibrium, the budding of flowers and the arrival of pollinators were perfectly temporally synchronized. With global warming this synchronization is disrupted, adding a new environmental pressure on both populations to quickly readjust. The challenge is that the environment has not yet stabilized.

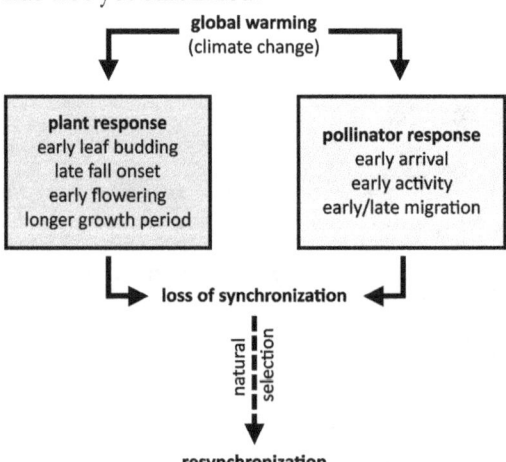

As long as global warming continues, there is no established set point for the population to re-synchronize. Those plants and pollinators that have the most genetically diverse gene pool stand a better chance of surviving the ongoing genetic drift caused by this catastrophic event.

DDT caused natural selection in pest population

DDT was a commonly used pesticide in the United States and is still used in some developing countries. Although initially effective at killing crop pests, its effectiveness was lost as pests evolved resistance to its toxicity. This evolution was due to natural selection in favor of individuals with favorable mutations that conferred resistance to DDT.

Due to its toxicity on most insects and animal species, DDT also had negative impact of natural flowering plants that relies on insect pollinators. Decreased plant productivity has a rippling affect throughout local ecosystems as primary consumers that rely on the fruits and seeds suffer food shortage. Furthermore, biological magnification cause dramatic increase in DDT concentration in top predators at the higher tropic levels as they consumed DDT-contaminated organisms at the lower levels.

Antibiotics and the emergence of new microbial diseases

Antibiotics are being prescribed and misused, lead to a drop in their effectiveness at destroying bacteria. Antibiotic resistance in bacteria population is driven by natural selection that favors those individuals that inherit the beneficial mutations against the antibiotics.

Heterozygous advantage and the sickle cell allele

The emergence of the sickle cell allele is a good example of how the heterozygous advantage preserved variation in the population. The sickle cell allele and the normal allele show incomplete dominance pattern of expression. Individual with 2 copies of the sickle cell allele have the condition and display the lowest relative fitness caused by poor binding and transport of oxygen. Although the sickle cell alleles seems to confer a competitive disadvantage, in areas where malaria is prevalent, individuals who are carriers are protected from severe infection of the malaria parasite because they express an intermediate phenotype that lowers their incidence of infection.

Peppered moth evolution driven by the industrial revolution

Prior to the industrial revolution, the white-bodied peppered moth experienced the highest relative fitness compared to the black-bodied moths because their light-colored body provided camouflage from predatory birds against the light-colored lichen on the trees of the English forests. By the end of the 19th century, industrial revolution cause the black-bodied peppered moth population to increase dramatically. This change is attributed to the increases in soot from coal burning factories that killed off many of the light-colored lichen, causing trees to darken. Darker trees provide camouflage protection for black-bodied, but not the light-bodied peppered moths.

SUMMARY OF KEY CONCEPT

- In addition to natural selection, artificial selection, genetic drift, gene flow, and sexual selection can all cause a population to evolve.
- Genetic drift has the greatest impact on small populations.
- Natural selection relies on existing phenotypic variation in the population
- Changes in the environment provide the selective pressure for natural selection.
- Some phenotypic variation can positively or negatively affect fitness
- Sickle cell anemia, natural selection on peppered moth, and DDT resistance are examples of natural selection.
- Humans can impact variation on other species through human activity or through active breeding programs.
- All forms of selection lead to a reduction of genetic variation in the population.

CHECK YOUR UNDERSTANDING

1. Evolution requires a population that

   (A) is homogenous.
   (B) is phenotypically heterogenous.
   (C) has sex selection.
   (D) is opened.

2. Variation in the gene pool of a population is effected by

   (A) genetic drift.
   (B) gene flow.
   (C) sex selection.
   (D) all of the above.

3. In a closed population, the primary source of phenotypic variation among individuals is

   (A) mutation.
   (B) gene shuffling.
   (C) genetic drift.
   (D) gene flow.

4. Gene shuffling in eukaryote typically requires

   (A) meiosis
   (B) crossing over
   (C) fertilization
   (D) all of the above

5. Selection processes that rely on the individual differences are:

   (A) natural selection.
   (B) bottleneck effect.
   (C) migration.
   (D) founder effect.

6. Natural selection can be driven by all of the following except

   (A) competition.
   (B) mate choice.
   (C) migration of some members of the population to a new area.
   (D) predation.

7. Which of the following is a selection process that requires two preexisting populations?

(A) founder effect
(B) bottleneck effect
(C) gene flow
(D) sex selection

8. In disruptive selection on a population of birds ranging in the pigment spectrum from white to black feather, the favored individuals are

(A) white only.
(B) black only.
(C) gray.
(D) black and white.

9. In stabilizing selection on a population of plants with varying height, the favored individuals are

(A) tall.
(B) short.
(C) medium height.
(D) tall and short.

10. The phenotypic difference commonly observed between males and females of a given species is primarily due to

(A) females being choosy about which males they select for mating.
(B) males being territorial.
(C) males displaying aggressive behavior to impress females.
(D) dominant males killing unrelated young males.

11. Genetic drift can be caused by all of the following except

(A) migration.
(B) flooding.
(C) immigration
(D) all of the above can lead to genetic drift.

12. All of the following processes can act to preserve genetic variation in the population except

(A) genes with multiple alleles that display a complete dominance pattern of expression.
(B) genes with multiple alleles where hybrid individuals display the highest fitness.
(C) gene with two alleles that display a complete dominance pattern of expression, and where the recessive phenotype confers a significantly greater fitness than the dominant phenotype.
(D) Alleles of a gene that is responsible of coat pigmentation in a prey population confer improved fitness at the lower phenotype frequency, and reduced fitness at higher frequency.

ANSWER & EXPLANATION

1. (B) is correct. (A) is wrong because a homogenous population lacks genetic variability, and therefore all individuals are genetically identical and have the same fitness. Such a population will have a gene pool that either goes extinct, or remains unchanged. Although (C) and (D) can cause a population to evolve, they are not absolutely required since it can also occur via other processes.

2. (D) is correct because each of these processes can drive evolution by reducing genetic variation. Both genetic drift (A) causes random removal of alleles, gene flow (B) can either introduce or remove alleles, and sex selection is a type of natural selection that favors certain alleles over others.

3. (B) is correct. Although mutations is responsible for variation in the gene pool, it is gene shuffling that is responsible for phenotypic differences between the individuals that make up the gene pool.

4. (D) is correct.

5. (A) is correct. Only natural selection, sex selection, and artificial selection rely on the phenotype. (B), (C), and (D) are driven by chance events that are not based on phenotypic variation.

6. (C) is correct.

7. (C) is correct. Gene flow is the exchange of genes between two populations.

8. (D) is correct because disruptive selection favors both of the extreme phenotypes.

9. (C) is correct because stabilizing selection favors the intermediate phenotype.

10. (A) is correct. Although each of the options can lead to sexual dimorphism, they are all tied to the fact that females are choosy. Males are territorial (B) to reduce competition for females (or female choice). Dominant males become dominant by impressing females when they physically out compete other males in their display of mating rituals, and agonistic or territorial behavior toward other males.

11. (C) is correct. Genetic drift is the random loss of alleles that occurs as a result of the founder effect, the bottleneck effect, or accidents. Immigration (C) is a type of gene flow that often introduces new alleles into the population. Migration (A) is often involved in the founder effect. Flooding (B) can cause catastrophic death that can result in the bottleneck effect.

12. (C) is correct because it is an example of allelic advantage which causes the homozygous recessive individuals to have significantly greater fitness. The can result in the rapid loss of the dominant allele, rather than its preservation. (A) is genetic preservation through diploidy. (B) is heterozygous advantage which maintain both alleles since the heterozygous, which has both alleles, has the highest fitness. (D) is an example of frequency-dependent selection.

07 Speciation

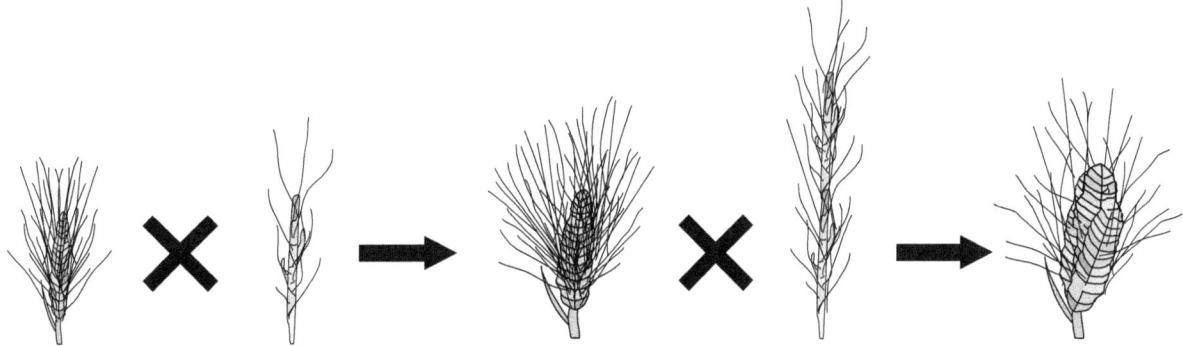

ALL STUDENTS MUST BE ABLE TO ANSWER THESE QUESTIONS

1. What are the factors that can influence speciation rates?
2. What are the factors that can influence extinction rates?
3. What is the molecular clock hypothesis and how is it used to infer relatedness between species?
4. What are the five major extinction events and the possible causes of each?
5. What are the 4 phases of speciation?
6. How do scientists determine when a speciation event is occurring?
7. How are pre- and post-zygotic barriers to reproduction distinguished?
8. What are the 4 pre-zygotic, and 4 post-zygotic isolation mechanisms?
9. What are the differences between allopatric, peripatric, sympatric, and parapatric speciation?
10. How is niche specialization involved in sympatric speciation?
11. How is punctuated equilibrium distinguished from gradualism?
12. How can polyploidy cause instant speciation in self-fertilizing species?

ALL STUDENTS MUST BE ABLE TO PERFORM THE FOLLOWING TASKS

1. Analyze a provided set of data to determine the degree of relatedness between species.
2. Explain how the modern bread wheat evolved through a series of allopolyploidy events.
3. Evaluate a provided set of data to determine if it confirms or contradicts the Grants' findings.

 Big Idea 1C

The Earth is constantly changing. These changes disrupt the ecosystem balance, causing the organisms that were previously well adapted to lose their competitive advantage. If this happens, these organisms must quickly re-adapt to the new environment or risk going extinct. Small, homogenous populations tend to be more vulnerable to environmental changes and are at greater risk of extinction.

MASS EXTINCTION AND CLIMATE CHANGE

Mass extinction occurs when more than 50 percent of the existing taxa simultaneously disappear over a short time period. The fossil record shows that Earth has gone through the following 5 major mass extinction events:

1. Ordovician event (443 million years ago)
2. Devonian event (360 million years ago)
3. Permian event (250 million years ago)
4. Triassic event (200 million years ago)
5. Cretaceous event (65 million years ago)

By correlating the fossil record and geologic data, scientist conclude that each of these mass extinction events coincided with some dramatic change in the environment. The figure shows the fluctuation in Earth's biodiversity over geologic time, as new species appeared and existing species died out. The temporal variation in the number of genera, particular around the 5 major events, is heavily influenced by changes environmental.

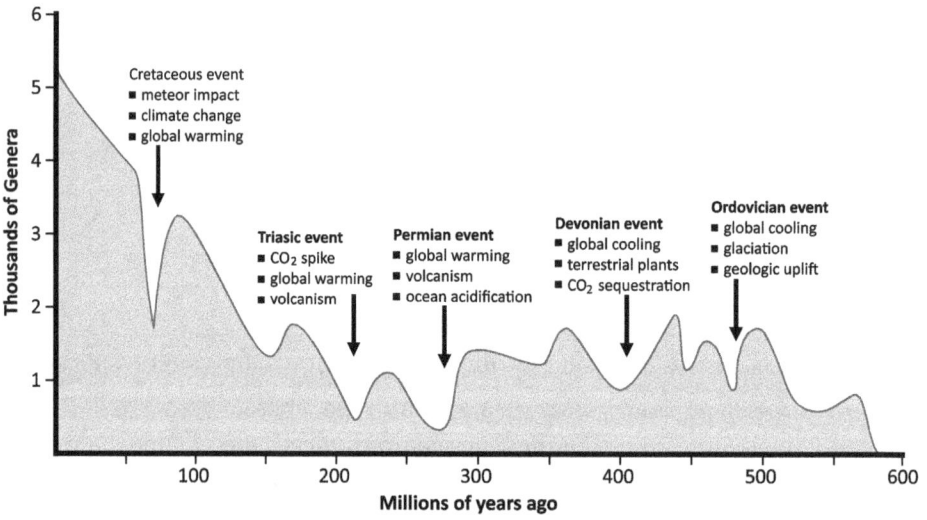

Throughout Earth's geologic history, the dominant global climate transformed several times from frozen glacial ice to warm tropical forest to hot dry deserts. If these changes occur rapidly, many species have little time to adapt and can die out. This sort of large-scale loss of biodiversity approaches a pattern expected for a bottleneck effect, rather than for natural selection, since very few individuals or species can withstand these dramatic shifts in the environment. For example, during Ordovician extinction, 85% of the existing species were wiped out. The primary geologic change may have been the drifting apart of the Gondwana landmass to form Africa, Antarctica, South America, Australia, and India. These changes also coincided with tectonic uplift to create mountain ranges, and a global cooling period that brought with it glaciation and a drop in sea levels. The lower global temperature, coupled with the loss of aquatic habitat may have driven most species to extinction.

A similar pattern of climate change is associated with each of the extinction events. Other events that may have caused the climate change included extraterrestrial radiation, loss of ozone, CO_2 sequestration, extraterrestrial bombardment by asteroids, increase in atmospheric O_2 levels, and volcanism.

DATING EVENTS

To understand the pattern of change in Earth's geologic history, scientists must determine when events occurred. The following methods are used to date events (see L3):
- Relative dating of fossils (biostratigraphy and fluorine dating
- Absolute dating of fossils with radioisotopes
- Genetic homology analysis using the molecular clock hypothesis

The molecular clock hypothesis

The pattern of species diversification that follows an extinction event, can be inferred by comparing the DNA of living species. Following a mass extinction event, many new niches open up for the few surviving species. With the availability of excess resources, these surviving populations rapidly expand their geographic ranges, split, and begin adapting to different ecological niches. This leads to genetic divergence as subgroups become reproductively isolated from each other. The time of divergence between two species can be determined through genetic homology analysis using the molecular clocks technique.

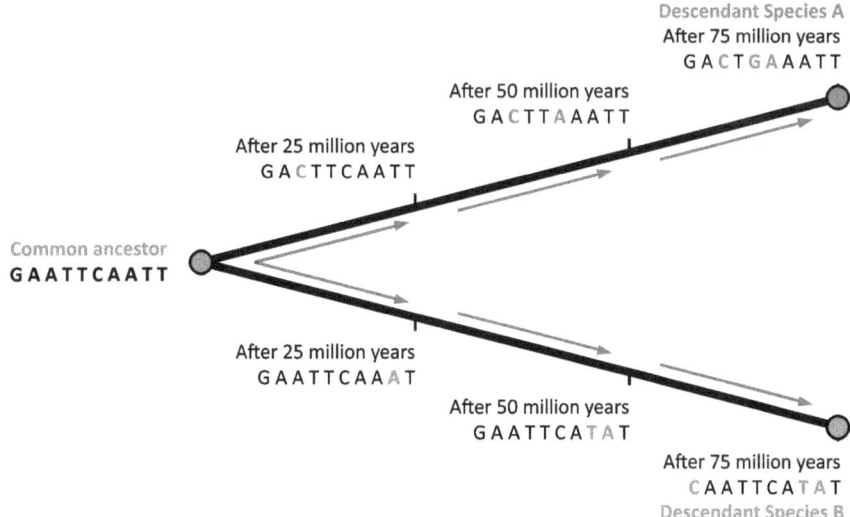

How does the molecular clock work?

According to the **molecular clock hypothesis**, mutations accumulate in an organisms' genome at a constant rate. By counting the number of genetic differences between two organisms, scientists can calculate how far back in time the two diverged from their most recent common ancestor. According to the previous figure, following the splitting of the ancestral species into the A and B lineages, there is an accumula-

GENETIC HOMOLOGY
Recall that the marsupial mammals once had an extensive global distribution prior to the expansion of the more competitive placental mammals. In the figure to the right, with the exception of the marsupial kangaroo, all of the other mammals are placental. As should be expected, the graph shows that human is more closely related to each of the placental mammals than to kangaroo.

tion of unique random mutations per lineage at a rate of 1 mutation per genome per 25 million years.

With this understanding, scientists perform genetic homology analyses to determine patterns of relatedness between taxa. The figure below shows the correlation of time of divergence to genetic homology analysis of the cytochrome c gene of different species. The graphs illustrates that humans are most closely related to chimpanzee and least related to the marsupial kangaroo. Furthermore, it provides approximate divergence times of roughly 10 and 100 million year ago, respectively.

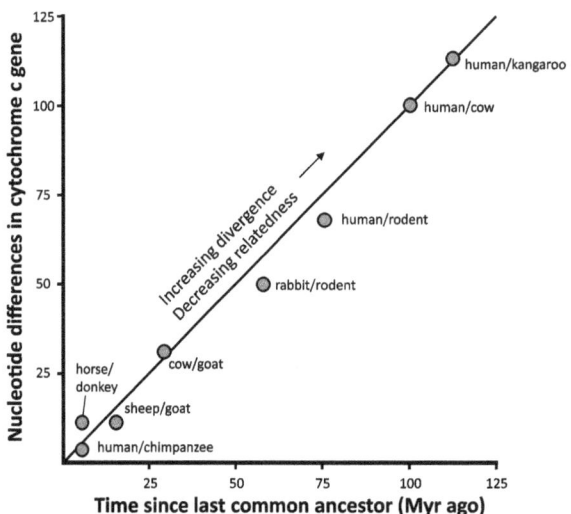

MECHANISMS OF SPECIATION

A species is an interbreeding population of organisms that can naturally produce viable (healthy) and fertile offspring. If a population splits into two isolated gene pools, the subgroups will begin to diverge into 2 separate species. The process of speciation leads to the emergence of new biological species that can no longer interbreed through natural processes. It occurs through the following 4 phases:
1. Original populations splits into 2 or more subpopulations;
2. Subpopulations become isolated from each other with separate gene pools;
3. Genetically isolated subpopulations undergo genetic divergence by accumulating unique mutations;
4. Genetic divergence leads to subpopulations becoming reproductively isolated from each other and can no longer naturally interbreed to produce viable hybrid offspring.

Reproductive barriers

There are **pre-zygotic** and **post-zygotic barriers** that ensure the reproductive isolation of two populations. The following 4 pre-zygotic barriers prevent the formation of a **zygote** (fertilized egg), by blocking copulation between different species:
-   **Habitat isolation** is where there is a geographic barrier such as a mountain and river that physically separates subgroups.
-   **Temporal isolation** is where the two subgroups have different mating times or seasons.

- **Behavioral isolation** is where the subgroups adapt incompatible innate behaviors such as mating and courting rituals.
- **Mechanical isolation** is where a morphological difference between the subgroups makes copulation impossible. Mechanical isolation is often due to large differences in body size, or sex organs.

Post-zygotic barriers prevent the hybrid zygote from developing into a viable and fertile adult. For example, crossing the reproductively isolated donkey and horse can produce a viable hybrid mule, but mule in infertile. The 4 post-zygotic barriers include:

- **Gamete isolation** is where morphologic difference between the gamete cells (sperm and egg) prevents the proper fusion of their haploid nuclei.
- **Reduced hybrid viability** is when the hybrid zygote fails to develop fully, and dies in the embryonic stage.
- **Reduced hybrid fertility** is where the hybrid is viable but infertile.
- **Hybrid breakdown** is where the hybrid is viable and fertile but the offspring from the second or later generation has a developmental defect that causes the lineage to breakdown.

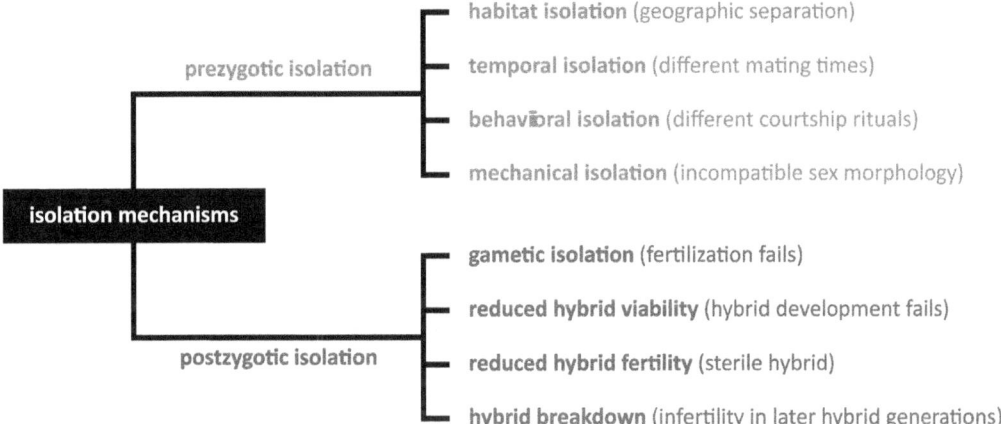

Speciation and geographic isolation

Speciation can occur with or without geographically isolation. Mechanisms that require geographic isolation are referred to as either allopatric or **peripatric speciation**. In **allopatric speciation**, a population splits into two subgroups that become separated by a geographic barrier such as a river. In peripatric speciation, a smaller subgroup is geographically isolated from the main population. In both instances, genetic divergence occurs if the gene pools remain isolated. In peripatric speciation, the small subgroup experiences a more intense genetic drift relative to the larger group.

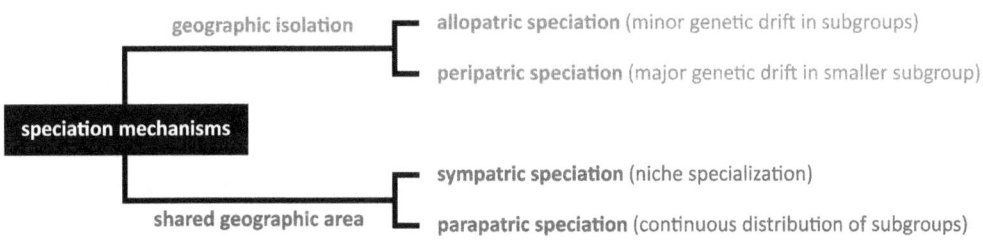

NOTES

SYMPATRIC SPECIATION
In the sympatric speciation example presented in the figure to the left the gray species can interbreed with both the black and white species. However, the white and black are reproductively isolated from each other and cannot interbreed.

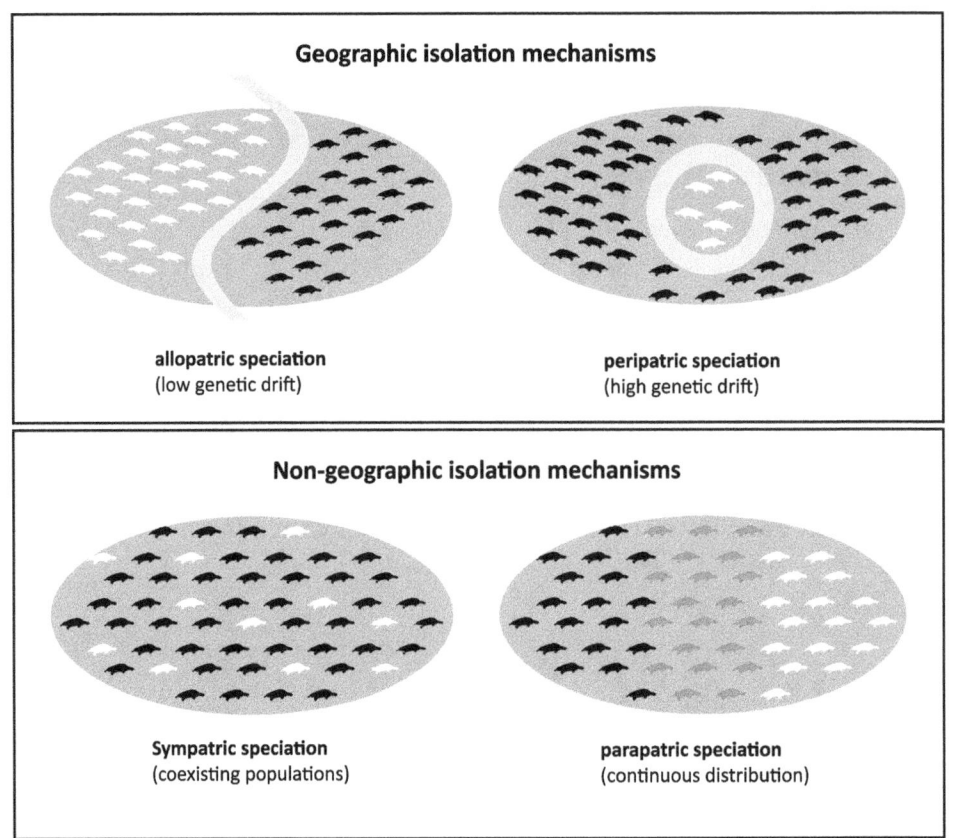

Speciation events without geographic isolation are rare. They typically occur when organisms of a given subgroup begin specializing in a vacant **ecological niche** within the same area as the larger group. In **sympatric speciation**, a subgroup of individuals from the larger population adapts to an alternative niche.

Sympatric speciation in flies

The divergence of the apple maggot flies from hawthorne maggot flies is a possible example of an ongoing sympatric speciation event. The hawthorne maggot flies lay their eggs exclusively on an apple relative called hawthorne. When apples were introduced into the US in the 19th century, a subgroup of hawthorne flies began laying their eggs on apples. This subgroup of flies was the founders of the apple maggot flies. The hawthorne and apple maggot flies each tend to mate and lay eggs on the fruit that forms part of its unique niche. This is a kind of behavioral isolation that has separated the gene pools of the two subgroups, potentially leading to future post-zygotic isolation. Note that the diverging subgroup experiences some other reproductive barrier such as derived innate behaviors that prevents mating and copulation.

In **parapatric speciation**, two or more populations occupy continuous geographic area with no specific geographic barriers between them. Mating is more likely between adjacent geographic neighbors. Subpopulations that are located further apart have gene pools that are isolated from each other, not by a geographic barrier, but by intermediate subgroups of the same species. This causes distant subgroups to diverge from each other.

Punctuated equilibrium versus gradualism

The rate of evolution in a population depends on the selective pressure caused by changes in the environment. **Gradualism** and **punctuated equilibrium** are 2 opposing views about how evolution occurs. Gradualism suggests that a descendent species (species C) evolves through slow and steady divergence from an ancestral species (species A), while punctuated equilibrium argues that species remain in a relatively stable state for much of their evolutionary history, experiencing rapid evolution over short periods when there is a sudden and dramatic change in the environment. In the figure, via punctuated equilibrium, the ancestral (species A) goes through 2 phases of rapid evolution to become the descendant species (species C).

NOTES

PUNCTUATED EQUILIBRIUM
Based on the fossil record, scientist believe that the horse evolved from a forest dwelling, deer-like ancestor through a series of abrupt changes that are in the pattern of punctuated equilibrium.

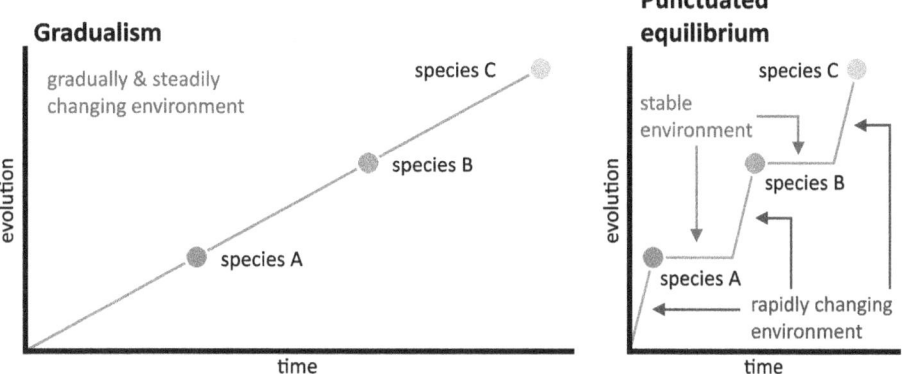

The following figure show the hypothetical evolution of a single ancestral butterfly species via gradualism and punctuated equilibrium. Notice that gradualism leads to incremental changes in pigmentation via intermediate species, while punctuated equilibrium involves with abrupt and dramatic changes.

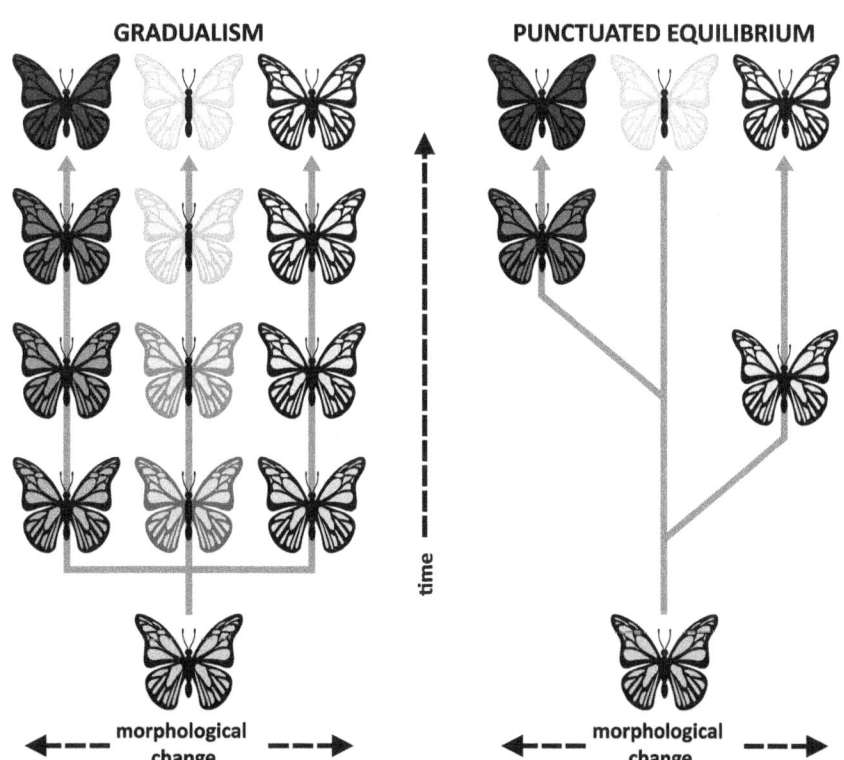

POLYPLOIDS
Polyploids contain more than 2 sets of chromosomes. Most eukaryotic species are diploid, meaning that their nuclei contain 2 copies of each chromosome.

Some specialized animal cells, such as human muscle cells, are polyploid. These polyploids that occur in otherwise diploid organisms are referred to as endopolyploids.

HOMOLOGOUS PAIRS
The term homologous pair refers to chromosomes in diploid cells that contain the same sets of genes. Each gamete cell that forms the zygote is haploid, meaning that they carry one of each chromosome. When the gamete cells fuse during fertilization, there are two copies of each chromosomes, thus all chromosomes in diploid cells are part of a homologous pair.

EVOLUTION DEPENDS ON MUTATIONS

Mutation is the ultimate source of all variation in the population. Millions of years may be needed to accumulate enough mutations before a population has sufficient variation for speciation. This does not mean that speciation always occur gradually as punctuated equilibrium can lead to rapid evolution.

Sources of mutations

1. Exposure to DNA altering environmental factors
   - Damaging chemicals called carcinogens that can damage the DNA
   - Radiation (UV, X-Ray, radioactive substances)
2. Error in internal cellular processes that can alter the DNA
   - DNA replication error
   - DNA repair mistakes

RAPID EVOLUTION IN PLANTS VIA POLYPLOIDY

Most eukaryotes are diploid, meaning that they have 2 copies, or homologous pairs, of each chromosome. For example, human somatic cells have 23 homologous pairs or 46 chromosomes each. However, it is not uncommon for some species of plants to have more than 2 copies of each chromosome. These polyploid species gain their higher ploidy states to become triploid (3 copies each) or tetraploid (4) through whole genome duplication. Once an organism gains a new ploidy state, it becomes reproductively isolated from members of its original species and is instantly a new species, making polyploidy the fastest mode of speciation. However, to be sustainable, a polyploidy event must lead to the establishment of a founder population from which the new species can expand. Consequently, speciation by polyploidy can only work with self-fertilizing species such as flowering plants where both male and female reproductive organs are present in each individual. Because animals require interbreeding to reproduce, they cannot undergo speciation by polyploidy. With self-pollinating flowers, a polyploidy event needs only occur in one individual that than self-pollinates to produce seeds to establish the starter population.

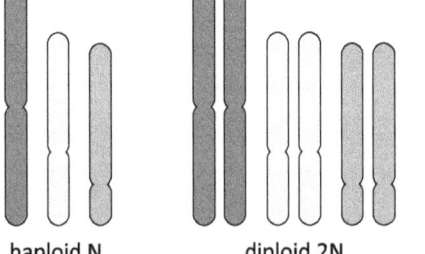 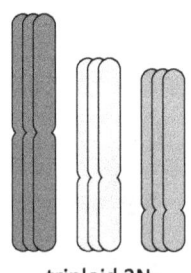 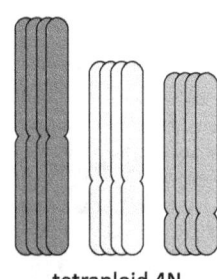

haploid N   diploid 2N   triploid 3N   tetraploid 4N

Autopolyploidy versus allopolyploidy

In autopolyploidy, 1 parent species gives rise to a new species. The figure shows chromosomal doubling (2n=6 to 4n=12) via meiotic non-disjunction of sister chromatids, followed by self-fertilization. In allopolyploidy, 2 individuals from different species (2n=4 and 2n=6) produce a hybrid (2n=10) following gamete fusion, mitotic non-disjunction, and self-fertilization.

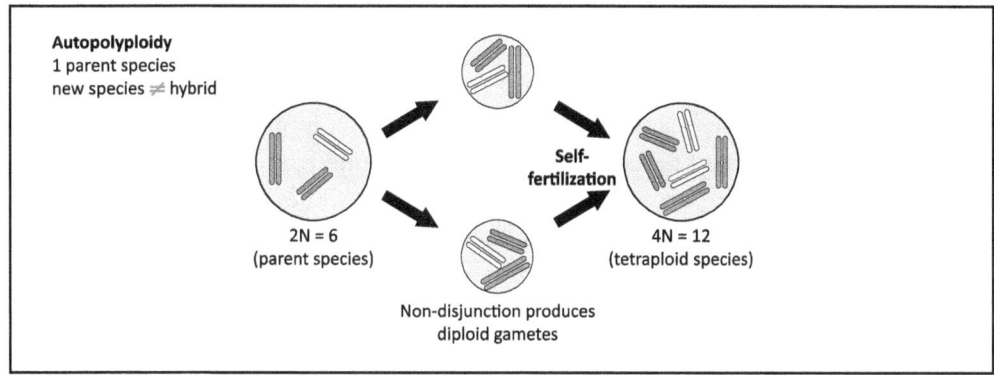

NOTES

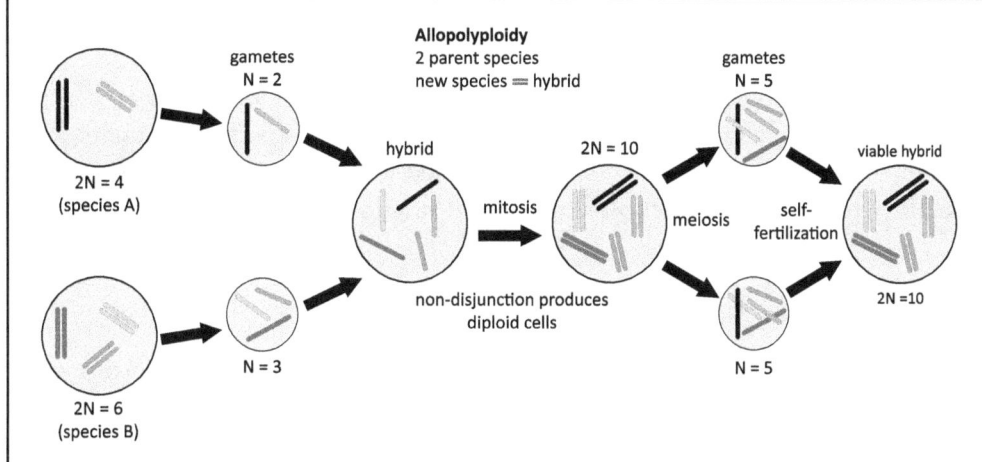

CASE STUDY AND EXAMPLES

Evolution of the modern bread wheat

The evolution of the modern common bread wheat occurred by polyploidy beginning 11,000 years ago. Initially a self-fertilizing hybrid species was produced through the hybridization of a true breeding einkorn species with a true breeding wild wheat species. This was followed by chromosomal doubling (self-pollination) from 2N = 14 to 2N = 28 to form fertile wild memmer. The final event was the hybridization of wild memmer with a wild wheat relative, followed by chromosomal doubling to modern common bread wheat (2N = 42).

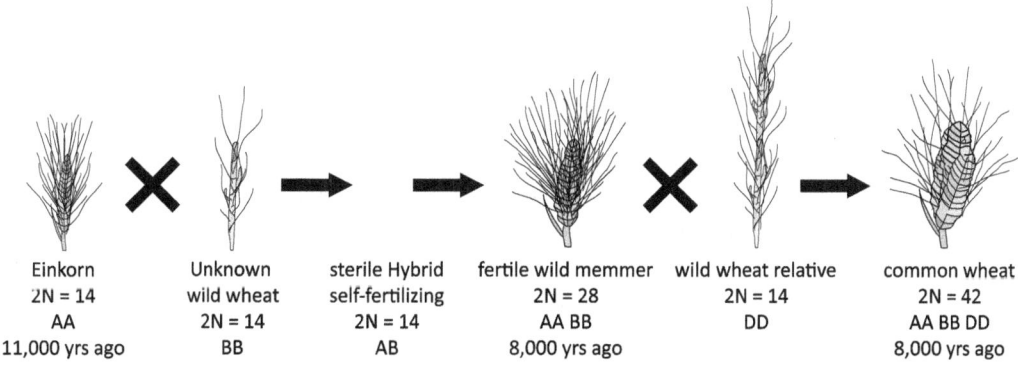

The human catastrophe

Burning of fossil fuels and climate change

The burning of fossil fuel causes the release of large volumes of greenhouse gases into the atmosphere. These gases trap heat in the biosphere, raising the average global temperature. Furthermore, when absorbed by the oceans, they act to increase the acidity (lower pH) of aquatic ecosystems. The increased global temperature and decreased aquatic pH, constitute a large scale alteration in the global environment that is pressuring species to either adapt or go extinct. On the geologic time scale, these changes are happening so rapidly that they are possibly ushering in a sixth major mass extinction event.

Human expansion

As human expand their geographic range by building communities, roads, highways, and shopping malls, and cutting down forests, other species lose their habitats and are displaced. This trend called urban sprawl, coupled with the burning of fossil fuels, is accelerating the rate of extinction and contributing to the 6th major mass extinction event.

Peter and Rosemary Grant

Peter and Rosemary Grant's observation of Galapagos finches provide one of the strongest evidences that evolution is still occurring today. The Grants were initially interested in understanding how inter- and intraspecific competition can affect an ecosystem's community.

To conduct a reliable investigation, they first established the following 5 criteria for the test group:

- A closed population where gene flow between the observed population and outside populations was not possible.
- A population that was small and manageable enough for all members to be accounted for and tracked throughout their lifetime, and included in collected data.
- A population that was large enough so that the collected data is statistically valid.
- A geographic area and species that allows for easy accounting of all newborns.
- A geographic area with strong cyclical selective pressure due to natural cyclical changes in the environment that can affect resource availability and, consequently, influence competition.

The Grant selected the two small islands of Daphne Major and Genovesa and the larger Santa Cruz Island of the Galapagos because the seem to meet the 5 study conditions described above. They specifically focused on understanding why the beaks sizes of Darwin's ground finches varied between the islands. On the island of Daphne Major, the medium ground finches have larger beaks than those on Genovesa, but smaller than the subpopulation just 10 km away on Santa Cruz Island.

The Grants hypotheses

The Grants hypothesized that were are 3 possible sources of the variation.

1. Variation was due to genetic drift, particularly due to the founder effect caused by 3 separate founder populations on each island.
2. The medium beak finches of Daphne Major were hybrids of the large and small beak finches from Santa Cruz and Genevosa, respectively. Gene flow from both to Daphne Major may have facilitated the hybridization of the two subgroups.
3. Variation in beak size was driven by natural selection coupled with allopatric speciation conditions. Geographically isolated subpopulations were adapted to their specific local niches, particular local food supply.

Field study (experimental) design

For over 30 years, they conducted field observations of finches by capturing birds and measuring their beak sizes throughout the months, years, and decades. They also kept an account of the number of newborns, their phenotypes, and their survival rates. They than correlated this data to changes in the local environment that were caused by cyclical wet El Niño and dry La Niña weather conditions.

The grants findings and conclusions

Neither genetic drift nor gene flow and hybridization could completely account for the variation. The most influential factor was the size of available food. On Daphne Major, the medium beak finches prefer smaller tender seed that are plentiful during the wet seasons. The larger, drier seeds are more difficult to crush, so these birds only resort to large seeds during the dry La Niña years when there are fewer or no small seeds available. Although large beak birds also feed on small seeds, they are less effective at securing them. The small beaks birds are unable to crush larger seeds and therefore feed exclusively on small seeds.

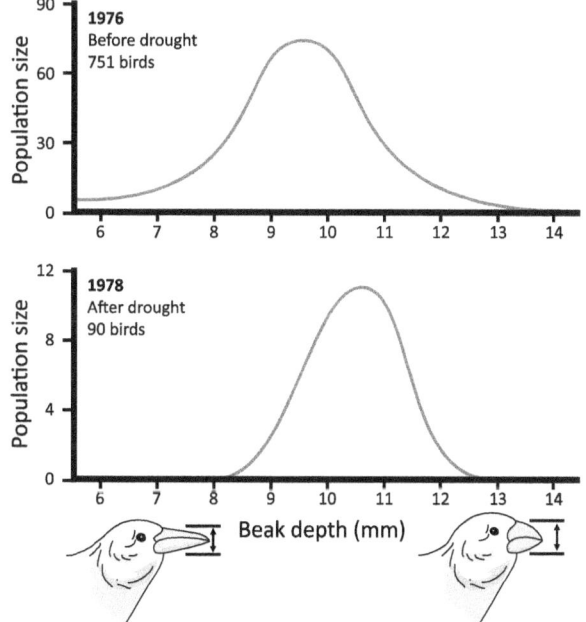

Directional selection on Galapagos finches

1976 Before drought 751 birds

1978 After drought 90 birds

Beak depth (mm)

During the wet season, small beak birds have the highest fitness and large beak birds have the lowest fitness. This switches during the dry season when the reproductive success of all birds is dramatically lowered, with small beak finches experiencing the most extreme negative selective pressure. The figure highlights these findings. It shows that there is ongoing directional selection from small to large beaks when the weather changed from wet to dry. Predictably, when the weather changes from dry to wet, there should be a directional shift toward the small beak birds.

SUMMARY OF KEY CONCEPTS

- A changing environment drives evolution.
- Speciation and extinction have occurred throughout Earth's history.
- Speciation occurs when two population undergo genetic divergence and become reproductively isolated from each other.
- The rate of extinction is accelerated by ecological stress due to a rapidly changing environment.
- The rate of speciation accelerates after intense ecological stress that reduces biodiversity and opens up previously occupied ecological niches to survivors.
- Pre- and post-zygotic barriers acts to reproductively isolate populations.
- Speciation can occur over multiple generations by punctuated equilibrium and gradualism, or suddenly by polyploidy.
- Human activity is accelerating climate change and displacing species from their natural habitats. We are possibly ushering in a 6th major mass extinction event.
- Peter and Rosemary Grant's observation of Galapagos finches provides strong evidence that evolution is still occurring today.

CHECK YOUR UNDERSTANDING

1. Changes in the global climate and/or environment has been caused by

   (A) increased volcanism.
   (B) meteor impact.
   (C) diversification of photosynthetic autotrophs.
   (D) all of the above.

2. Which of the following statements about the evolution of Earth is incorrect?

   (A) Early Earth had a reducing atmosphere.
   (B) Climate change has occurred multiple times throughout Earth's history.
   (C) Large scale and abrupt global environmental changes is correlated with the major mass extinction events.
   (D) The evolution and diversification of organisms capable of cellular respiration preceded the Earth's transformation to an oxidizing atmosphere.

3. According to the molecular clock hypothesis

   (A) The mutations rate is the same for all species.
   (B) As species converge, they accumulate similar mutations and nucleotide sequences.
   (C) As subgroups become reproductively isolated, they begin to diverge as each acquires unique mutations.
   (D) All of the above statements are correct.

4. Which of the follow statements describes two populations that belong to the same biological species?

   (A) Lions and tigers have been able to produce fertile offspring in zoos.
   (B) A relatively large great Dane and the miniature chihuahua are different breeds of dogs.
   (C) The bonobo and the chimpanzee are great apes like humans that can interbreed to produce fertile viable offsprings, and occupy habitats that are separated by the Congo River.
   (D) Three categories of domesticated cats that are common in urban settings are the housed pet, stray cats, and feral cats.

5. Which of the following species pairs would you expect to have the highest genetic homology?

   (A) dog and wolf
   (B) horse and donkey
   (C) dolphin and whale
   (D) flying squirrel and flying phalanger

6. Which of the following statements about speciation is incorrect?

   (A) Speciation always requires a split in the ancestral population.
   (B) Speciation requires that the gene pools of diverging subgroups remain closed to each other.
   (C) Speciation requires that diverging subgroups become physical separated from each other.
   (D) Speciation requires that the gene pools of diverging subgroups acquire unique mutations.

7. Which of the following is an example of pre-zygotic reproductive barriers?

   (A) The Congo river separates the bonobo from the chimpanzee.
   (B) The gorilla's sperms is unable to successfully fertilize the baboon's eggs.
   (C) A cross between the horse and the donkey produces an infertile mule.
   (D) Hybridization of the cotton species G. barbadense, and G. hirsutum produces apparently viable and fertile hybrid offspring, but their F2 progenies die as seeds or early development.

8. A certain species of grass that grows in areas contaminated by heavy metals from local mines, has undergone speciation to produce a strain that can tolerate the toxic overflow. Tolerance levels in grass populations increase with their proximity to the contaminated site, as does a shift in their flowering season, with adjacently located population overlapping. Which speciation mechanisms is described above?

   (A) allopatric speciation
   (B) peripatric speciation
   (C) sympatric speciation
   (D) parapatric speciation

9. The various species of Galapagos finches are all descended from a single mainland species that diverged as subgroups began occupying different niches defined by the subgroups' food sources on shared islands. Which speciation mechanisms is described above?

   (A) allopatric speciation
   (B) peripatric speciation
   (C) sympatric speciation
   (D) parapatric speciation

10. Which of the following can occur as a result of human-driven global warming?

   (A) gradualism on surviving local species
   (B) an evolutionary shift on local species that is predicted by punctuated equilibrium
   (C) extinction of all primates
   (D) None of the above.

11. Which of the following statements about polyploidy is incorrect?

(A) Autopolyploidy requires hybridization between two parent species.

(B) Polyploids have more than 2 copies of each chromosome.

(C) Polyploid speciation is common in plants, but not animals.

(D) All of the above statements are correct.

12. Which of the following statements about the Grants research is incorrect?

(A) They hypothesized that genetic drift, geographic isolation, and niche specialization was responsible for variation in the beak sizes of Galapagos finches.

(B) El Niño and La Niña weather conditions posed a significant problem for their experimental design.

(C) They discovered that genetic drift, gene flow, and hybridization were not solely responsible for the observed variation in beak size.

(D) During the El Niño years, small beak birds had the highest fitness.

ANSWERS & EXPLANATIONS

1. (D) is correct.
2. (D) is correct. Option (A) refers to the absence of oxygen in the early Earth atmosphere.
3. (C) is correct. Option (A) is incorrect because the mutation rates in prokaryotes is higher than eukaryotes. Prokaryotes have higher rates of mutation as a result of higher rates of DNA replication, less advanced DNA repair mechanisms, and increased viral-mediated lateral gene transfer. (C) is incorrect because convergence of does not lead to species sharing their acquired mutations; it only leads to similar, but independently derived adaptations.
4. (D) is correct because they are all domestic cats that can naturally interbreed. Feral cats were born and raised in the wild and have had no human contact. Stray cats were pets that were abandoned or lost to the wild, but have had human contact. Option (A) describes two species that are geographically isolated in the wild - tigers are native to Asia and Lions are native to Africa. Also, their hybrids are typically infertile. (B) is incorrect because, even though they are both considered dogs, they are clearly mechanically isolated. (C) is incorrect because bonobo and chimpanzee are geographically isolated by the Congo river.
5. (A) is correct because dogs are directly descended from ancestral wolves and can still be hybridized with wild wolves to produce fertile and viable offspring. (B) is incorrect because the hybrid mule is infertile (reduced hybrid fertility). (C) is incorrect because whales and dolphins are at least mechanically and behaviorally isolated. (D) is incorrect because squirrel is a placental and flying phalanger is a marsupial mammal.
6. (C) is correct because speciation does not always require a geographic barrier between diverging subgroups. Recall that in sympatric species, subgroups can occupy the same area but must occupy different niches. The Galapagos finches, for example, adapt to different food sources.
7. (A) is correct because it represents geographic isolation. (B) represents gametic isolation, (C) is reduced hybrid viability, and (D) is hybrid breakdown.
8. (D) is correct because the information describes a continuous distribution of grass populations that can interbreed with adjacent, but not distant populations that are temporally isolated.

9. (C) is correct because the different finch species occupy the same island but are adapted to different food sources.

10. (B) is correct because the global warming is occurring so rapidly that effected species do not have time to gradually adapt. Consequently, natural selection that approaches, is causing a rapid shift in the surviving species as expected for punctuated equilibrium.

11. (A) is correct because autopolyploidy requires only one parent species that undergoes chromosomal duplication.

12. (B) is correct because El Niño and La Niña conditions were needed to provided the strong cyclical selective pressure on the finch populations.

08 Endosymbiotic theory

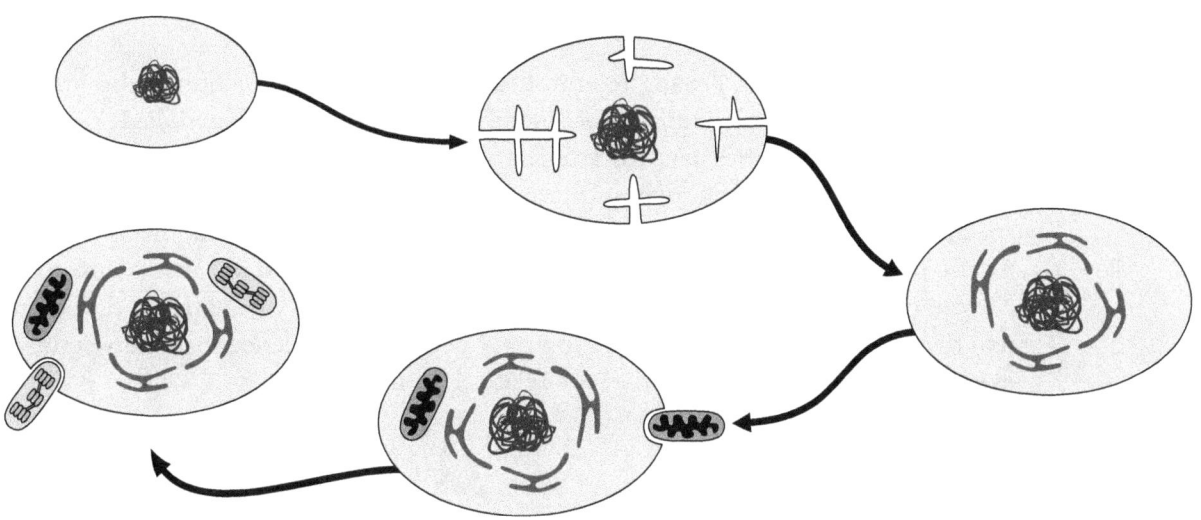

ALL STUDENTS MUST BE ABLE TO ANSWER THESE QUESTIONS

1. What is the endosymbiotic theory?
2. How did Kwang Jeon's observation proof that endosymbiosis may have repeatedly occurred throughout Earth's history?
3. What are the three types of symbiotic relationships and how do they differ?
4. What are the hypothetical series of events that lead to the evolution of eukaryotic cells?
5. How was membrane invagination important to the evolution of eukaryotic cells?
6. What are six evidences that endosymbiosis led to the formation of mitochondria and chloroplasts?
7. How has endosymbiosis and lateral gene transfer from the bacteria domain impacted the phylogenetic tree?

---

1. All Students must Be Able to perform the following tasks
2. Apply Kwang Jeon's observation to justify the claim that endosymbiosis has repeatedly occurred.
3. Use illustration to support your descriptions of the endosymbiotic evolution of eukaryotes.

   Big Idea 1B1b3-4

The **endosymbiotic theory** explains how modern eukaryotic cells may have taken the giant evolutionary leap from a single cell organism to multicellularity. According to this theory, both mitochondria and chloroplasts are descended from formerly free-living prokaryotes that became **phagocytized** (engulfed) by another cell about 1.5 billion years ago.

KWANG JEON'S OBSERVATION OF ENDOSYMBIOSIS

In 1966, microbiologist Kwang Jeon noticed an unexpected infection of his amoeba colony by a strain of x-bacteria. Most of the amoeba became sick and died. However, after several months, the surviving amoeba and their descendants were healthy despite having x-bacterium thriving within their cytoplasm. When treated with antibiotics to specifically kill the x-bacteria (antibiotics do not normally kill amoeba), the host amoeba also died. Jeon hypothesized that the amoeba had evolved to become symbiotically cooperative with the x-bacteria so that they now required an x-bacteria produced protein to continue living. This observation provided strong evidence that endosymbiotic events have occurred and that they are relatively common.

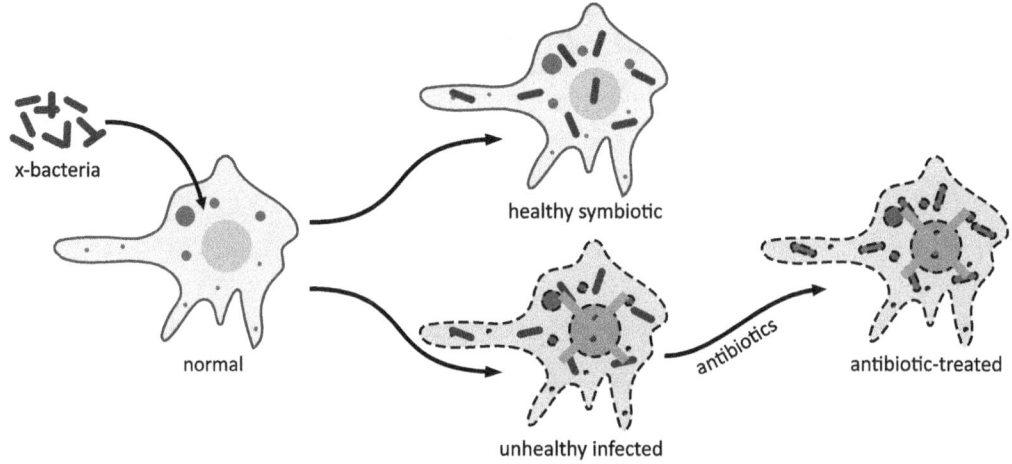

ENDOSYMBIOSIS IN EUKARYOTIC EVOLUTION

The hypothetical series of events that lead to the evolution of eukaryotic cells include:
1. A precursors prokaryotic cell increased in size and developed membrane invagination to increase surface area to volume ratio for enhanced nutrient uptake.
2. Invaginated membranes pinched off to form the early **endomembrane systems** that later becomes a complex of membrane structure that forms the nuclear membrane, ER, Golgi apparatus, and vesicles. This cell was now the primitive eukaryote (prokaryotes do not have endomembrane systems).
3. An **aerobic prokaryote** became engulfed by this early eukaryote and the two cells (endosymbiont and host) evolved a **symbiotic relationship** that benefited both cells. In this relationship, the host provides nutrients, water, protection, and the endosymbiont produces ATP through aerobic respiration. The endosymbiont underwent further evolution to become the modern mitochondria.
4. Another evolutionary lineage branched off when one of the aerobic eukary-

otes engulfs a photosynthetic **cyanobacterium** that becomes a second endosymbiont. This photosynthetic endosymbiont underwent further evolution to become the modern chloroplast. This second lineage give rise to plants and other lineage give rise to the non-photosynthetic eukaryotes.

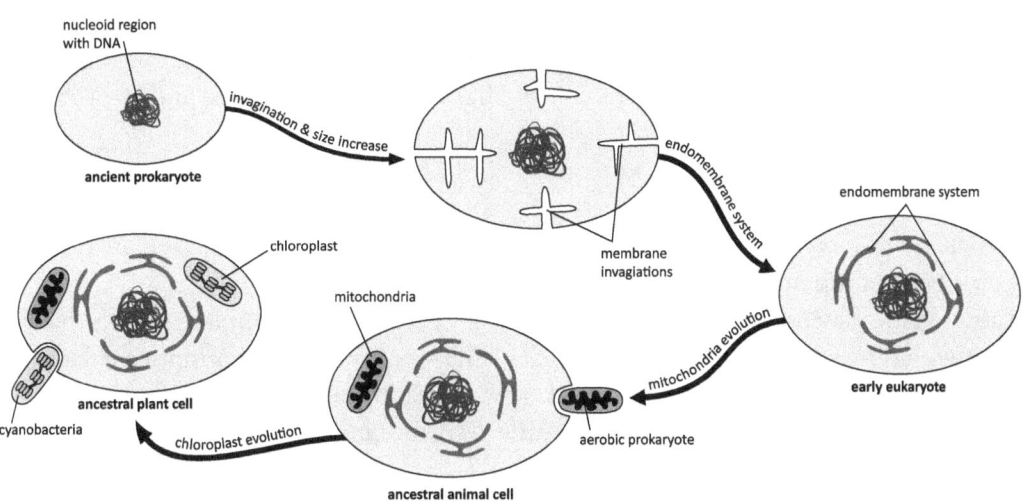

Evidence for endosymbiosis

Evidence in support of the endosymbiosis theory include:

- Mitochondria and chloroplasts have inner and outer membranes, suggesting that they underwent successive endosymbiosis events that preserved the membrane of the original bacteria endosymbiont.
- The outer membranes of mitochondria and chloroplasts contain transport proteins called porins and lipids called cardiolipin found in bacteria membrane.
- Chloroplasts and mitochondria are similar in size to other prokaryotes.
- DNA in both mitochondria and chloroplasts are more similar to the bacterial circular DNA in structure and genetic composition.
- The ribosomes of mitochondria and chloroplast are more similar to bacteria cell ribosomes than to the cytoplasmic eukaryotic ribosomes.
- Both mitochondria and chloroplast reproduce in a similar manner to bacteria through the process of binary fission that divides the parent cell into two equal daughter cells.

ENDOSYMBIOSIS AND THE PHYLOGENETIC TREE

The following figure shows how the evolution of eukaryotes through endosymbiosis fits into the phylogenetic tree. Although eukaryotes are one of the 3 main branches of the tree, the evidence for endosymbiosis suggests that there has been lateral gene transfer of genetic information from the bacteria domain to the eukaryotic domain after the 3 lineages split. It is also safe to assume that there may have been lateral transfer between the Achaea domain and eukaryotes and bacteria as well.

SUMMARY OF KEY CONCEPTS

- Eukaryotes have internal membrane-bound organelles, including the ER, Golgi apparatus, vesicles, and nucleus that make-up the endomembrane system.

- They also have linear DNA packed into structures called chromosomes rather than circular DNA.

- Prokaryotes lack internal membrane-bound organelles and nucleus. They have a circular DNA.

- Mitochondria and chloroplast are more similar to prokaryotes (circular DNA, division by binary fission, similar membrane bound proteins and lipids, ribosomes, genes, and size) than to eukaryotes in which they are contained.

- Scientists believe that eukaryotes evolved from a series of processes including two endosymbiosis events.

- An early prokaryote increased in size; membrane invagination increased surface area to volume ratio for better nutrient absorption.

- Infolded membrane pinched off internally to form an early endomembrane system that included the nuclear envelop, ER, Golgi, and vesicles. This became the first primitive eukaryote.

- Aerobic bacteria was engulfed by the early eukaryote, established internally as an endosymbiont and eventually evolved into the mitochondria. This became the ancestor of all modern eukaryotic heterotrophs.

- Some early eukaryotic heterotrophs subsequently engulfed photosynthetic cyanobacteria that later evolved to become the chloroplast.

- Both mitochondria and chloroplasts have a double membrane with the inner membrane possibly preserved from the original engulfed prokaryote.

CHECK YOU UNDERSTANDING

1. Which of the following statements is a correct description of the endosymbiotic theory?

   (A) It explains how the organic soup give rise to eukaryotic cells.

   (B) It describes how macrophages are able to phagocytize amoeba.

   (C) It describe how the chloroplast and mitochondria may have evolved in eukaryotic cells.

   (D) It explains the consequences of prokaryotic phagocytoses of eukaryotes.

2. Kwang Jeon's observation and investigation showed that

   (A) natural selection favors amoeba cells with inherited characteristics that allowed them to form a mutualistic relationship with the x-bacteria.

   (B) Antibiotics only affected the bacteria endosymbiont, not the host.

   (C) A majority of the infected amoeba could live with an x-bacteria endosymbiont.

   (D) The x-bacteria endosymbiont became a new organelle in the amoeba host.

3. Which of the following provides the likely evolutionary sequence (first to last) for the following eukaryotic organelles?

   (A) chloroplast > mitochondria > nucleus

   (B) golgi > chloroplast > mitochondria

   (C) rough ER > mitochondria > chloroplast

   (D) mitochondria > vacuole > chloroplast

4. Which of the following statements in incorrect?

   (A) Aerobic bacteria can produce ATP an oxidizing atmosphere.

   (B) Cyanobacteria was likely the first life form.

   (C) Mitochondria and chloroplasts both have an inner membrane that is believed to have been preserved from the original endosymbiont.

   (D) The genetic material of mitochondria is packaged into 4 linear DNA strands.

ANSWERS & EXPLANATIONS

1. (C) is correct.

2. (A) is correct. Those amoeba cells that survived and thrived must have had inherited characteristics that allowed them to develop a symbiotic relationship with the x-bacteria.

3. (C) is correct. Recall that the endomembrane system (nucleus, ER, Golgi, and vesicles) evolved from membrane invagination and infolding, followed by mitochondria and chloroplasts.

4. (D) mitochondria and chloroplasts have a single circular chromosome that is similar to that of bacteria.

BIG IDEA 2

Biological systems

- THERMODYNAMICS
- ENERGY IN LIVING SYSTEMS
- MATTER IN LIVING SYSTEMS
- MEMBRANES
- CELL COMPARTMENTALIZATION
- ENERGY CAPTURE
- DYNAMIC HOMEOSTASIS

09 Thermodynamics

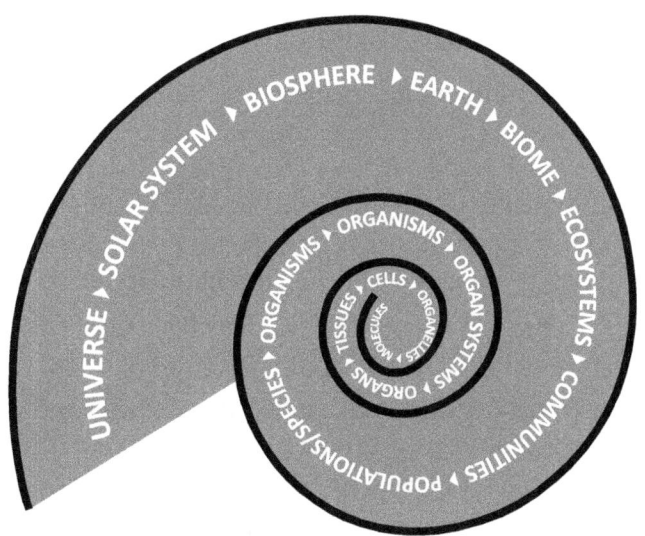

ALL STUDENTS MUST BE ABLE TO ANSWER THESE QUESTIONS

1. What is thermodynamics?
2. How does thermodynamics influence the relationship between the system and surrounding?
3. What are the levels of organization in biological systems?
4. Why is the Earth considered an closed system (although it is actually an open system)?
5. What are the three laws of thermodynamics?
6. How are each of the laws of thermodynamics relevant to biological systems?
7. Hoes does energy enter and flow through ecosystems?
8. How do biological systems used energy as heat (q) and energy as work (w)?
9. What is the relationship between enthalpy (H), Gibbs free energy (G), temperature (T), and entropy (S)?
10. Which metabolic processes are primarily used by biological system to absorb and release energy?
11. How do biological system use coupling mechanisms to ensure internal organization?

ALL STUDENTS MUST BE ABLE TO COMPLETE THE FOLLOWING TASKS

12. Use examples to explain why the laws of thermodynamics are relevant to biological systems.
13. Use provided information to determine is ΔE and/or ΔG is positive, negative or zero, and whether the biological process is endergonic or exergonic.

 Big Idea 2A1a-b

Thermodynamics is the study of how heat flows within and between systems, and this influences the movement of matter and energy. For example, the movement of warmer air toward colder air causes weather winds, and variation in localized water temperatures produces the ocean currents.

SYSTEMS AND SURROUNDINGS

In **thermodynamics**, the temperature and energy of two defined parts of the universe is compared. These parts can be as grand as the whole planet versus the rest of the universe, or as specific as a cell's organelle versus the rest of the cell. In both instances, the smaller part is contained within the larger part and is referred to as the **system** and the **surrounding**, respectively. In other instances two systems that share a surrounding can be compared. For instance, a population of lions and another of gazelles that coexist in the Serengeti ecosystem are separate systems what a share surrounding.

Biological systems can exist at a number of different levels of the biological hierarchy, such as the biosphere, biomes, ecosystems, species, populations, individual organisms, organ systems, organs, tissue, cells, organelle, macromolecules, genes, and more.

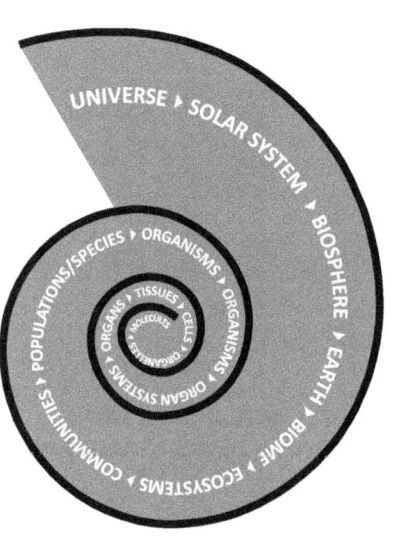

Levels of organization
- Universe
- solar system
- biosphere
- biomes
- ecosystems
- communities
- populations
- organisms
- organ systems
- organs
- tissue
- cells
- organelles —— nucleus
- molecules —— chromosomes
  —— DNA
  • gene

A specific level of this hierarchy that is being studied is defined as the system to distinguish it from the rest of the surrounding universe. Of course, biologists typically do not consider the entire universe when conducting an investigation. Instead they focus on the system and a defined local surrounding.

Example 1
When studying a population of buffalo (biological system), an ecologist might identify the surrounding as the local habitat, considering only other local buffaloes, other local species (predators and competitors), local resources (food sources, water supply), and local weather conditions, and may chose not to consider global weather patterns, or drought or oil spill in a distant geographic areas.

Example 2

If the biological system is a cell, all lower hierarchical levels such as organelles and molecules are subsystems of the cell. This means that anything inside of the cell is a part of the cell and anything outside of the cell are part of its surrounding. Cell A is part of cell B's surrounding and vice versa. Collectively the cells can interact to form higher organization of tissues that have emergent biological properties that are absent in the individual cell. For example, a single neuron is unable to function as both a photoreceptor of light and a motor neuron that stimulates a muscle to contract. The collective of all the neurons that form the tissues of the nervous system can integrate all of these functions to coordinate an appropriate response to an external stimulus. The larger environment of the tissue forms the surrounding of each cell (system) that it contains. The same goes for tissues (system) that form organs (surrounding), and organs (system) that form organ systems (surrounding), and organ system (system) that for the whole organism (surrounding).

Opened, closed and isolation systems

An **isolated system** is a system that is completely seal off, exchanging neither matter nor energy with the surrounding. As far as scientist can tell, the only truly isolated system is the universe as a whole. An **open system** exchanges both energy and matter with the surrounding, while a **closed system** is like a glass house that exchanges energy, but not matter.

Earth is an open system in that extraterrestrial bodies such as asteroids can enter and spaceships can leave. However, the amount of matter exchange with the rest of the universe is so insignificant that we will, henceforth, consider the Earth a closed system that exchanges energy in the form of radiation and heat, but not matter.

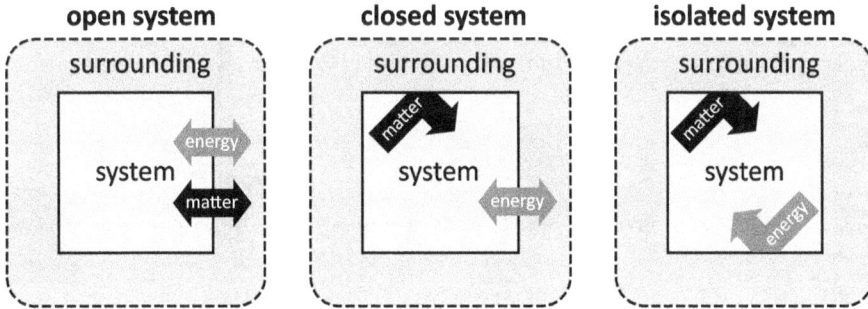

THE LAWS OF THERMODYNAMICS

The 3 laws of thermodynamics govern all biological processes and are one of the key unifying principles in biology.

The first law

The **first law** deals with the conservation of energy. It states that energy can be neither created nor destroyed. It can only be transferred from one substance to another or changed from one form to another. Two systems that are opened to each other will exchange energy until equilibrium is reached.

In the following example, system A (T = 100°C) releases energy to system B (T = 0°C) until they are in equilibrium with each other (T = 50°C).

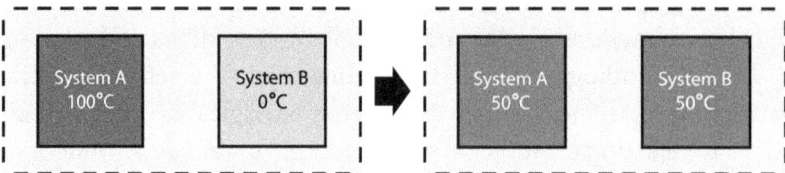

Relevance for biological systems

Living organisms receive energy through energy transfer from other sources. With the exception of **producers** (or **autotrophs**), all other organisms are **consumers** (**heterotrophs**) that get energy through feeding. Food contains energy that was stored in the covalent bonds of organic molecules such as carbohydrates, fats, and proteins. The energy is released through metabolic processes that breakdown the covalent bonds.

Producers such as plants are at the based of all food chains since they are the first entry point for energy into ecosystems. Through photosynthesis, they absorb energy from light and store it in glucose molecule. When an animal eats, it receives this stored energy. This energy is later released during **respiration** as glucose is metabolized. The energy is then recaptured by **oxidative phosphorylation** for the endergonic process of ATP production.

In ecology, organisms are grouped according to their specific energy source. Each group comprising of organisms that receive energy from a similar source is assigned to the same **trophic levels**. The first tropic level includes producers that source their energy from the sun. The second tropic level includes **primary consumers** that are **herbivores** such as cattle and deer that feed exclusively on producers. And the third (or higher) tropic levels include **secondary consumers** (or **tertiary consumer** or higher level) such as **carnivores** that feed exclusively on consumers of the lower trophic levels and **omnivores** that feed on lower level consumers and autotrophs.

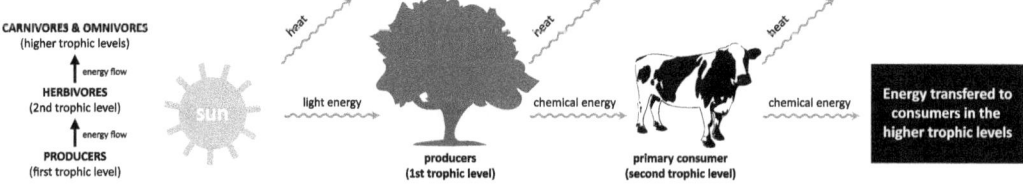

The second law

The **second law** deals with maximum entropy, or maximum disorder in biological systems. It states that an isolated system will naturally increase in **entropy (S)** over time by becoming more disorderly until a state of thermodynamic equilibrium is reached. At this point, it has achieved its maximum possible entropy state, meaning that it is maximally disorder.

Disorder is often counter intuitive. For example, if a drop of food coloring is added to a glass of water, you will immediately observe a seemingly disordered dispersion of the color that will diffuse throughout the liquid until the entire solution becomes

TROPHIC LEVELS
Organisms are assigned to trophic levels based on the highest level in which they feed. An omnivore that feeds on both the 1st and 2nd trophic level is assigned to the 3rd trophic level. And a hyena that feeds on the 2nd and 3rd levels is placed in the 4th trophic level.

homogenous. Intuitively, it may appear that the system began in chaos or disorder and ended with homogenous order. However, the reverse is correct. It began with order, and driven by thermodynamic forces (diffusion), became increasingly disordered until equilibrium was achieved (maximum entropy). Simply put, a system is more ordered when its component parts are locally concentrated rather than equally distributed. Thermodynamic equilibrium or maximum entropy is the point when equal distribution of "stuff" is achieved through diffusion.

Relevance for biological systems

Just as a room filled with stuff tends to become naturally disorder, a biological system such as a cell tends to become disordered over time. There is a threshold limit to disorder (different from maximum entropy) that each biological system can withstand before permanently losing functionality and dying. Biological systems have evolved special features to resist the thermodynamic drive toward this threshold limit (and toward maximum entropy). The most obvious feature is the cell membrane that limits the diffusion of component parts from the cell to the surrounding, slowing down the drive toward the entropy threshold. Every biological system has some sort of barrier toward diffusion that allows it to build up an ideal concentration of material and energy. Other examples of biological barriers include:

- Geographic barriers that separate 2 populations of the same species
- Behavioral barriers that separate 2 populations that recently diverged
- Reproductive barriers that separate 2 different species
- Organelle membranes that separate each organelle from the cytoplasm
- The physical structure of chromosomes that separates chromosomes from each other
- Non-coding parts of DNA called introns that separates the genetic parts of DNA called exons

No biological barrier can absolutely prevent the drive toward the thermodynamic threshold. Coupled with energy intake, biological barriers only slow down the drive and allow biological systems to propagate life. In time, every biological system, including ecosystems, will reach the threshold. This can happen by the gradual accumulation of disorder such as with old age, rapid accumulation of disorder such as in accidental death from a fetal wound or forest fire, targeted disruption of order due to sickness caused by pathogens, or an internal flaw in the biological system that prevents it from maintaining internal order (inherited fatal medical condition).

The third law

The **third law** deals with minimum entropy. It states that the entropy of a system correlates directly with its temperature, such that as the temperature of a system decreases, so does its level of disorder.

Relevance for biological systems

The amount of energy in biological systems correlates directly with the biological system's level of disorder. Simply put, a frozen ecosystem is less disordered than an ecosystem on fire. This explains why the initiation of life on Earth required a cooling event. In the early Earth, high amounts of free energy created a highly chaotic environment that prevented the organization of matter and energy into living systems. A cooling event reduced the amount of free energy, making the environment sufficiently stable for the organic soup to form, and for life to emerge.

HEAT AND WORK

Energy can be neither created nor destroyed, but it can change form. Whenever a system undergoes a transformation there is a change in its energy (Δ**E**). For example, a redwood forest has considerable energy stored in its tissue of the organisms that inhabit this ecosystem, particularly within the tissue of the dominant tree species. During a forest fire, a significant portion of this stored energy is lost. The difference in energy in the forest before and after forest fire is $\Delta E = E_{after} - E_{before}$, which in this instance is negative.

In biological systems, energy is used to do **work** (**w**) by moving matter, or it is exchanged as **heat** (**q**) so that internal thermal equilibrium is maintained. The following are specific examples of how biological systems use energy:

- A predator move its body to chases a prey by muscle contraction (w), which produces body heat (q) due to respiration. ($\Delta E < 0$)
- Proton pumps push hydrogen ions across the inner mitochondria membrane (w) against the concentration gradient of hydrogen ions. ($\Delta E < 0$)
- Plants absorb sun energy and use the energy to build glucose and ATP molecules (w), to transport water and minerals (w), and release heat (from respiration) by transpiration (q). ($\Delta E = 0$)
- Fish swims by moving its fins against the surrounding water (w), which produces heat (q) from respiration that is exchanged with the surrounding water. ($\Delta E < 0$).
- A bear moves various body parts (w) to eat and digest honey. ($\Delta E > 0$)

GIBBS FREE ENERGY

The total energy of a biological system (**enthalpy, H**) includes the energy used for organization (negative **entropy change, -S**), heat (**temperature, T**), and **Gibbs free energy** (**G**).

$$H = G - TS$$

Heat measured as temperature (T)

Warm-blooded organisms such as mammals have special metabolic process to convert the chemical energy stored in the bonds of organic compounds into heat. Much of the solar radiation that enters the ecosystem is also converted to heat that is than absorbed by organisms. For example, cold-blooded organisms such as reptiles heat their bodies by lying in the sun.

Energy used for organization (-ΔS)

The energy of systems is measured as energy comparison before and after the system changes. The change in the energy that a biological system uses for internal organization is a negative magnitude to entropy change (-ΔS) because it counters entropy. It includes the energy that is used to build structural components (proteins, lipids, and carbohydrates) and to perform work (muscle contraction, vesicles, cytokinesis).

The change in Gibbs (ΔG) is determined using the following equation,

$$\Delta G = \Delta H - T\Delta S$$

where ΔG is the change in Gibbs free energy, ΔH is the change in total energy (or enthalpy change), T is temperature, and ΔS is entropy change. ΔG can be used to determine if biological events release or absorb energy.

- $\Delta G < 0$, spontaneous or exergonic process; energy released to surrounding
- $\Delta G > 0$, endergonic process; energy absorbed from surrounding.
- $\Delta G = 0$, equilibrium.

Exergonic and endergonic reactions are coupled

No reaction is truly spontaneous because activation energy (AE) is needed to initiate each reaction. This benefits biological systems by keeping molecules in a stable state, allowing organisms to store the energy and release it only when needed. Protein catalysts called **enzymes** lower the activation energy needed to speed up metabolic reactions. These enzymes do not affect the molecule's Gibbs free energy.

Aerobic respiration is coupled to oxidative phosphorylation

The energy absorbed during photosynthesis is stored in glucose. When glucose is metabolized during aerobic respiration, the energy is released (exergonic). This exergonic process is coupled to oxidative phosphorylation that captures the released energy and stores it ATP (endergonic). Thus, while glucose is releasing energy, ATP is absorbing it into its high energy phosphate bonds. ATP is the most common use fuel for the various endergonic metabolic processes of cells.

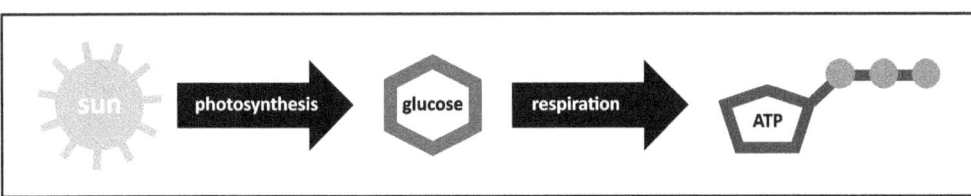

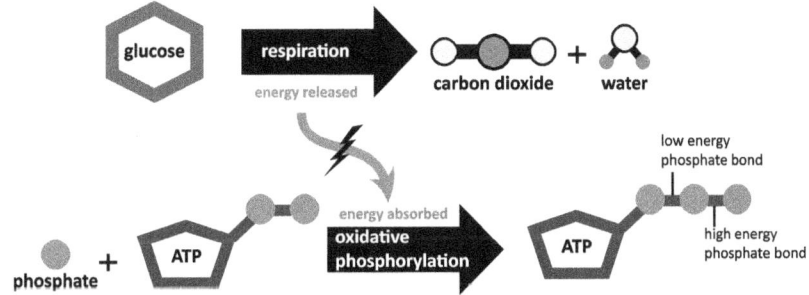

SUMMARY OF KEY CONCEPTS

- The universe has a natural tendency toward disorder.
- Disorder is uniformity; an equal distribution of energy and matter through space.
- Energy is needed to counter the drive toward thermodynamic equilibrium (maximum disorder or maximum entropy).
- To maintain order, all living organisms need energy to power biological mechanisms that act to counteract entropy.
- The loss of energy or order will cause living organisms to die.
- Living systems follow the second law of thermodynamics by maintaining a balance between order (decreased entropy) and disorder (increased entropy).
- Maintaining this balance is energy demanding, requiring living systems to take in more energy than is used up.
- Exergonic reactions release captured free energy to living systems for the maintenance and increase of order.
- Endergonic reactions do not occur spontaneously because they require energy input from the surrounding. Many endergonic reactions take in energy by absorbing heat from the surround (endothermic).
- Oxidative phosphorylation is coupled to respiration.
- Both mitochondria and chloroplasts have a double membrane with the inner membrane possibly preserved from the original engulfed prokaryote.

CHECK YOU UNDERSTANDING

1. Which of the following statements about the Serengeti plains is correct?

    (A) If both lions and hyenas live in the plains and are fierce competitors, the lions are part of the surrounding of the hyena.

    (B) The grasses of the plains are the primary producers, while the lions and hyenas reside in the third trophic level or higher.

    (C) The Serengeti plains are at the ecosystem level of biological organization.

    (D) All of the above are correct.

2. A systems that has a non-porous transparent boundary is considered a

    (A) closed system

    (B) opened system

    (C) isolated system

    (D) semi-closed system

3. 10 mL of solution A (conc. = 10 g/ml sucrose) is placed in a dialysis tube that is permeable to water, but not sucrose. The tube is than sealed and immersed into a beaker contain 10 mL of Solution B (conc. = 5 g/ml sucrose). What would you expect to observe at equilibrium?

    (A) [solution A] > 10 g/ml

    (B) [solution B] > 10 g/ml

    (C) volume of solution A > 10 mL

    (D) volume of solution B > 10 mL

4. In an experiment, 10 caterpillars consumed 100 grams of collard greens. If the greens store 7 kilocalories per gram of total mass and caterpillar store 9 kcal/g, what is the average mass of each caterpillar? Assume that only 10% of the energy that enters a trophic level is passed on to the next trophic level.

    (A) 128 g

    (B) 78 g

    (C) 0.8 g

    (D) None of the above statements is correct.

5. Which of the following statements is an accurate assessment?

    (A) If the cell membrane of a bacterial cell was made porous by a certain drug, the bacteria will rapidly reach thermodynamic equilibrium with its surrounding and die.

    (B) Sunscreen lotions provided an additional protective boundary that blocks UV radiation from entering the skin cells of the sunbather.

    (C) The ozone layer is a boundary component of the Earth that specifically blocks out harmful radiation.

    (D) All of the above are accurate statements.

6. While grazing, a zebra failed to notice the pack of wild dogs racing toward its direction. While the dogs fed on the gazelle's remains, a crowd of vultures anxiously waited for a chance at the leftovers. Which of the following statements about work and energy is correct?

   (A) $\Delta E_{dogs} > 0$ (after chase, but before feeding)

   (B) $\Delta S_{gazelle} > 0$

   (C) $w_{dogs} > 0$

   (D) All of the above are correct.

7. The process of oxidative phosphorylation

   (A) converts glucose to ATP

   (B) release energy from glucose

   (C) harnesses the energy released by respiration during glucose breakdown, and uses this energy to produce ATP molecules.

   (D) All of the above are correct.

ANSWERS & EXPLANATIONS

1. (D) is correct. Since the lions are external to the hyena, and vice verse, they inhabit the same local environment, and are competitors, option (A) must be correct. (B) is correct because grasses are the primary producers in plains ecosystem, and the lions and hyenas are carnivores that are minimally in the third trophic level - the second trophic levels contains herbivores, and the first, producers. (C) is correct because the Serengeti plains is a grassland ecosystem.

2. (A) is correct because transparent systems can let in energy as light, but since it is non-porous, it restricts the exchange of matter.

3. (C) is correct. At equilibrium, [sucrose] = 10 ml (10g/ml + 5 g/ml) ÷ 20 ml = 7.5 g/ml. Both options (A) and (B) are eliminated because the exceed the predicted equilibrium concentration. Since only water can move across the membrane, solution A will be diluted down to 7.5 grams by gaining water from solution B, causing its volume to become greater than 10 mL.

4. (C) is correct. 100 g greens contain 700 kcal of energy. The caterpillars receive 10% of this energy, or 70 kcal. Since the 9 kcal of energy is stored per gram, 70 ÷ 9 = 7.8 g. Since there are 10 caterpillars, 7.8 g / 10 = 0.8 g.

5. (D) is correct.

6. (B) is correct. The gazelle has greater disorder (entropy) after death, therefore ΔS is positive. Option (A) is wrong because the process of chasing releases energy to the environment and causes ΔE to be negative. It is only after the dogs feed that they will have a net gain of energy. (C) is wrong because the dogs performed work by chasing the gazelles.

7. (C) is correct.

10 Energy in living systems

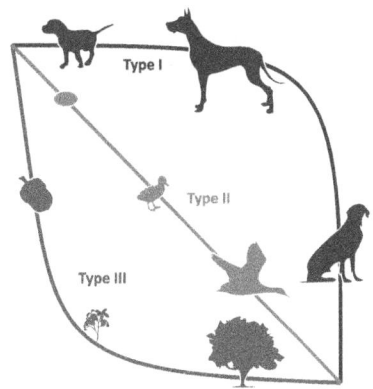

ALL STUDENTS MUST BE ABLE TO ANSWER THESE QUESTIONS

1. Why do organisms need energy, and what are 5 ways that they use excess consumed free energy?
2. What is the consequence of organisms experiencing a loss of internal order or free energy flow?
3. How is the natural drive toward disorder and entropy offset in living systems?
4. What is the coupling of endergonic and exergonic processes used to help maintain order in biological systems?
5. What are the primary mechanisms used by plants, ectotherms, and endotherms for maintaining optimal internal temperatures?
6. Why does reproduction require additional energy beyond what is needed for growth and maintenance of organization?
7. How do k- and r-strategists differ in their metabolic and reproductive rates, and size?
8. How is seasonal reproduction typically synchronized with the amount of available free energy?
9. How do annuals, biennials, and perennial differ in their reproduction and life-cycle?
10. How does mass-specific metabolic rates compare to whole-body metabolic rates, and what are their relationship to body size?
11. How does the amount of available free energy effect population size and the ecosystem in general?

ALL STUDENTS MUST BE ABLE TO COMPLETE THE FOLLOWING TASKS

1. Use a provided set of data to explain how biological systems use energy for organization, growth, and reproduction.
2. Use examples to demonstrate that energy related pathways in biological systems are sequential.
3. Use a provided set of data to determine the survivorship pattern of organisms.
4. Use the historical trends in human population growth to explain the relationship between population size and available free energy.

 Big Idea 2A1

ORGANIZATION, GROWTH, AND REPRODUCTION

The energy that organisms consume is used for maintaining organization, growth, and reproduction. For example, following implantation in the uterine wall, the fertilized egg receives energy in nutrients that diffuse from the mother's blood. The embryo uses the energy to progressively transform into a functional multicellular organism. This involves various stages of embryonic development through which the cell divides (growth and reproduction) and builds specialized tissue and organs (growth and organization).

While organisms are busy absorbing energy, entropy is relentlessly taking it back. This loss of energy to entropy can reverse effort at organization. However, healthy organisms consume considerably more energy than they lose to entropy, allowing them to counter the effect.

Energy intake occurs through sequential processes

Energy enters biological systems at different points. In producers, it enters through the process of photosynthesis that converts captured light energy into chemical energy as glucose molecules. Consumers receive their energy directly from glucose. However, most endergonic biochemical pathways within cells cannot directly use the energy stored in glucose. Glucose must be metabolized to release the energy that is than used to make ATP through the process of oxidative phosphorylation. Metabolism of glucose occurs through anaerobic respiration (fermentation and glycolysis) and aerobic respiration (Krebs and Calvin cycles). Prokaryotes lack mitochondria and can produce 2 ATP only via the less efficient process of fermentation. Eukaryotic cells use aerobic respiration to produce about 36 ATP molecules.

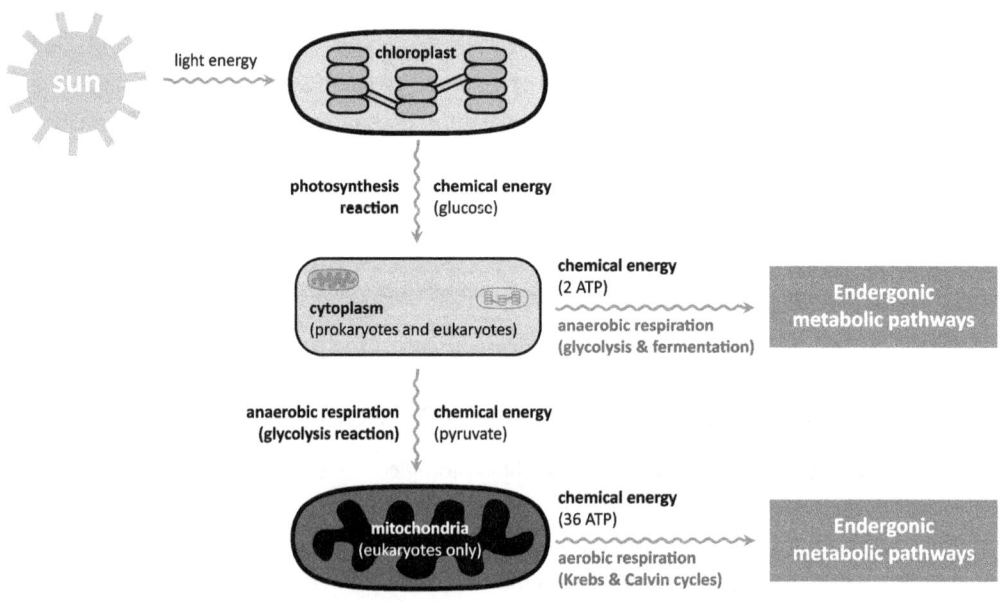

What is the excess energy used for?

- During time of excess food resources, energy can be stored in fats and carbohydrates and reserved for when energy sources are scarce.
- Energy can be used to build structural and functional molecules such as proteins, lipids, and fibers that facilitate internal organization. One such molecule is the phospholipid – the basic subunit of all cell membranes. As the primary barrier between cells and their surroundings, the cell membranes is critical to all living systems.
- Energy is constantly used to power all of the life-sustaining processes such as endergonic biochemical reactions, muscle contraction, and nerve impulse propagation.
- Cold-blooded organisms can absorb energy directly as heat and use it to help maintain internal thermal equilibrium.
- Warm-blooded organisms can also absorb energy directly as heat. However, through heat releasing exothermic reaction, they can metabolize fats and carbohydrates to maintain internal thermal equilibrium.

Free energy is used for regulating body temperature

Endotherms such as mammals and birds are warm-blooded animals that rely on internal mechanisms to maintain body temperatures. Two important heat-producing mechanisms used by endotherms are:

- Heat-releasing exothermic reactions that convert chemical energy into heat
- Heat released by rapid reflex contraction of muscles (shivering) in response to low temperature.

Having an internal heat source means that endotherms can remain metabolically active for longer periods in regions with sharp temperature differences between the day and night, or the seasons. This explains why most nocturnal animals are endotherms, and why they are also more successful than ectotherms in regions with cold winter seasons.

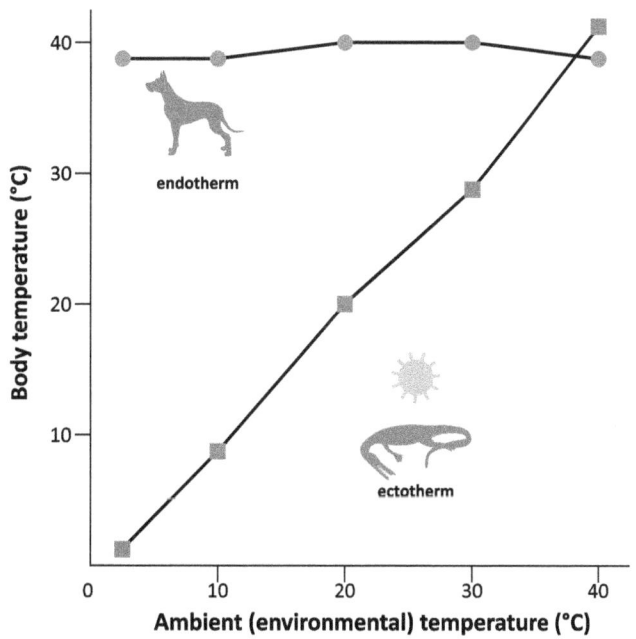

However, this comes at a cost of increased energy demand. A warm-blooded animal such as a lion must feed regularly, while a cold-blooded animal like the anaconda need to feed only once every few weeks.

Ectotherms such as reptiles, amphibians, and fish are cold-blooded animals whose body temperature equilibrates to that of the surrounding. Consequently, in order to maintain optimal thermal condition, ectotherms must adjust their behavior to exchange heat with their surrounding. For example, lizards will typically lie in the sun to reabsorb heat, or position their bodies in the wind or water to cool down.

Plants do not regulate their internal temperature, as do animals. However, they have the following internal mechanisms to protect against the damaging effects of frost:
- Production of chemicals that lower the freezing point of plant tissues
- Prior to the onset of winter, plants transport water out of their cells to avoid cells damage by ice crystals.

Free energy is used for reproductive success

k-strategists versus r-strategists

Species with higher reproductive rates such as grasses and mice are referred to as r-strategists (or r-selected) species. These organisms are typically small in size and consume a relatively small amount of energy per individual. They also reach maturity quickly, reproduce and die shortly after. They tend to display a Type III survivorship pattern in which most individuals die without having reproduced.

K-strategists such as elephants produce fewer offspring, provide more parental care, and have a lower mortality rate. They also require lots of energy per individual, take a more time to mature, and have a longer life expectancy. Unlike r-strategist, k-strategist can reproduce more than once in their lifetime. They tend to display Type I or II survivorship pattern in which most individuals lives to their maximum life span.

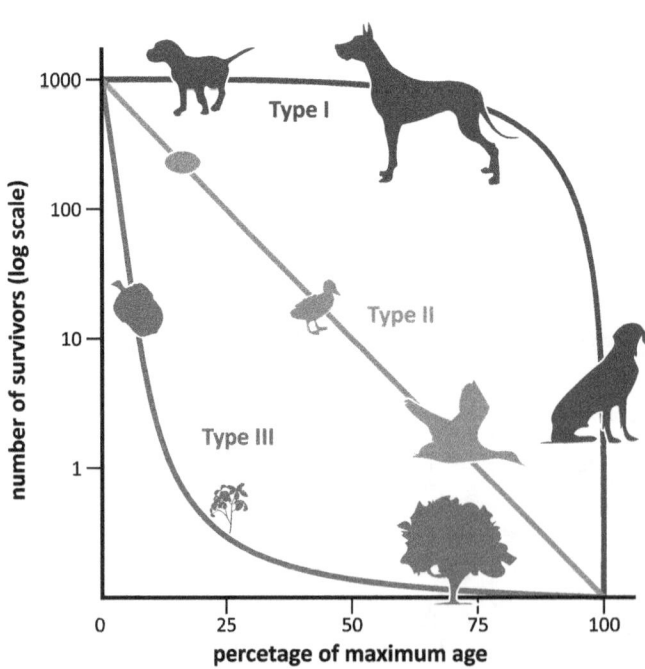

Seasonal reproduction

Plants and animal reproduce seasonally to coincide with the availability of free energy, nutrients, and water supplies. Because the care of offspring is energy demanding, parents must have access to more free energy than is needed for their own internal maintenance and growth. Consequently, peak reproduction occurs after winter months when a rise in temperature is coincided with increased plant productivity, which in turn increases the food supply and consumable free energy within the entire ecosystem.

Annuals, biennials, and perennials

Annuals, biennials, and perennials are adapted to life cycles of one, two, or more years, respectively with intermittent winter dormancy when growth is paused. Biennials require vernalization during the dormancy period before they are able to flower. Dormancy slows down the plants growth and prevents seed germination from occurring before winter has passed since if early germination was to occur the seedling will die from the winter frost.

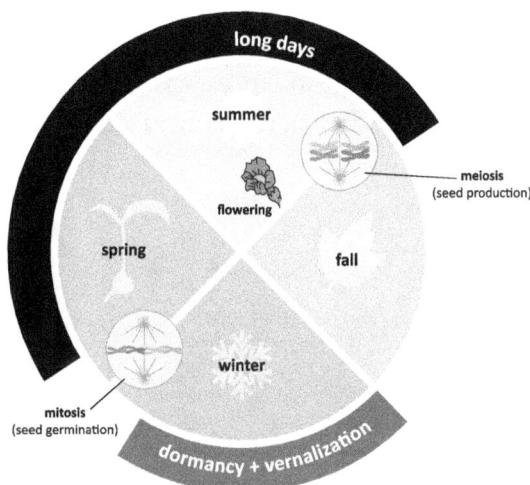

METABOLIC RATES RELATE TO BODY SIZE

As the mass of organisms increase, they require greater organization, which increases their energy demand. This explains why k-strategists use more energy than do r-strategists. The energy that an organism consumes can be used calculate its metabolic rate. There are two types of metabolic rates:

- Whole-animal metabolic rate is the amount of energy that the organism consumers per hour. Whole-animal metabolic rate correlates directly with the organism's mass, meaning that as an organism's mass increases, so does its whole-animal metabolic rate.
- Mass-specific metabolic rate is calculated by dividing the whole-animal metabolic rate by the organism's body mass. Mass-specific metabolic rate measures how efficient a specific organism is with energy. It is inversely correlated with the organism's size, meaning that the larger organisms have lower mass-specific metabolic rate and greater energy efficiency.

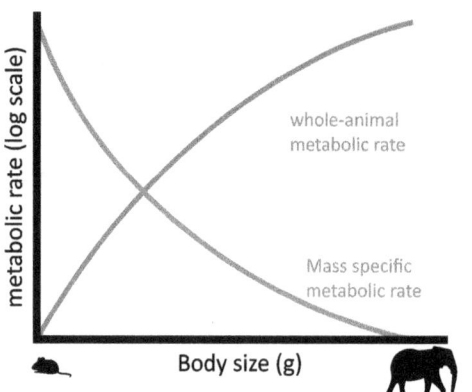

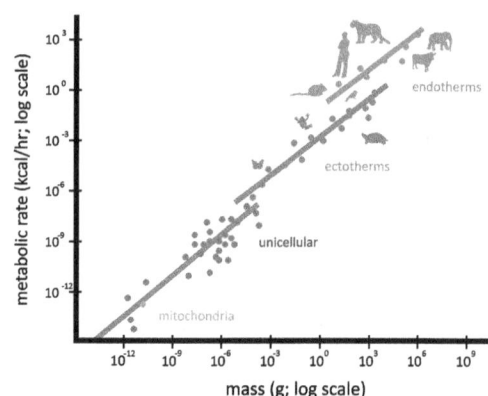

POPULATION SIZE DEPENDS ON AVAILABLE ENERGY

All animals receive a majority of their energy from the food that they consume. For example, the historical trends in human population growth correlate well with past improvement in agriculture technology that increased food supply. Today, humans are rapidly expanding their range through urban sprawl, deforestation, and overfishing, resulting in a reduced supply of food and energy to other species. Consequently, the population these organisms are declining, resulting in a current mass extinction event.

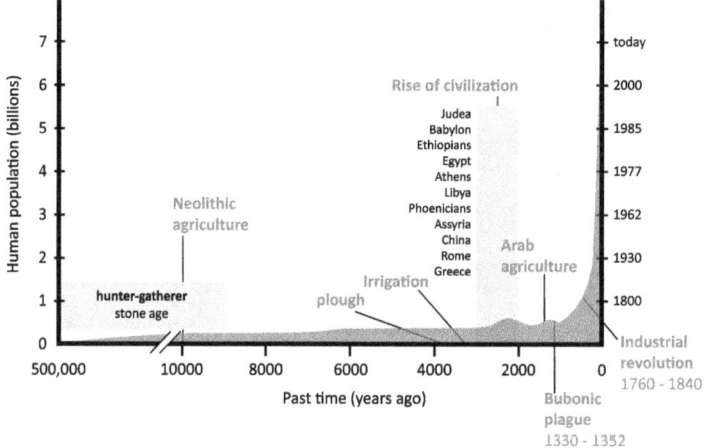

FREE ENERGY AND ECOSYSTEM STABILITY

Ecosystems are biological systems that are structured by trophic levels, with all organisms in a given trophic level receiving their energy from a similar source. Producers located in the first tropic level receive energy from sunlight, and store about 1% of solar energy they received. Each of the higher tropic level receives about 10% of the energy from the level below.

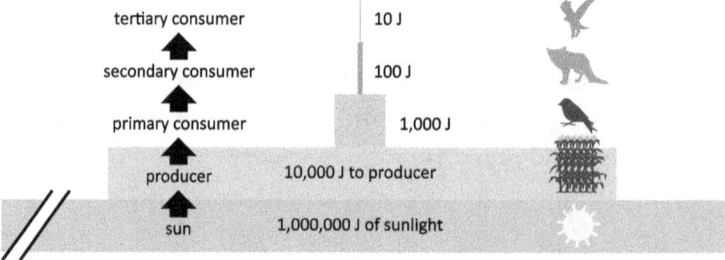

If access to sunlight is obstructed, primary productivity by producers drops, causing a disruption of energy flow to the higher level. This initiates a rippling effect throughout the ecosystem, possibly reducing the number of trophic levels and/or the sizes of each.

If the disruption is instead at the top trophic level the consequences can also be severe. For example, if the eagle in the figure is eliminated (-10J), the population of foxes in the third tropic level will increase. Foxes will than prey more heavily on the small birds and severely reduce their population. Because each trophic level is intertwined, these disruptions can have unpredictable consequences, and may lead to a complete collapse and destruction of the ecosystem.

SUMMARY OF KEY CONCEPTS

- Living organisms have different strategies for regulating body temperature and metabolism. Endotherms take in excess energy and use some for metabolic processes that increase body temperature (heat by metabolism). Ectotherms take in less energy, but have behavioral strategies to exchange heat with the surroundings (heat and cool by behavior)
- Plants that are adapted to cold environments produce low freezing point solutes, and have mechanisms to transport water out of cells to intercellular spaces to protect against freezing.
- The larger the organism, the more free energy is consumed (higher energy/individual; lower whole-animal metabolism). Smaller organism need more energy per unit mass (lower energy/individual; higher whole-animal metabolism) .
- Larger organisms are more energy efficient.
- Growth and reproduction require intake of excess energy. Loss of energy and dehydration causes death.
- Food supply directly correlates with energy supply and population size.
- Disruption in energy supply at one tropic level (particularly at the producer level) can have detrimental and unpredictable impacts on the ecosystem.

NOTES

CHECK YOUR UNDERSTANDING

1. Organisms use consumed energy for

    (A) growth
    (B) reproduction
    (C) maintenance of organization
    (D) All of the above.

2. Organisms use the excess energy than they consume for

    (A) to build energy storage molecules such as fats and carbohydrates.
    (B) synthesizing structural and functional molecules that are involved in internal organization.
    (C) producing heat.
    (D) All of the above.

3. Endotherms are organisms that

    (A) have heat producing processes that can maintain thermal homeostasis within a range of ambient temperature.
    (B) heat their bodies primarily through behavioral adjustments.
    (C) have the highest energy efficiency.
    (D) consume a small amount of food relative to their body size.

4. Which of the following would you expect to displays a Type III survivorship?

    (A) frogs
    (B) oysters
    (C) pine tree
    (D) All of the above.

5. Which of the following is likely an r-strategist?

    (A) eucalyptus tree
    (B) lions
    (C) polar bear
    (D) None of the above.

6. Which of the following statements in incorrect?

    (A) Annuals have a one year life cycle and are typically the first plants to bud in the spring.
    (B) Vernalization prevents seed germination.
    (C) Dormancy stops or slows down growth.
    (D) Both germination and dormancy continues until the passing of winter.

7. When comparing a whale and a dolphin,

    (A) the whale's whole-body metabolic rate is lower than the dolphin's.

    (B) The whale is more energy efficient than the dolphin

    (C) The dolphin's mass-specific metabolic rate is lower than the whale's.

    (D) The dolphin and whale have similar metabolic rates because they are both marine mammals..

8. During the period of 1330 to 1352, the change in human population can be attributed to

    (A) an increase in free energy due to improvements in agricultural technology.

    (B) a decrease in free energy due to a drop in crop productivity as new crop pathogens emerged.

    (C) outbreak of wars in Europe that reduced the European population.

    (D) None of the above.

9. What percent of the total energy that is absorbed by plants is transferred to the 4th trophic level?

    (A) 10%

    (B) 1%

    (C) 0.1%

    (D) 0.01%

10. Which of the following trophic levels likely contains the fewest number of individuals?

    (A) first

    (B) second

    (C) third

    (D) Additional information needed.

ANSWERS & EXPLANATIONS

1. (D) is correct.

2. (D) is correct.

3. (A) is correct. Option (B) is referring to ectotherms, not endotherms. (C) is wrong because endotherms use more energy per body mass than do ectotherms. Compared to endotherms, ectotherms require very little energy intake for maintenance of internal order, growth, and reproduction.

4. (D) is correct. All three organisms produce lots of offspring with very few surviving past adolescence.

5. (D) is correct because none of the organisms reach maximum maturity quickly, and each live has a relatively long life expectancy.

6. (B) is incorrect because vernalization prevents flower buds from forming, not germination.

7. (B) is correct because, although whales have a higher whole-body metabolic rate, they have a lower mass-specific metabolic rate. They used less energy per unit mass of body than do dolphins.

8. (D) is correct. The decline in global human population was due to human death from the bubonic plague.

9. (D) is correct.. First trophic level receives 10%, the second receives 1%, the third receives 0.1%, and the fourth receives 0.01%.

10. (D) is correct because in an a redwood forest where each of the dominant producer species is a large tree, the first trophic level might have very few individuals relative to the higher trophic level. This does not change the fact that the first trophic levels always has the most energy because each producer is large enough to store much greater amounts of energy. In a grassland ecosystem where the dominant producer is grass, the first trophic level should have the most individuals.

11 Matter in living systems

 Big Idea 2A3a

In addition to energy, biological systems exchange material with their environment. Material in the form of atoms and molecules are absorbed used for growth, reproduction, and the maintenance of internal order, or released as waste products.

Unlike energy that flows linearly through the trophic levels and leaves the ecosystem as heat, matter as a finite resource must be recycled. Biological systems rely on their surrounding to provide a constant supply matter in forms of elements and molecules.

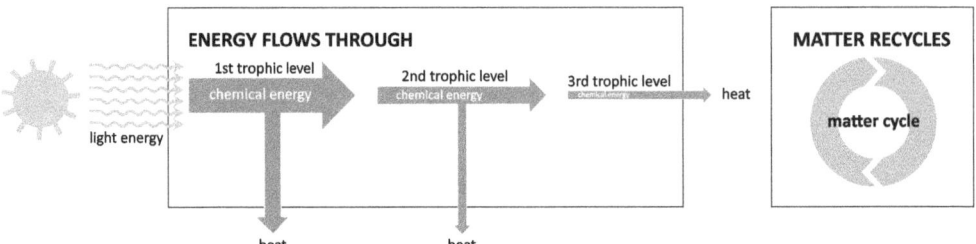

Organisms use the absorbed matter to build organic compounds that support life processes. All compounds require carbon, hydrogen, oxygen, phosphorus, and nitrogen, which are supplied through the following four biogeochemical cycles:

- Carbon cycle (supplies C and O)
- Water cycle (supplies H and O)
- Phosphorus cycle
- Nitrogen cycle
- The Carbon cycle

Carbon dioxide (CO_2) is the primary source of carbon atoms. It is absorbed from the atmosphere (or water in aquatic ecosystems) through the leaves of plants and combined with water (H_2O) during photosynthesis to produce glucose ($C_6H_{12}O_6$) and oxygen (O_2). Both plants and animals used the glucose and oxygen during respiration to release stored energy (produces CO_2 and water as by-products) that is than coupled to oxidative phosphorylation for ATP production.

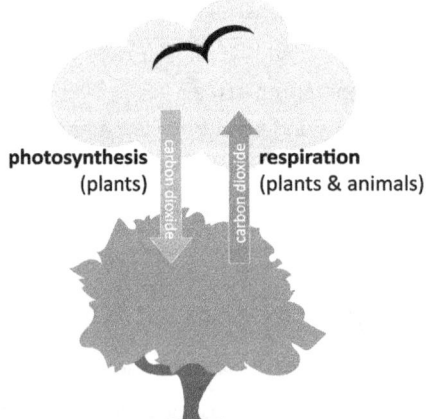

Not all of the absorb carbon is coupled to ATP production. Some are used for producing storage carbohydrates (plant **starches** and animal **glycogen**), structural carbohydrates (plant **cellulose** and animal **chitin**), and proteins, lipids, and nucleic acids (DNA and RNA).

THE NITROGEN CYCLE

Nitrogen (N_2), the most abundant gas in the atmosphere, is needed to build proteins and nucleic acids. Most organisms are not equipped to directly absorb nitrogen from the air. Plants called **legumes** receive their nitrogen supply directly from the symbiotic **nitrogen-fixing bacteria** that live in their roots, while other plant absorbed nitrogen that is deposited by nitrogen-fixing bacteria living in the soil, or by decomposition of dead organisms. **Denitrifying bacteria** release nitrogen from the soil and return it to the atmosphere. It can also be artificially added to the soil in fertilizer.

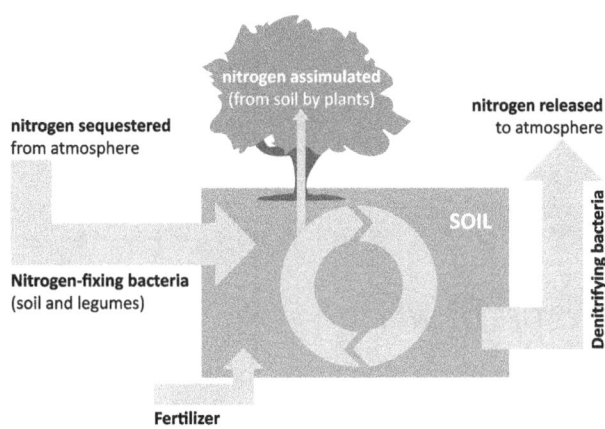

nitrogen assimulated
(from soil by plants)

nitrogen released
to atmosphere

nitrogen sequestered
from atmosphere

SOIL

Nitrogen-fixing bacteria
(soil and legumes)

Denitrifying bacteria

Fertilizer

THE PHOSPHORUS CYCLE

Most phosphorus in the environment is locked up in rocks and minerals or decomposed **humus** and plant residue. Some are artificially added to soil as fertilizer. As water flows over these phosphorus deposits, it is dissolved and can than be absorbed by plants. The plants assimilate the absorbed phosphorus into nucleic acids and certain types of lipids.

THE WATER CYCLE

The water cycle moves water between various water reservoirs, and in so doing has a major impact on ecosystem weather conditions. The key water reservoirs are:
- Clouds
- Ice (snow and glaciers)
- Surface bodies of liquid water such as oceans, seas, lakes, and rivers
- Underground aquifers of liquid water

Most clouds form as a result of water condensation following evaporation from liquid reservoirs or plant transpiration. Water is also released to the atmosphere directly from ice reservoirs through by sublimation.

Water returns to Earth by precipitation, which includes rain, snow, hail, and sleet. Snowmelt also returns water to liquid reservoirs, where water can be further distributed by both surface and groundwater flow. Terrestrial distribution delivers the water to plants for uptake. Plant use water to deliver nutrients, to transport macromolecules, for photosynthesis, and for cooling.

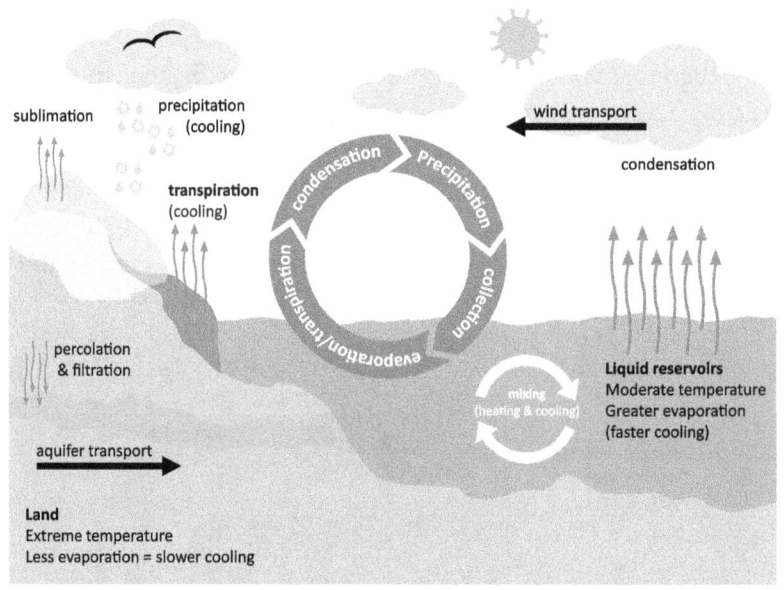

EMERGENT PROPERTIES OF WATER

In biology, an emergent property is a unique characteristics that arises only after the component parts of a biological system are assembled to form the completed biological system. For example, a cell can only display the characteristics of a cell when it is assembled with all of its parts, including organelles, cytoplasm, cell membrane. Likewise, water has emergent properties that are critical to life that arise with the component parts of water molecules - hydrogen and oxygen - are assembled together in the correct structure to form water molecules.

Polar properties

The most important emergent property of water is that it is a highly polar molecule, meaning that it has a relatively large charge difference the negative pole (oxygen end) and positive pole (hydrogen ends). This polarity is due to the presence of a strong electronegative oxygen atom (O) that is covalently bonded to 2 weak electronegative hydrogen (H) atoms that creates the partial charges at the poles.

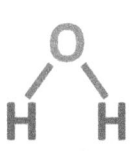 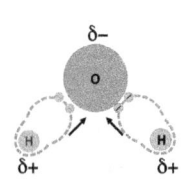 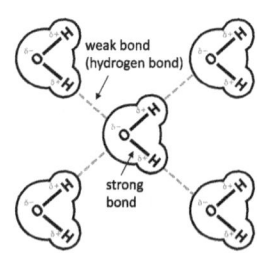

Cohesive and adhesive properties

The polarity of water allows it to form **hydrogen bonds** with other substances. Think of hydrogen bonds as a sticky intermolecular attractions (also called **Van der Waals interaction**) between its hydrogen atoms of water and nitrogen (N), oxygen (O) or fluorine (F) atoms of other polar molecules. H-bonds are responsible for **cohesion** between water molecules and **adhesion** between water and other polar substances.

Universal solvency, capillary action, and surface tension

Universal solvent property
Due to its adhesive properties, water is considered a **universal solvent** can adhere to and dissolve most substances to some degree. However, dissolving other polar substances, particularly those that form H-bonds with water. Substances that dissolve well in water therefore referred to as hydrophilic (water loving) substances.

Surface tension and capillary action properties
The movement of water through the vessels (xylems) of plants occurs by capillary action. Capillary action relies on cohesion between water molecules at the surface of water to establish a strong surface tension, and adhesion between water molecule and the vessels walls to lift the fluid up to the leaves.

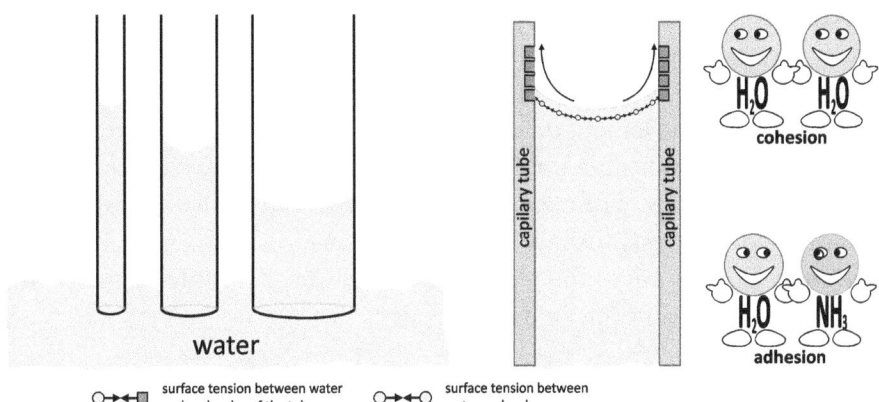

Heat capacity (thermal storage) & temperature moderation property

H-bonding and the cohesive property are directly responsible for the high **specific-heat (c)** of water that allows it to store heat (thermal storage). Specific heat is a measure of the relative amount of energy a substance can absorb before changing phase (solid → liquid → gas). For water c =4.2 J/g° C, meaning that 4.2 J of energy is required to raise the temperature of 1 gram of water by 1°C. This ability to store so much energy is due strong H-bonds that are able to withstand lots of energy before breaking and causing water to become a gas.

Biological systems benefit from the moderate global temperature that is kept within a livable range, thanks in large part to the ability of water to store lots of energy from the sun during the day (prevents daytime overheating), and to slowly release the energy during the nights (prevents nighttime freezing). This intermediate thermal conductivity makes water ideal for moderator of global temperature.

substance	SH (J/g° c)
iron	0.4
gold	0.1
ethanol	2.4
wood	1.7
ice	2.1
water	4.2
steam	2.0

Heat of vaporization & fusion
Heat of vaporization is the energy needed to break the hydrogen bonds that hold water molecules together so that it becomes a gas. **Heat of fusion** is the energy required to weaken the H-bonds so that ice melts. Both occur by the heating effect and require the absorption of free energy from the surroundings. The reverse occurs by the cooling effect.

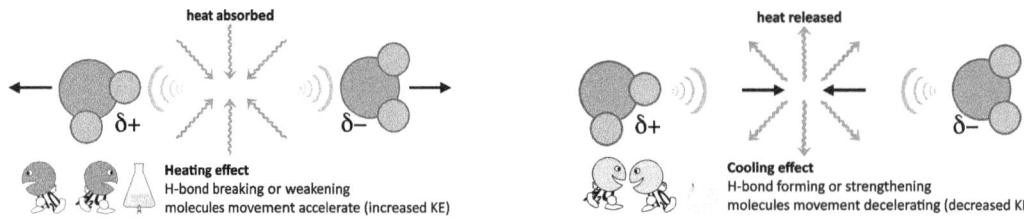

Ice is a good insulator

Ice is less dense than liquid water due to air becoming trapped between the water molecules as they liquid crystallize, resulting in a greater increase in volume than mass (decreased density). Ice therefore float on liquid water and can form a solid cover over aquatic ecosystem in regions that freeze in the winter. This ice cover also serves as an insulator of the aquatic ecosystem from the cold frigid air (think igloos). As the ice melts, it forms cold water that is denser than the ideal higher temperature water of the aquatic ecosystem beneath. Therefore, while the floating ice insulates from cold regions air above, cold water sinks, and warm water rises, protecting aquatic ecosystems from freezing. This movement of water base on temperature difference of the solid or liquid state is also responsible for the oceans currents that play a major role in moderating global temperature. For example, rising warm water causes local air temperature to rise at the equator, establishing a polarity between the air temperature at the poles and the equator. This drives the global air current due to winds caused by the movement of warm air toward the colder polar airs.

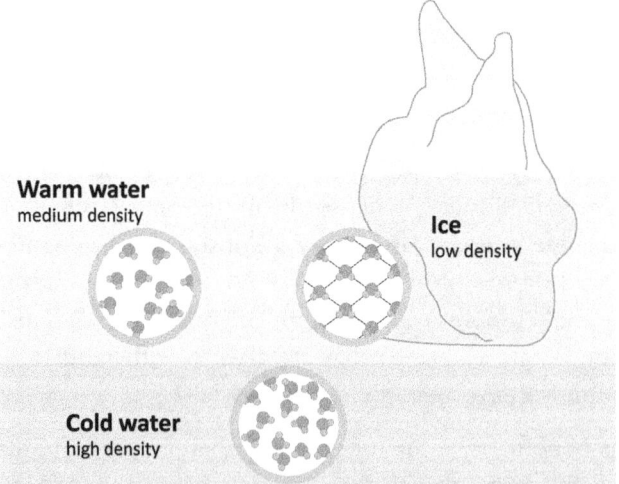

SUMMARY OF KEY CONCEPTS

- Polar compounds have partially positive ($\delta+$) and partially negative ($\delta-$).
- Van der Waals forces are responsible for intermolecular bonds that hold similar molecules (cohesion) and different substances (adhesion) together.
- Water is a polar molecule that forms hydrogen bonds, which are the strongest of the Van der Waals interactions.
- Van der Waals interaction give water its emergent properties: polarity, cohesive, adhesive, universal solvent, capillary action, surface tension, high specific heat, temperature moderator, insulator, density (ice < warm water < cold water).

CHECK YOUR UNDERSTANDING

1. The energy of an ecosystem

    (A) is finite.

    (B) is recycled.

    (C) must be release through controlled burning of managed forest so that the energy not build-up to dangerously levels.

    (D) typically enters as light and leaves as heat.

2. A primary natural reservoir for carbon is

    (A) carbon containing rocks.

    (B) fossil fuels.

    (C) the oceans.

    (D) none of the above.

3. The primary process through which carbon enters the trophic levels of ecosystems is

    (A) aerobic respiration.

    (B) fermentation.

    (C) photosynthesis.

    (D) oxidative phosphorylation.

4. The carbon absorbed by organisms is directly used for all of the following except,

    (A) photosynthesis

    (B) respiration in animals

    (C) building storage carbohydrates.

    (D) ATP production

5. Which of the following biogeochemical cycles is involved in oxygen recycling?

    (A) carbon cycle

    (B) water cycle

    (C) both

    (D) neither

6. All of the following are ways that nitrogen is introduced into ecosystems except

    (A) direct fixation into soil.

    (B) direct absorption by legumes.

    (C) fertilizer.

    (D) direct absorption through leaf stomata.

1. Which of the following statements about the water cycle is incorrect?

   (A) Precipitation returns water from the clouds to the surface.
   (B) Ambient terrestrial temperature experiences less fluctuation near large bodies of water than near dry, inland desert areas.
   (C) Most evaporation of water occurs in dry ecosystems.
   (D) Water is delivered to plants by percolation, aquifers, and surface flow.

2. All of the following is directly caused by an emergent property of water except?

   (A) Water dissolves hydrophilic substances.
   (B) The insect commonly known as water strider can walk on the surface of water.
   (C) Water quenches your thirst.
   (D) When only one corner of a paper towel was placed in a beaker of water and left there, the entire paper towel soon after became wet.

3. When heated in a pressurized container which of the following will occur?

   (A) Water will absorb more energy because the pressure will provide an additional counter force to the heat and will help hold the H-bonds together.
   (B) The boiling point of water will decrease because the pressure will increase the amount of force acting to break the hydrogen bonds.
   (C) There will be no observed effect on the heat capacity of water.
   (D) The heat capacity of water will decrease.

4. Which of the following lists the states of water in increasing order of density?

   (A) steam, warm water, cold water, ice
   (B) steam, ice, warm water, cold water
   (C) ice, warm water, cold water, steam
   (D) None of the above.

ANSWERS & EXPLANATIONS

1. (D) is correct. (C) is somewhat true in that forest are often controlled burned in order to reduce the build-up biomass from fallen leaves and other plant litter that is a wild-fire hazard. However, it is a artificial process.

2. (C) is correct because aquatic plants receive carbon from dissolved carbon dioxide.

3. (C) is correct.

4. (D) is correct. Oxidative phosphorylation is coupled to the breakdown of glucose during respiration, however it does not directly involve the use or release of carbon. Respiration (option B) in animals can directly metabolizes carbon from consumed glucose to produce carbon dioxide and water, and to release energy to oxidative phosphorylation.

5. (C) is correct.

6. (D) is correct.

7. (C) is correct.

8. (C) is correct. Option (A) is due to water's universal solubility. (B) is due to surface tension. (D) is due to capillary action.

9. (A) is correct.

10. (B) is correct.

12 Membranes

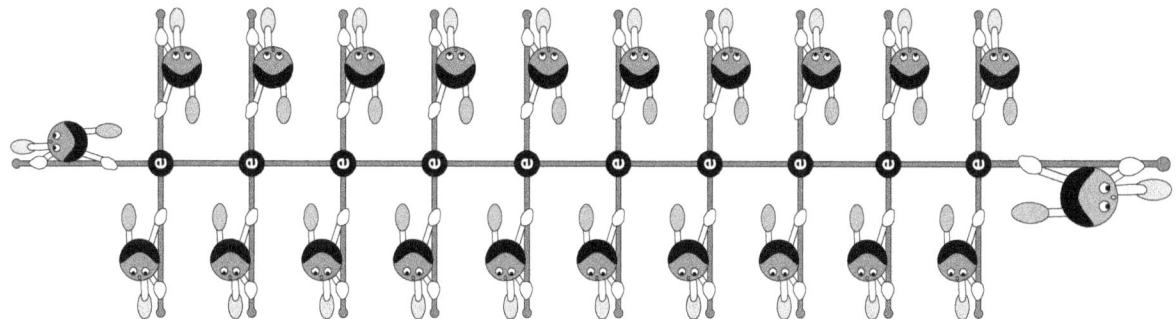

ALL STUDENTS MUST BE ABLE TO ANSWER THESE QUESTIONS

1. How does electronegativity influence the polarity and hydrophobicity of molecules?
2. How are hydrophobic, hydrophilic, and amphiphilic substances different in their properties and structure?
3. What are the 3 classes of lipids? What are their distinguishing structural and functional features?
4. What are the similarities between soap and phospholipids?
5. Why do micelles and phospholipid bilayers form?
6. What are steroids? What role do they play in stabilizing the cell membrane?
7. Which particles are able to simply diffuse across the cell membrane? Which cannot?
8. What are 4 challenges that selective permeability pose for the cell?
9. What are the 9 membrane bound structures that help the cell overcome these challenges?
10. Which properties determine whether a membrane bound protein is peripheral or integral?
11. What are the 2 general differences between active and passive transport?
12. What is the similarity and difference between simple diffusion, osmosis, and facilitated diffusion?
13. What are the 2 factors that effect the rate of diffusion across a membrane?
14. What happens to animal or plant cells placed in a hypertonic, hypotonic, or isotonic solution?
15. How does coupled transport work?
16. Under which conditions are endocytosis and exocytosis used for membrane transport?

ALL STUDENTS MUST BE ABLE TO COMPLETE THE FOLLOWING TASKS

1. Use models to investigate the function and properties of the cell membrane, including selective permeability, and active and passive transport.

 Big Idea 2B1-2

A cell is a biological system with a membrane to separate it from the surrounding. With this membrane boundary, cells are able to internally concentrate energy and matter, selectively secrete waste, and restrict the diffusion of resources and internally produced organizational molecules such as proteins, carbohydrates, and lipids. All of this is made possible by the hydrophilic and hydrophobic nature of the cell membrane that directly impacts its semipermeability.

HYDROPHOBIC VERSUS HYDROPHILIC

Hydrophobic substances are uncharged particles and non-polar molecules that tend to repel water molecules. **Hydrophilic** substances are charged particles (ions) and polar molecules that are attracted to water (cohesion) due to **Van der Waals interactions** with the δ+ and δ- poles of water molecules. Hydrophobic and hydrophilic substances do not mix well with each other. This means that nonpolar substances (fats and oils) are poor solvents for polar substances (water and alcohols), and vice versa. Instead, nonpolar hydrophobic substances are soluble in other hydrophobic **solvents**, and polar hydrophilic substances are soluble in other in polar solvents.

An easy way to determine where on the spectrum of hydrophobicity a particular substance lies is to add a drop of water to a surface that has been coated with the substance (substrate). A highly hydrophilic substance will cause water to spread more evenly over the surface so as to maximize their Van der Waals interaction., while when added to a highly hydrophobic substance, spherical water beads form that minimize interaction.

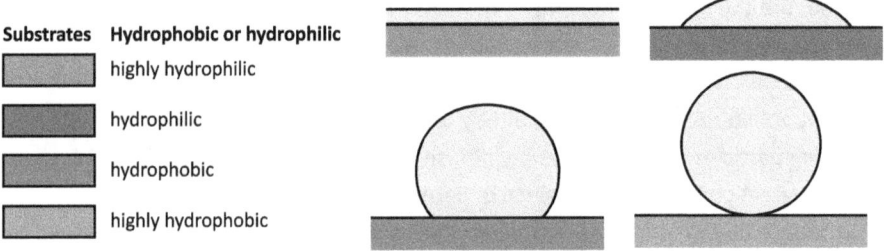

Adhesion of water to different surfaces

Structure of hydrophobic and hydrophilic substances

The shape of particles and molecules are determined in part by how much the nuclei of the substances atoms attract electrons. The strength of this attraction (**electronegativity**) correlates with where the specific type of atom (**element**) lies on the periodic tables. Excluding the **Nobel gases**, the periodic trend is that electronegativity increases across and up the table from the bottom left atom (Fr) to top right element (F).

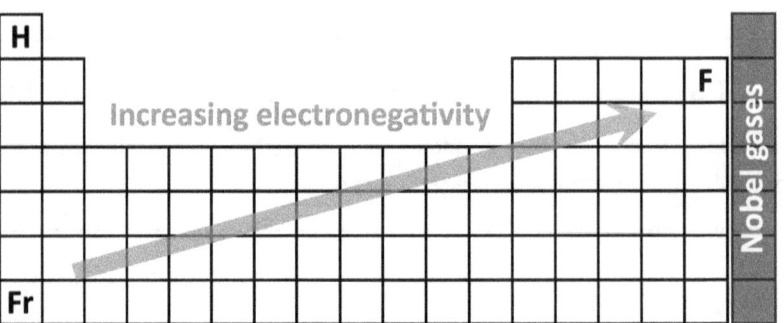

Hydrophobic substances

Hydrophobic substances do not mix well with water. They are nonpolar and have no charge, and are therefore not attracted to the δ- or δ+ poles of water. The absence of a charge can be a result of once of the following:

- Symmetric distribution of **electronegativity** among their atoms so that they lack partially charged poles. The illustration below is a analogy shows the symmetrical distribution of H atoms around carbon. Notice that the electronegativities of H atoms cancel out each other so that the molecule lacks polarity.

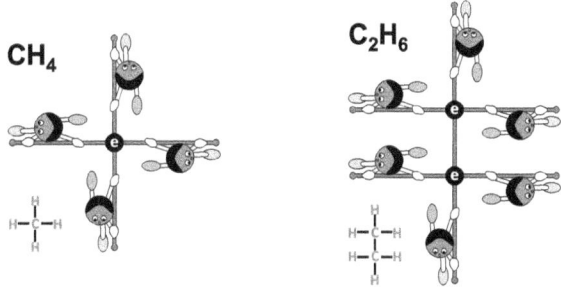

- The substance may be a single uncharged atom that have neither gained nor lost electrons to form an ion.

Hydrophilic substances

Hydrophilic substances are attracted to water because:

- The substance is a fully charged particle (ion) that have either gained (anion) or lost (cation) an electron.
- The substance is a molecule that has an asymmetrical distribution of electronegativity among its atoms, causing partially negative (δ-) and partially positive (δ+) poles to form. In the illustrative analogy below, notice that one of the hydrogens have been replaced with a highly electronegative chlorine atom (Cl) or hydroxyl functional group (-OH) so that the molecule is no longer symmetrical balanced. The chlorine and hydroxyl ends are δ-.

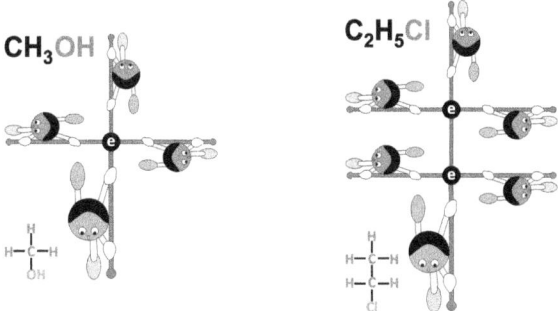

Amphiphilic substances

Some large hydrocarbons, such as phospholipids and soap, contain both hydrophobic and hydrophilic region. These substance are classified in a separate group and are referred to as amphiphilic substances. It is the amphiphilic nature of phospholipids that allows cell membranes to form. In the following illustrative analogy, notice that soap molecule has a long tail with a symmetric distribution of hydrogen atoms, except

NOTES

POLARITY
Single atoms are never polarized because polarization requires a molecule with at least two atoms. However, when an atom loses or gains an electron, it becomes fully charged with a positive or negative charge, respectively.

from the left-most trailing hydrogen that is offset by the leading negatively charged carboxyl (COO-) head. The influence of the COO- head on the trailing hydrogen is significantly reduced with the length of the tail - the tail's hydrophobicity is directly correlated with number of carbon atoms in the tail.

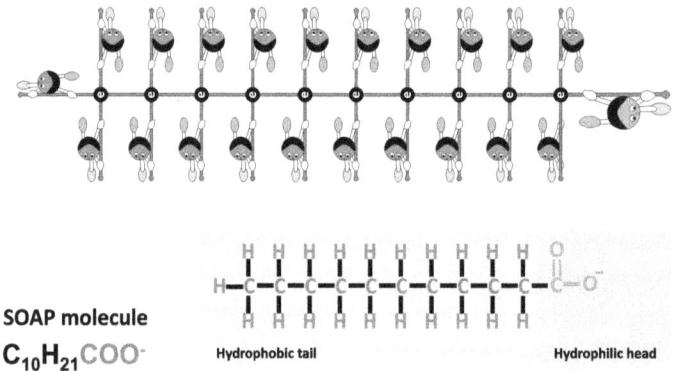

SOAP molecule

$C_{10}H_{21}COO^-$

Hydrophobic tail   Hydrophilic head

LIPIDS

Lipids are special groups of macromolecules that play a major role in the formation of cell membranes. The 3 major classes of lipids are fats, phospholipids, and steroids

Fats are synthesized through a **dehydration reaction** that combines 1 triglycerol with 3 fatty acids. They are used for energy storage, as cushion for organs, and to insulate the body from the cold.

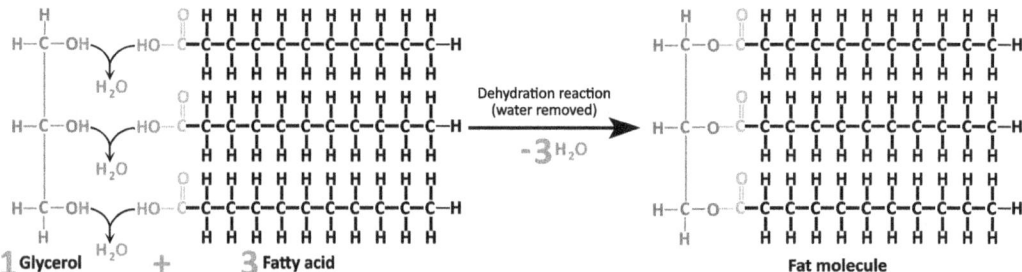

Phospholipids

Phospholipids are synthesized through a dehydration reaction that combines 1 glycerol, 2 fatty acids, and 1 phosphate-choline. Similar to fats, the fatty acid tails are hydrophobic, while the phosphate-choline heads are hydrophilic. The primary function of phospholipids is to form the lipid bilayer of the cell membrane (separates cell from surroundings), and the organelle membranes (allows cells to compartmentalize).

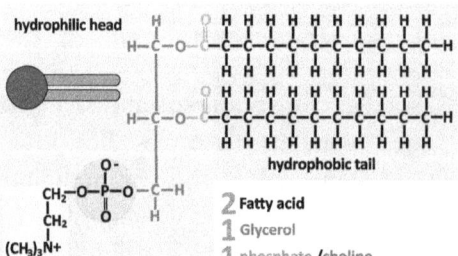

Phospholipid bilayers and micelles

When placed in water, the hydrophobic tails of phospholipids are attracted to each other resulting in the formation of a one layer spherical structures called **micelles**, or large double layer sheets that can fuse to form large bilayer spheres.

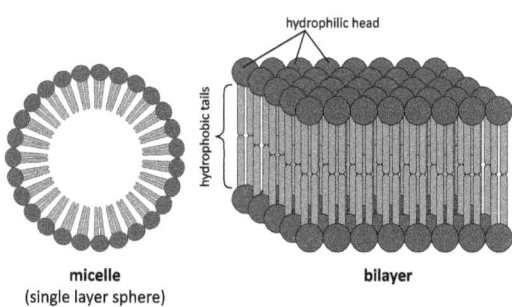

micelle
(single layer sphere)

bilayer

Steroids

Steroids are cyclic lipid compounds that have a 5-carbon and three 6-carbon rings fused together. The most important steroid is cholesterol, which is used for the following critical functions:

- It is a key component of the cell membrane of animal cell where it stabilizes the membrane by restricting phospholipid movements to reduce fluidity.
- It is used as a precursor molecule for building other steroid hormones such as testosterone and estrogen.

Cholesterol and membrane integrity

Membranes that contain many saturated phospholipids are more tightly packed. This creates a more viscous and rigid membrane and prevents the movement of membrane bound proteins. Membranes that contain more unsaturated phospholipids are more fluid since the bend in the unsaturated fatty acid prevents tight packing of phospholipid. Cholesterol fills the bends and makes the unsaturated membrane more viscous, while still allowing some movement of membrane proteins.

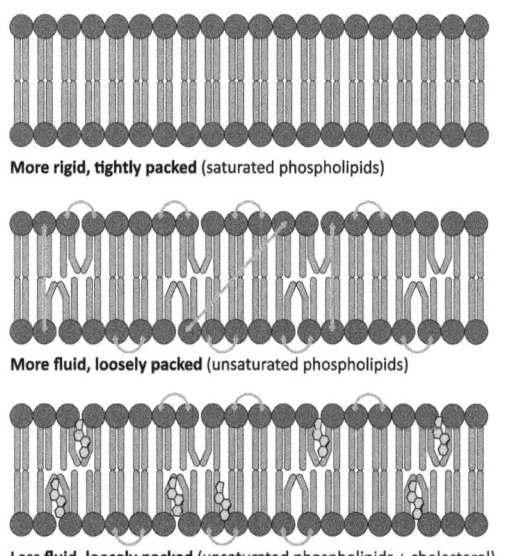

More rigid, tightly packed (saturated phospholipids)

More fluid, loosely packed (unsaturated phospholipids)

Less fluid, loosely packed (unsaturated phospholipids + cholesterol)

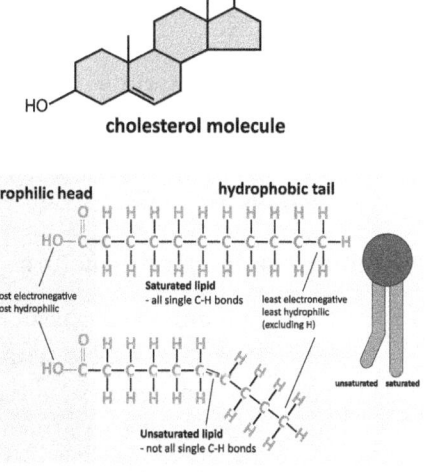

cholesterol molecule

unsaturated fats are bent

SELECTIVE PERMEABILITY OF THE PHOSPHOLIPID BILAYER

The cell membrane is **selectively permeable**, meaning that some particles can pass through by simple diffusion, while other particles are restricted. Simple diffusion is an exergonic process that does not require energy. It is the movement of particles from an area of high concentration to an area of low concentration. Particles that can cross by simple diffusion are small-uncharged polar molecules, and small hydrophobic molecules. Particles that are prevented from crossing include, large-uncharged polar molecules, and charged particles (ions). Because of this selective permeability, the phospholipid bilayer is a good cellular barrier to the external environment.

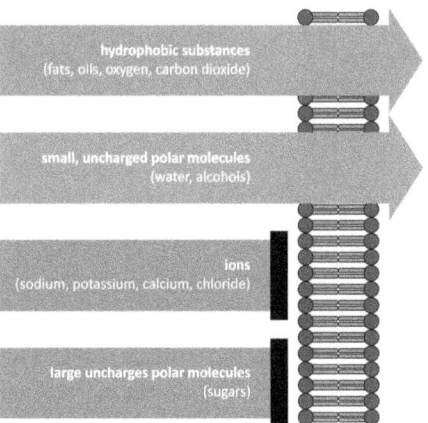

Challenges posed by selective permeability

Selective permeability is beneficial, but it does present some challenges:

1. It prevents the exchange of restricted substances with the surroundings such as the following:
   - Wastes that need to be excreted
   - Some large polar molecules that need to be secreted and sent to other tissue. For example, glucose produced by chloroplasts must leave the photosynthetic leaf cells and be transported to root storage cells.
   - Some large polar molecules that need to be imported. For example, the glucose needed for respiration must enter the mitochondria by crossing the outer and inner mitochondria membranes.
   - Critically needed exchange of charged particles with the surrounding. For example, the movement of sodium and potassium ions across the neuron is critical to the propagation of nerve impulses.

2. It severely limits or stops cell-cell communication. This blocks the cell's ability to respond to changes in the external environment

Membrane-associated structures

The following membrane-associated structures help the cell resolve challenges:

- **Transport proteins** (or **channel proteins**) facilitate the movement of charged particles (ion channels) and large polar molecules across the membrane. Channel proteins can be permanently opened, or can be gated channels that only open when a specific stimulus instruct them to do so.
- **Protein pumps** that actively move particles across the membrane and against

their concentration gradient from high to low concentration.

- **Aquaporin proteins** are special channels that increase the rate of water movement across the membrane.
- **Ligand receptors** are receptor proteins that are typically associated with channel proteins or some other transmembrane protein that can transmit a communication signal into the cell. Ligand receptors bind to a signaling molecule (the ligand) to receive message from another cell or from the external environment. For example, a neuron receives the signal from another neuron when a neurotransmitter binds to its ligand receptor on Na/K channels, causing these channels to open so that sodium (Na+) and potassium (K+) can cross and depolarize the membrane.
- **Enzymatic proteins** that are embedded in the membrane are involved in cell communication and catalytic reactions. Some are peripherally attached to the membrane where they act as second messengers, receiving signals from ligand receptors and propagating that signal to other second messengers. Others are also involved in catalyzing various types of chemical reactions, such as ATP synthase in oxidative phosphorylation to produce ATP, or electron transport proteins in the electron transport chains (ETC).
- **Structural lipids** such as cholesterol that provide structural support for the membrane, and junction and extracellular matrix (ECM) proteins.
- Finally, membrane structures called **glycolipids** and **glycoproteins** are displayed on the extracellular side of the cell membrane and play a role in cell recognition. Immune cells can distinguish "self" cell from pathogens based on the specific types of glycoproteins and glycolipids that cells display.

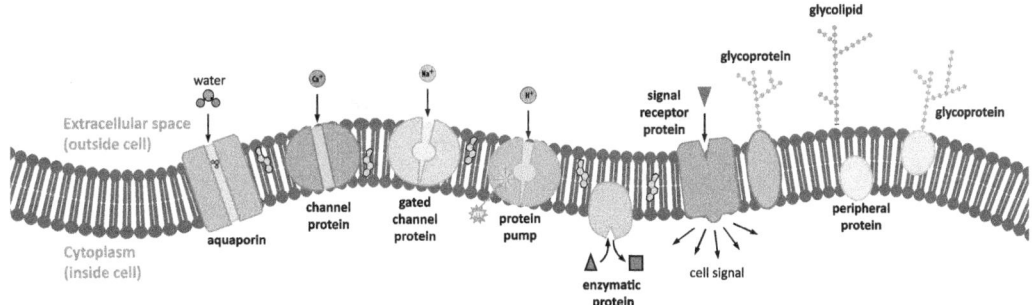

Hydrophobicity of membrane bound proteins

Cell membrane proteins use Van der Waals forces to interact with the phospholipids of the bilayer. **Peripheral proteins** (yellow) are strictly hydrophilic and can only interact with the hydrophilic heads of the phospholipids. Because they lack an affinity for the sandwiched hydrophobic tails, peripheral proteins do not transcend the membrane and can only temporarily bind to the surface of the membrane.

Proteins that have both hydrophobic and hydrophilic regions are called **integral membrane proteins** or **transmembrane proteins**. These proteins can interact with both the hydrophilic heads and hydrophobic tails, and therefore can transcend the full depth of the membrane. They tend to be permanently attached to the membrane with the hydrophilic regions exposed to extracellular matrix or cytoplasm, and their hydrophobic regions in middle of bilayer.

When comparing two solutions, the hypertonic solution is the one that has the higher solute concentration, and the other is the hypotonic solution. Movement of water always goes from the hypotonic (high water potential) to the hypertonic solution (low water potential). When the solute concentration and osmotic pressure is balanced, the solutions are at equilibrium and become isotonic. (Note that both osmotic pressure and solute concentration influence the water potential. This is discussed in more details in Investigation 4.)

When animal cells are placed in a hypotonic solution, they tend to burst their volume increase from water import. Plant cells are protected from lysis by their cell wall, which counters the rising osmotic pressure and limits osmosis to prevent the membrane from overstretching.

ACTIVE AND PASSIVE TRANSPORT

Organisms rely on 2 types of membrane transport for absorbing needed material and removing waste, and balancing their water and electrolyte content:

- ■ **Passive transport**
- ■ **Active transport**

Passive transport (osmosis, simple and facilitated diffusion)

The transport of substances across the membrane from an area of high concentration to low concentration is driven by entropy and occurs by **diffusion**. Water, however, diffuses from an area of low concentration of particles (high water potential) to high concentration (low water potential), a process called **osmosis**. All entropy driven movement of particles (diffusion and osmosis) are classified as passive transport because they do not require an energy input. Those particles that are restricted from crossing the membrane (ions, polar molecules) are help across by membrane proteins (aquaporin, channel proteins, pumps), a process known as **facilitated diffusion**.

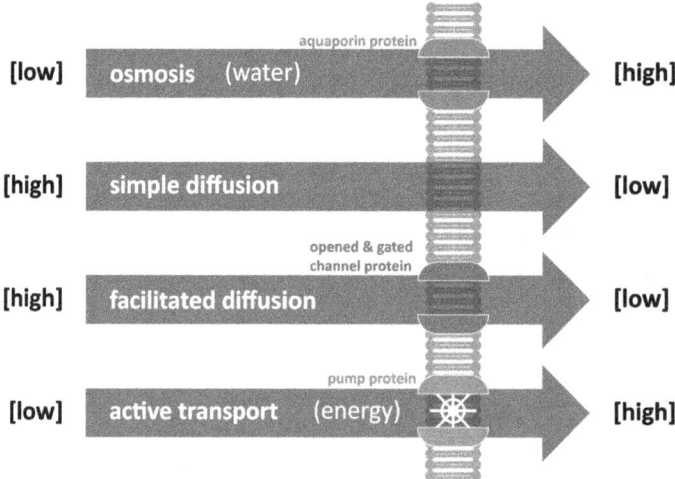

Active transport from [low] to [high]

The movement of a particle against its concentration gradient from [low] to [high] is counter to entropy. This type of membrane transport is classified as active transport because it requires energy input. Active transport uses pump proteins that are coupled to a exergonic process such as ATP dephosphorylation.

CELL RESPONSE TO OSMOSIS

■ The number of transport proteins present in the membrane can influence the diffusion rate.

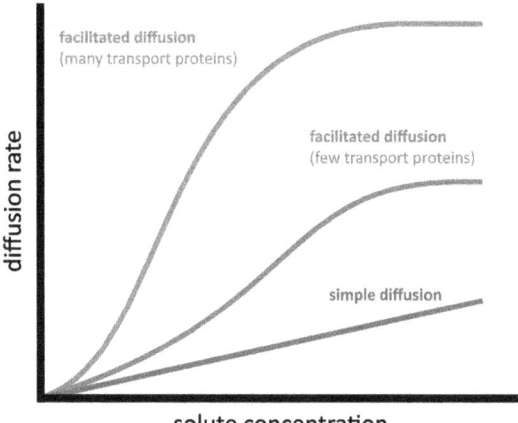

■ When two solutions are separated by a semipermeable membrane, water will move toward the more concentrated solution until equilibrium is achieved.

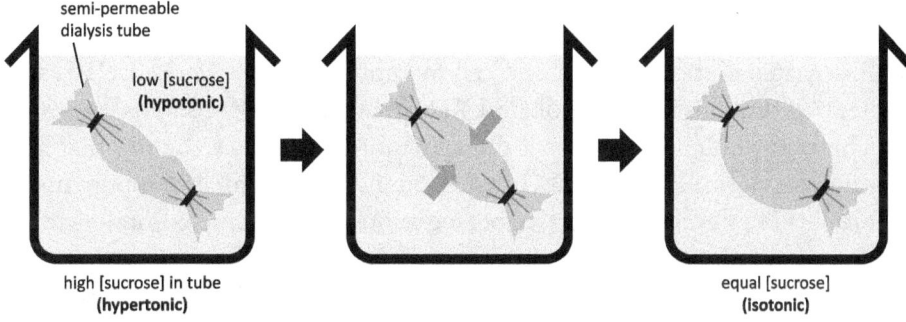

■ When a red blood cell is placed in an **isotonic solution** there is no net movement of water across the cell membrane because the cell is at equilibrium with isotonic solutions.

■ When a red blood cell is placed in a **hypotonic solution**, water moves into the hypertonic cell and causes the RBC to lyse.

■ When a red blood cell is placed in a **hypertonic solution**, water moves out of the hypotonic cell and causes the RBC to shrink.

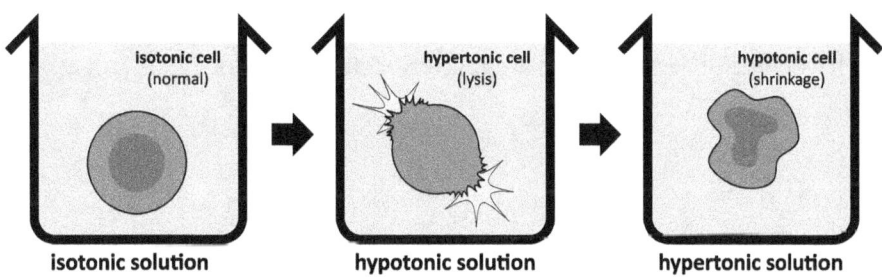

Phagocytosis is the import of large particulate matter such as whole bacteria cells and food. White blood cells use this mechanism to defend against bacteria infection.

Pinocytosis is the import of fluid that contains restricted material such as large molecules. It is initiated when a signal molecule binds to targeted membrane receptors to trigger the infolding of cell membrane. The infolded cell membrane than pinches off to form cytoplasmic vesicles that contain the imported fluid.

■ Protein transporters are classified as either **uniporters**, **symporters**, or **antiporters**. Uniporters, such as aquaporin proteins, are specialized to carry only 1 type of particle across the membrane. Symporters, such as the sodium-glucose co-transporter, can carry 2 different particles simultaneously in the same direction. Antiporters such as the voltage-gated sodium-potassium channels, carry 2 different particles, one at a time, in opposite directions.

■ Some transport proteins are coupled to another transport protein. These types of transport protein are called coupled transporters. 1 transporter of the pair is a endergonic (requires energy) such as a protein pump, while the other is exergonic, such as protein channel. For example, the Na/K pump, uses energy from ATP to build a concentration (or charge) gradient across the membrane, while the exergonic transport, such as the Na/glucose channel, uses the energy stored in the gradient to move particles across membrane.

■ Some large molecules are transported across the membrane through **endocytosis** (import) and **exocytosis** (export) mechanisms. Endocytosis can occur either by **phagocytosis** (cell eating) or **pinocytosis** (cell drinking). All endocytosis and exocytosis mechanisms use vesicles that fuse with (export) or pinch off (import) the phospholipid bilayer.

SUMMARY OF KEY CONCEPTS

■ Cell growth and homeostasis requires import of water, energy, and material.

■ Associated components of the cell membrane include various classes of embedded proteins, cholesterol, and glycoproteins and glycolipids.

■ The phospholipid bilayer is a fluid membrane with a structure that supports selective permeability. Embedded proteins are essential components of the bilayer. They enhance selective permeability and facilitate communication with the surroundings, including with other cells.

■ The viscosity and fluidity of the bilayer is reduced by cholesterol molecules that restrict the movement of phospholipids.

■ Passive transport is the diffusion of particles across the cell membrane from high to low concentration. Active transport requires energy to move substance from low to high concentration.

■ Examples of membrane proteins include Na/glucose co-transporter, Na/K antiporter, Na/K pump. Some protein transporters are coupled.

■ Osmosis is a passive process for the movement of water from an area of low to high solute concentration. During osmosis, water moves from the hypotonic to the hypertonic solution until equilibrium is reached. At equilibrium, the solutions are isotonic to each other.

■ Movement of large particulate matter occurs by endocytosis and exocytosis.

Check you understanding

1. Which of the following statements about electronegativity is incorrect?

    (A) Bromine has stronger attraction for electrons than does iron.
    (B) When the electronegativity between two atoms of a molecule is large, they will form a non-polar covalent bond.
    (C) Electronegativity decrease moving down the period table.
    (D) Electronegativity influences the polarity of molecules.

2. A hydrophobic substance

    (A) works well as detergents.
    (B) tend to be non-polar.
    (C) dissolve well in water.
    (D) tend to have an asymmetrical distribution of electronegativity among their atoms.

3. Soaps work well for washing away substances that are both hydrophilic and hydrophobic because

    (A) soaps are amphiphilic.
    (B) soaps interact well with water and hydrophilic substances.
    (C) soaps interact well with hydrophilic substances.
    (D) All of the above.

4. Lipids are macromolecules that

    (A) store energy.
    (B) are critical to the formation of cell membranes.
    (C) help enhance the structural integrity of cell membranes.
    (D) All of the above.

5. Which of the following statements about the cell membrane is true?

    (A) Membranes are made of a double layer of phospholipids with the hydrophilic head connected on the interior.
    (B) The membrane becomes more fluid as the number of saturated phospholipids increase.
    (C) Membranes that contain a relatively large number of unsaturated phospholipids require cholesterol for stability.
    (D) Saturated phospholipids have a bent shape and are less rigid.

6. Which of the following membrane-bound structures is responsible for moving charged particles across the membrane in the direction of the concentration gradient?

(A) aquaporins

(B) protein pumps

(C) channel proteins

(D) glycolipids

7. Which of the following membrane-bound structures is responsible is involved in cell recognition/communication?

(A) glycolipids

(B) protein pumps

(C) channel proteins

(D) None of the above.

8. Which of the following structures is involved in active transport?

(A) aquaporins

(B) protein pumps

(C) channel proteins

(D) All of the above.

9. Which of the substances can use simple diffusion to enter the cell?

(A) sodium ion

(B) steroid hormones

(C) sugar

(D) None of the above.

10. What happens to a plant cell that is placed in concentrated salt solution?

(A) Salt ions will move into the cell so that its electrolyte (salt) concentration is equal to the solution.

(B) Water will move from the cell into the salt solution, causing the cell to shrink.

(C) Water will move from the solution into the cell, causing the it expand.

(D) The cell will lyse.

ANSWERS AND EXPLANATIONS

1. (B) is an incorrect statement because large differences in electronegativity causes unequal sharing of electrons, and polarization of covalent bonds.

2. (B) is correct.

3. (D) is correct.

4. (D) is correct. Not that option (C) relates to the role of cholesterol in stabilizing membrane that contain lots of unsaturated phospholipids.

5. (C) is correct. Option (A) is incorrect because it is the hydrophobic tails that are attached at the interior. (B) and (D) are incorrect because saturated fats are straight and pack tightly to give the membrane a more rigid structure.

6. (C) is correct.

7. (A) is correct.

8. (B) is correct.

9. (B) is correct because steroid hormones are lipids and are therefore hydrophobic. Hydrophobic substances can use simple diffusion. Option (A) is a charged particle and (C) is a hydrophilic substance. Both require facilitated diffusion by a protein channel.

10. (B) is correct because the cell is hypotonic relative to the hypertonic solution. Water always move from the hypotonic to the hypertonic solution.

13 Cell compartmentalization

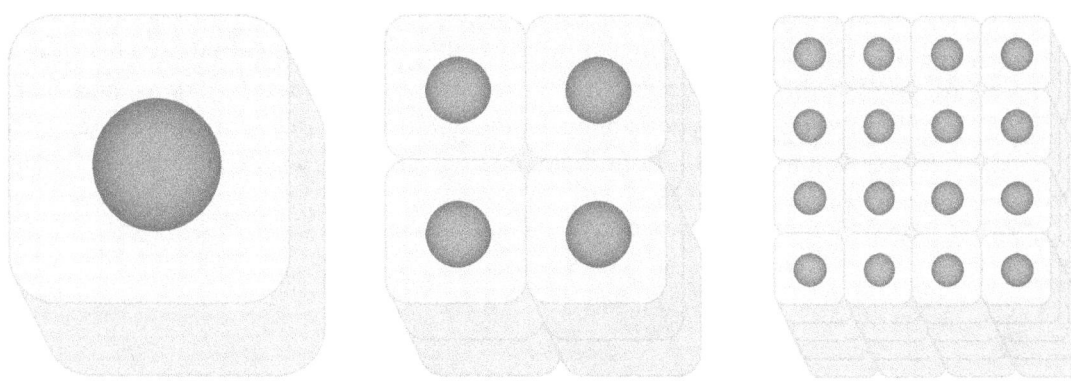

ALL STUDENTS MUST BE ABLE TO ANSWER THESE QUESTIONS

1. How does cell size influence membrane transport and membrane-bound reactions?
2. How is the SA/V ratio related to cell size?
3. What are the 2 evolutionary events that lead to the emergence of the endomembrane system?
4. How does cell compartmentalization minimize competing interaction? Provide specific examples.
5. What is the function of each organelles, and the endomembrane system in general?
6. How is the ER involved in the maintenance of the endomembrane system?
7. How is the endomembrane system used for exchanging resources between the various parts of the cells, and between the cell and its surroundings?
8. What are the similarities and differences between plants and animals cells?
9. What are the similarities and differences between prokaryotic and eukaryotic cells?

ALL STUDENTS MUST BE ABLE TO COMPLETE THE FOLLOWING TASKS?

10. Use examples, analogies, and models to explain how the endomembrane system and cell compartmentalization in general support the cell's function.
11. Use models to compare and contrast prokaryotes and eukaryotes.
12. When provided an illustration of a cell, distinguish the cell parts, including the organelles that make up the endomembrane system, and other organelles.

 Big Idea 2B3

To maximize internal organization, biological systems have barriers (or boundaries) separating them from their surroundings. For example, the Earth has its atmosphere as a barrier, ecosystems typically have geographic barriers, species have reproductive barriers, and cells have cell membranes. Each of these barriers arose through geologic change or through evolutionary adaptations.

In addition to a cell membrane, cells also have internal subsystems called organelles that are separated from the rest of the cell by internal membranes. These organelles are compartments specialized for different organizational functions, and that minimize competing interactions between metabolic processes. They also increase the surface area for membrane bound reactions such as the electron transport chains and oxidative phosphorylation.

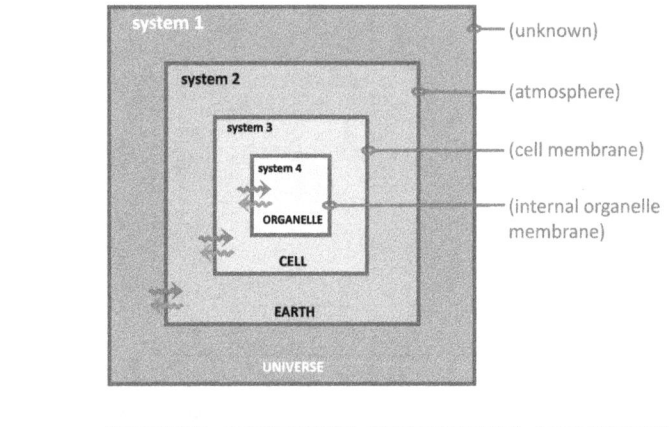

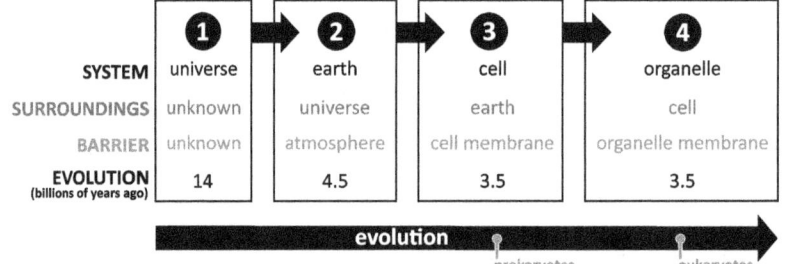

CELL SIZE

The size of the cell directly influence two import cellular processes: membrane transport, and membrane-bound metabolic reactions that are catalyzed by proteins embedded in the membranes.

Membrane transport

The size, measured as volume, of a cell is inversely related to the cell's surface area-to-volume ratio such that as volume decreases, the SA:V ratio increases. SA:V ratio directly impacts the cells efficiency at membrane transport. Since all imported and exported material and waste crosses the cell membrane, the smaller cell has more membrane relative to size (higher SA:V ratio) and is faster and more efficient at import/export. This competitive advantage lead to natural selection favoring smaller cells of larger ones. However, there is a threshold limit to how small a cell can become before the size advantage is lost, as too small of a volume can restrict and limit internal organization.

Reaction rates

In addition to transport efficiency, many important enzymes and protein must be associated with the cell or organelle membranes to function. For example, in the chloroplast, the **chlorophyll-protein complex** is embedded in the **thylakoid membrane** where it acts to absorb energy from light. For these types of metabolic reaction where membrane interaction is necessary, membrane surface area can impact reaction rates. The higher the SA:V ratio, the greater is the relative number of activated processes, and the more quickly it can be brought to conclusion.

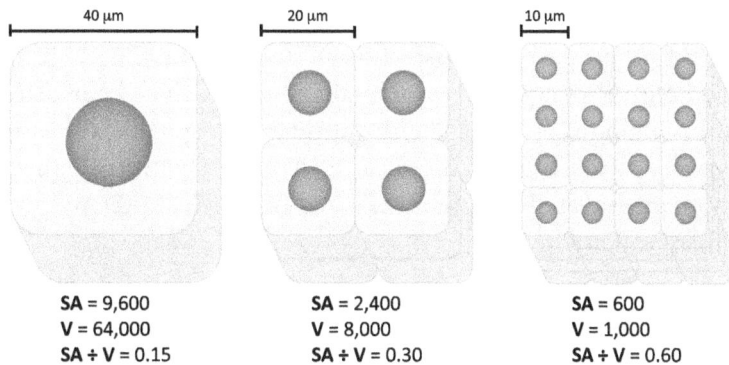

COMPARTMENTALIZATION

The various metabolic processes in the cell compete for energy and material, and can have negative consequences on each other. For example, the digestive **lytic enzymes** in lysosomes recycle material and digest food particles. In macrophages, lysosomes are also involved in the immune response. They **phagocytize** invading bacteria or infected cells into large vacuoles that are than fused with the lysosomes. The lytic enzymes and low pH of the lysosome then kills the engulfed cells, thereby protecting the whole organism for disease. The digestion does not harm the macrophage because lysosomes are internal compartments that isolate the lytic reaction from the rest of the cell. However, if the content of the lysosome were released into the cell's cytoplasm, the whole cell would die. This can happens when the cell initiates the automatic self-destruction process called **apoptosis** in order to prevent infection of other cells.

There are many instances where a metabolic process that is beneficial for one thing can be detrimental for the whole organism. To resolve this cells use organelles to compartmentalize reactions and separate them from each other. **Cell compartmentalization** to form the **endomembrane system** is a derived character of eukaryotes (absent in bacteria and achaea). It evolved prior to endosymbiosis as a result of the invagination of the cell membrane. It comprises of the cell membrane and most of the cells organelles. The specific organelles that are part of the endomembrane system are vesicles, vacuoles, lysosomes, Golgi apparatus, endoplasmic reticulum, and nuclear envelop. After the emergence of the endomembrane system, an endosymbiosis event gave rise to the mitochondria and chloroplast. Each of the organelles of the 2 sets of evolutionary events is specialized for a specific task that supports internal organization.

A specialization analogy

Instead of hiring on general practitioner physician that understand a bit about a broad range of medical conditions, a hospital decides to hire specialists such as brain surgeons, cardiologist, urologist, and anesthesiologists, each an expert of within the specific field of medicine. A heart disease patient is better served by an experienced cardiologist, rather than a less experience general practitioner.

Eukaryotic cells are like specialists, each capable of performing a specific life-sustaining task, relying on other cells for the rest. For example, a red blood cell does its part for the whole organisms by specializing in oxygen and carbon dioxide transport, while cardiac muscles do their part by contracting (pumping) the blood so that RBCs can deliver their contents. Prokaryotes are generalists, with each cell performing all of the necessary life sustaining tasks. Just like the specialist hospital that needs many more doctors since each specialist can treat only a small pool of patients with specific conditions, eukaryotes are multicellular (many cells per organism) and have greater levels of organization than prokaryotes.

The endomembrane system

The endomembrane system evolved prior to endosymbiosis as a result of the invagination of and subsequent pinching off from the cell membrane (see Lecture 8). All component of the endomembrane system are compartmentalized by their shared phospholipids that move from one to the other via vesicles that pinch off from and fuse with their membranes.

Nuclear envelop
- Contains a double bilayer membrane (4 lipid layers).
- Separates the contents and tasks of the nucleus from the cytoplasm.
- Contains the entire genomic DNA.
- Controls the cell's activity.

Endoplasmic reticulum (ER)
- The largest organelle, and an extension of the nuclear envelop.
- Produces, processes, and transports biochemical compounds.
- Produces membrane bound proteins and lipids that make up cell membranes.
- Receives all proteins that are destined for export.
- Divided into two compartments:

Rough ER (rER) have attached ribosomes
  - Receives newly synthesized polypeptides from ribosomes
  - Chemically modifies proteins and phospholipids
  - Folds polypeptide into correct 3D protein shape
  - Builds all of the membranes of the cell and its organelles
  - Communicates with sER

Smooth ER (sER) lack attached ribosomes
  - Site of budding transport vesicles that leave for the Golgi apparatus
  - Synthesizes lipids (oils, phospholipids, steroids)

- Metabolizes carbohydrate (glycogen breakdown in liver cells)
- Detoxifies drugs and poisons (high concentration of sER in liver)
- Ca^{2+} pumps of sER involved in muscle contraction (covered later)
- Sends vesicles to Golgi apparatus

Golgi apparatus
- Compartmentalized into cis (faces ER) and trans (faces cell membrane) Golgi.
- Receives most proteins destined for secretory vesicles.
- Modifies and synthesizes carbohydrates.
- Modifies proteins into glycoproteins
- Sends secretory vesicles to cell membrane

Vesicles
- Small spherical lipid bilayer compartments.
- Transports material between organelles.
- Transports imported material from the cell membrane.
- Transports exported material to the cell membrane.

Vacuole
- Giant vesicle-like structures
- Largest organelle in plant cells
  - Called central vacuoles (plant cells)
  - Animals cell vacuoles are used for endocytosis and exocytosis
- Stores nutrients, wastes, and pigments
- Regulates turgor pressure (water pressure)
- Balances the pH of the cell.

Lysosomes
- Vesicle-like structures that contain hydrolytic enzymes
- Digests and recycles of used cell parts.
- Fuses with phagocytized vacuole to digest imported food, and phagocytized bacteria and infected cells.
- Carries out apoptosis (cell suicide) to protect whole organism from disease.

NOTES

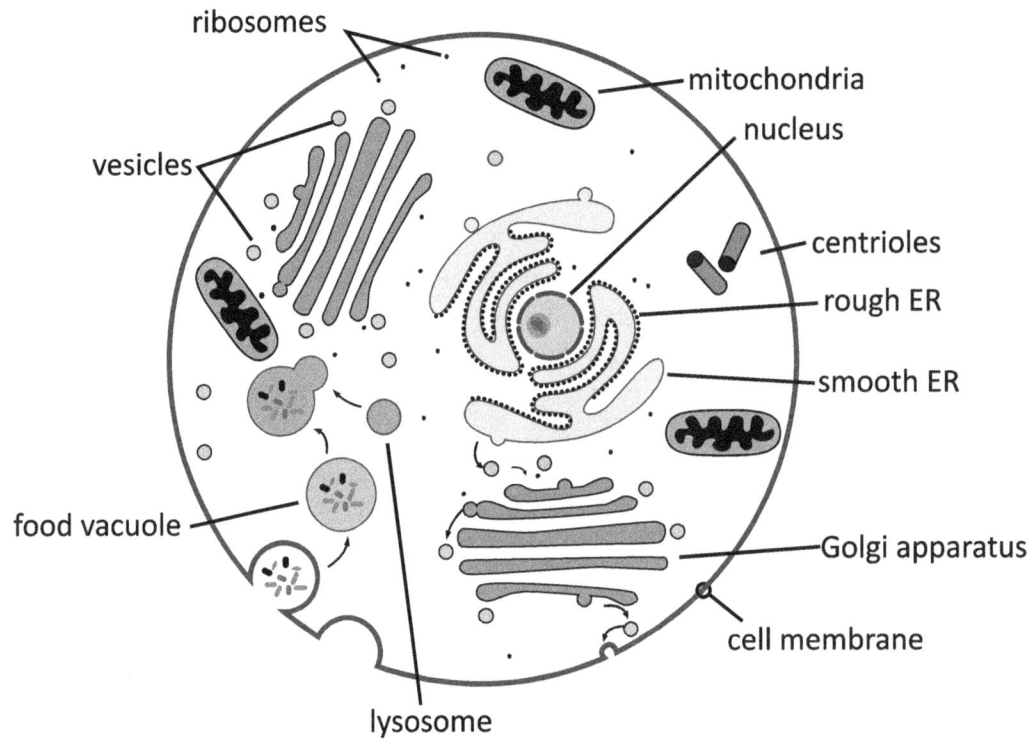

ribosomes

mitochondria

nucleus

centrioles

rough ER

smooth ER

vesicles

food vacuole

Golgi apparatus

cell membrane

lysosome

ANIMAL CELLS

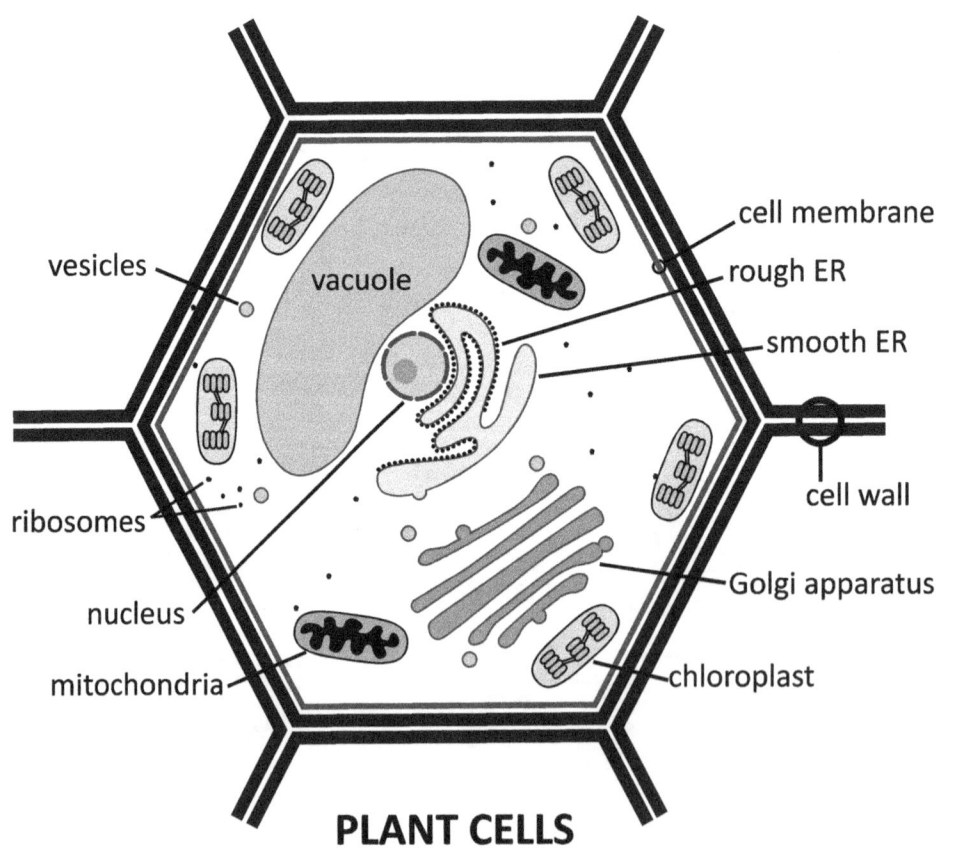

vesicles

vacuole

cell membrane

rough ER

smooth ER

cell wall

ribosomes

nucleus

Golgi apparatus

mitochondria

chloroplast

PLANT CELLS

SUMMARY OF KEY CONCEPTS

- Eukaryotes have internal membrane compartments called organelles.
- The cell membrane and most of the organelles are part of the endomembrane system.
- The endomembrane system makes up the majority of internal membrane structures. It includes, the nuclear envelope, rough ER, smooth ER, Golgi apparatus, vesicles, lysosomes, vacuoles, and cell membrane.
- Other organelles include the chloroplast and mitochondria that evolved via endosymbiosis.
- Prokaryotes (bacteria and achaea) lack organelles.
- The purpose of internal membranes is to increase membrane surface area for compartmentalized reactions and enzymes, and to create specialized subsystems (organelles).

CHECK YOUR UNDERSTANDING

1. Cells have evolved to be relatively small in order to

   (A) reduced the amount of matter and energy demand for growth, reproduction, and maintenance of order.

   (B) increase the amount of internal membrane that is can incorporate.

   (C) increase the efficiency of membrane transport.

   (D) maximize energy efficiency.

2. Natural selection have restricted cells from become infinitely small because cells that are too small

   (A) have a low rate of membrane import.

   (B) are limited in their ability to internally compartmentalize due to limited internal space.

   (C) form strong H-bonds with substances around them and become stuck.

   (D) All of the above.

3. If the length of a cubical cell is doubled, its volume will change by

   (A) $\frac{1}{2} \times$

   (B) $2 \times$

   (C) $8 \times$

   (D) $16 \times$

4. During the first few series of cell division following fertilization, the embryo increase cell count without increase its total volume. This process is called cleavage, and it produces equally sized daughter cells that collectively has the same volume as the fertilized egg. How many rounds of mitosis is needed to produce 64 daughter cells?

   (A) 2

   (B) 4

   (C) 6

   (D) 8

5. Which of the following statements accurately compares the 64 descendant daughter cells to the original fertilized egg?

   (A) The daughter cells has a total surface area that is 6.3% of the fertilized egg.
   (B) The cell volumes has decreased by 64 ×.
   (C) the SA/V ratio has increased by 4 ×.
   (D) All of the above.

6. Which of the following organelles is not part of the endomembrane system?

   (A) mitochondria
   (B) chloroplast
   (C) peroxisomes
   (D) all of the above.

7. Which of the following organelles is relatively abundant in cells of the prostate gland?

   (A) smooth ER
   (B) Golgi apparatus
   (C) lysosomes
   (D) vacuole

8. Which of the following organelles is relatively abundant in macrophages?

   (A) smooth ER
   (B) Golgi apparatus
   (C) lysosomes
   (D) ribosomes

9. Which of the following organelles is relatively abundant in intestinal mucosal cells that absorb fatty acids from ingested food, synthesize them into triglycerides, and secrete the triglycerides into the lymphatic vessels.?

   (A) endoplasmic reticulum
   (B) nucleus
   (C) chloroplast
   (D) ribosomes

ANSWERS AND EXPLANATIONS

1. (C) is correct.

2. (B) is correct.

3. (B) is correct because for if the initial length = x, the cell had an initial volume, $v_i = x^2$. If length is doubled ($l_f = 2x$), $v_f = (2x)^2 = 4x^2 = 4v_i$.

4. (C) is correct because the 1st division produces 2 cells, than 4, 8, 16, 32, and the 6th division produces 64 cells.

5. (D) is correct because

    $$v_f = 1/64 \, v_i = (l_f)^3$$
    $$l_f = (1/64 \, v_i)^{1/3}$$
    $$SA_f = 6 \, l_f^2 = 6 \, (1/64 \, v_i)^{2/3}$$
    $$SA_i = 6 \, v_i^{2/3}$$
    $$SA_f / SA_i = (1/64)^{2/3} = 0.0625 = 6.3\%$$

6. (D) is correct.

7. (B) because, as a glands that secretes enzymes and other proteins, it should have lots of secretory cells. Within these cells, secretory vesicles leave the Golgi and fuse with the cell membrane to release their products.

8. (C) is correct because macrophages require the lytic enzymes stored in lysosomes in order to digesting the particles and cells that they phagocytize.

9. (A) is correct because lipids are synthesized in the ER.

14 Energy capture

ALL STUDENTS MUST BE ABLE TO ANSWER THESE QUESTIONS

1. What are 2 ways that organisms release energy?
2. What is the energy source for chemosynthesis? How does it compare to photosynthesis?
3. What is the relationship between the structure and function of chloroplast? Mitochondria?
4. What is the relationship between the light reactions and the Calvin cycle?
5. How do the functions of the various chloroplast and mitochondrial membranes compare?
6. How is energy transferred to electron and ATP along the ETC of the chloroplast and mitochondria?
7. Why is ATP needed in the Calvin cycle?
8. What is the role of chlorophyll a and antenna pigments in energy capture and electron transfer?
9. How does the P650 and P700 chlorophyll a pigments differ in their function?
10. Why it is beneficial to have different light absorbing pigments in the photosystems?
11. What are the 5 steps of energy absorption and electron transport in the light reactions.
12. What is the link between energy, NADPH, ATP, and electrons?
13. Why is there a difference in the number of ATP produced from $FADH_2$ and NADH?
14. What are the major catabolic and anabolic pathways?
15. How do cells use catabolic processes? Anabolic processes?

ALL STUDENTS MUST BE ABLE TO COMPLETE THE FOLLOWING TASKS

16. Use a self-developed analogy to explain the light reaction.
17. Diagram and describe the structure of the mitochondria and chloroplast.
18. When provided an illustration of a cell, distinguish the cell parts, including the organelles that make up the endomembrane system, and other organelles.

 Big Idea 2A2

ENERGY AND NUTRIENT FLOW

Biological systems rely on the sun as a primary source of energy. Plants and other photosynthetic autotrophs harness solar energy and store it in the bonds of glucose molecules. This absorbed energy is the total energy that enters the ecosystem is called the gross primary production (or GPP). Plants release some during respiration for ATP production by coupled oxidative phosphorylation, and as waste. ATP provides the energy needed to power the different life sustaining processes such as growth (synthesis of new molecules) and tissue repair, reproduction, and maintenance of internal homeostasis. Due to energy inefficiency, each of these processes a loss of lots of energy as heat. What is left is the net primary production (NPP), which accounts for the total biomass that contain the energy available to the higher trophic levels.

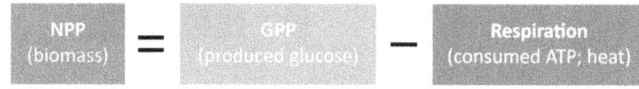

$$\text{NPP (biomass)} = \text{GPP (produced glucose)} - \text{Respiration (consumed ATP; heat)}$$

While photoautotrophs receive energy directly from the sun, heterotrophs get their energy by feeding on other organisms. Herbivores of the second trophic level feed exclusively on producers, and those in each of the higher trophic levels feed on the organisms in the level(s) below. As energy flows through, only about 10% of the gross production from one level is passed on the next

Photosynthesis versus chemosynthesis

Deep beneath the ocean floor where light never reaches, there are regions that are rich with life. In unique aquatic ecosystem, glucose production is not based on photosynthesis, but on chemosynthesis. Specially adapted archaea bacteria harness energy from inorganic hydrogen sulfide release from hydrothermal vents of underwater volvanoes. These bacteria are the primary producers there, and they pass their captured energy up to the organisms in higher trophic levels.

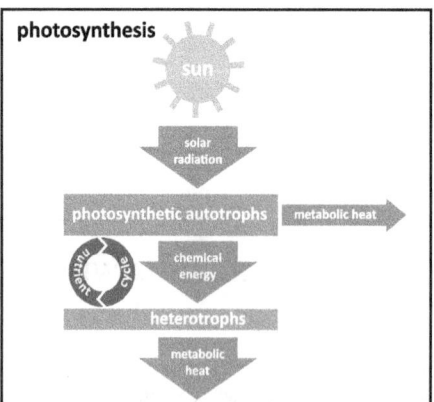

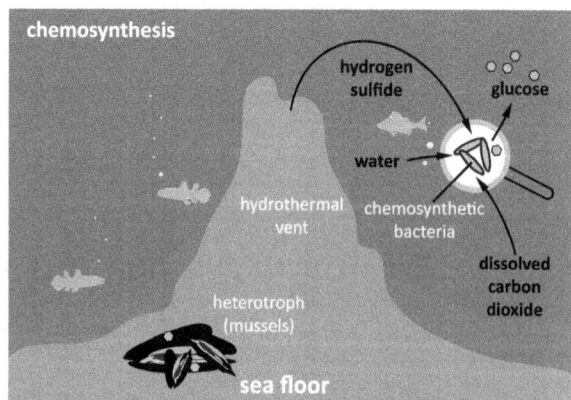

CHLOROPLAST AND PHOTOSYNTHESIS

Chloroplasts are the photosynthetic organelles of plants. It contains an outer and inner membrane, and internal structures with called thylakoids that are specialized compartments much like the cytoplasmic organelles. The thylakoids are separated from the rest of the chloroplast by a thylakoid membrane.

Inner and outer membranes

The outer membrane separates the chloroplast from the rest of the cell. It is porous and permeable to most small molecules and ions. The inner membrane is separated from the outer membrane by an intermembrane space and is less permeable to substances. To facilitate selective permeability, the inner membrane has embedded transport proteins that facilitate needed cytoplasmic proteins and glucose into and out of the chloroplast. The space within the inner membrane is called the stroma and is analogous of the cytoplasm of the cell.

Thylakoid membrane

The thylakoid membrane segments that chloroplast into the photosynthetic thylakoid compartments. Think of the thylakoids as the chloroplast's own mini-organelles that are specialized for capturing the light. The thylakoid space contains the lumen fluid that fills stacked discs called granum. Grana connect to each other via tunnel-like bridge structures called stromal lamellae that transport nutrients and other material needed to sustain the chloroplast.

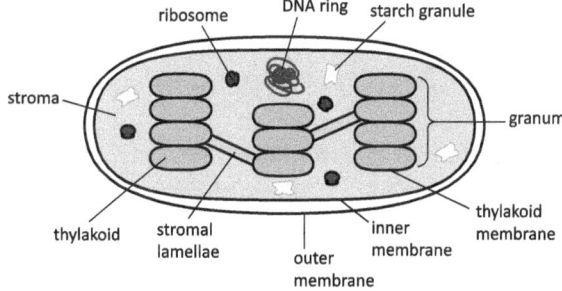

Photosynthesis reactions

Photosynthesis occurs through 2 reactions, light-dependent reaction that absorbs light energy, and Calvin cycle that synthesizes glucose with energy from ATP that is produced by oxidative phosphorylation, which is coupled to the light reaction.

The light dependent reaction

- Occurs in the thylakoid membrane.
- Requires transparent membranes so that light can reach the thylakoid.
- Reactants come from the cytoplasm (water) and Calvin cycle (ADP, P, $NADP^+$)
- Products are released to the lumen (oxygen and protons) and Calvin cycle (NADPH, ATP)
- Oxygen diffuses across the thylakoid, inner, and outer membranes to enter the cytoplasm.
- High [H^+] in the establishes a proton gradient that drives ATP production by

PROTEIN DEPENDENCY
Although the chloroplast and mitochondria have their own DNA, they still require proteins that are translated from the nuclear DNA. Special protein channels on the inner membranes selectively import the proteins.

ATP PRODUCTION
ATP is not directly produced by the light reaction. The energy that is captured from light is released to electrons derived from water molecules. As these energized electrons move through the ETC, some of their absorbed energy is released for oxidative phosphorylation, which is coupled to the light reaction.

ELECTRON DONORS
In photosynthesis, water is the electron donor and NADP+ is the final electron acceptor of the ETC. In respiration NADH and FADH$_2$ are electron donors and oxygen is the final electron acceptor of the ETC.

COVALENT BONDS
In order to form covalent bonds, the Calvin cycle requires electrons. Each covalent bond is held together by at least 2 electrons that are shared between the atoms connect by the bond. Although the atoms themselves provide some of the electrons, more are needed. These electrons are provided by water molecules.

oxidative phosphorylation (discussed later).
- $NADP^+$ is the final electron acceptor of ETC in photosynthesis.
- Reaction equation: $H_2O + NADP^+ + ADP + P \rightarrow O_2 + H^+ + ATP + NADPH$

Calvin cycle (or dark reactions)
- Occurs in the stroma.
- Reactants are come from the cytoplasm (carbon dioxide) and light-dependent reaction (ATP, NADPH)
- Carbon dioxide is imported to the stroma from the cytoplasm by diffusion across the inner and outer membranes.
- Products (glucose, ADP, P, $NADP^+$), which are released into the stroma, go to the light-dependent reaction.
- Glucose is exported by facilitated diffusion across the inner membrane to the cytoplasm.
- Reaction equation: $CO_2 + NADPH + ATP \rightarrow Glucose + ATP + NADP^+ + P$

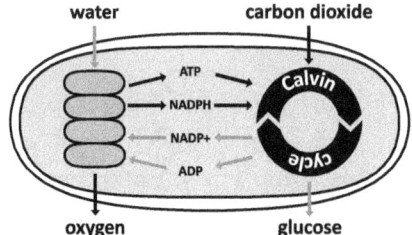

Energy and electron requirements

The ultimate purpose of photosynthesis is to convert solar energy into chemical energy in the form of glucose. The Calvin cycle produces glucose, but in order to do so, it needs the following:
- C, H, and O atoms that are provided by water (H_2O) and carbon dioxide (CO_2).
- High-energy electrons needed to form the covalent bonds between the C, H, and O atoms of glucose.
- Energy in the form of ATP to power the endergonic steps in the Calvin Cycle.

Photosystems II and I
The light reaction takes in all of the energy needed for photosynthesis and sends it to the Calvin cycle where the actual synthesis of glucose occurs. But, how does the energy get to the Calvin cycle? The capture and transfer of energy involves five key players:
- Light absorbing pigment molecules (**chlorophyll** *a* and **antenna pigments**)
- Light absorbing protein complexes called photosystem I, and photosystem II
- Electron transport chain (ETC
- NADP+ / NADPH cofactor electron carrying cofactor.

The photosystems are made of chlorophyll *a*, antenna pigments (chlorophyll b, lutein, xeaxanthin, β-carotene, and lycopene), and associated proteins. The two photosystems are embedded in the thylakoid membranes, and are the structural and functional light absorbing units of the chloroplasts. Each pigment absorbs a different wavelength of light. For example the lutein (P470) absorbs yellow light at a wavelength of 470 nm, and β-carotene (P500) absorbs orange light at 500 nm. This variability allows for a broader range of the light spectrum to be used, maximizing the amount of energy that can be absorbed.

NOTE: To better understand this section, it helps to refer to the figure below.

All of the energy collected in PSII and PSI are funneled from the antenna pigments down to their respective chlorophyll a (P650 or P700). Electrons from chlorophyll a (P700) of PSI become excited and are quickly snatched up by NADP⁺, which is reduced (gains 2 electrons) to NADPH. This creates an electron vacuum in P700. Simultaneously, a similar situation occurs at PSII that causes chlorophyll a (P650) to donate its electrons to the ETC. These donated P650 electrons fill the P700 vacuum, while P650 fills its vacuum by snatching electrons from water molecules, causing them to split into oxygen and protons (H⁺ ions).

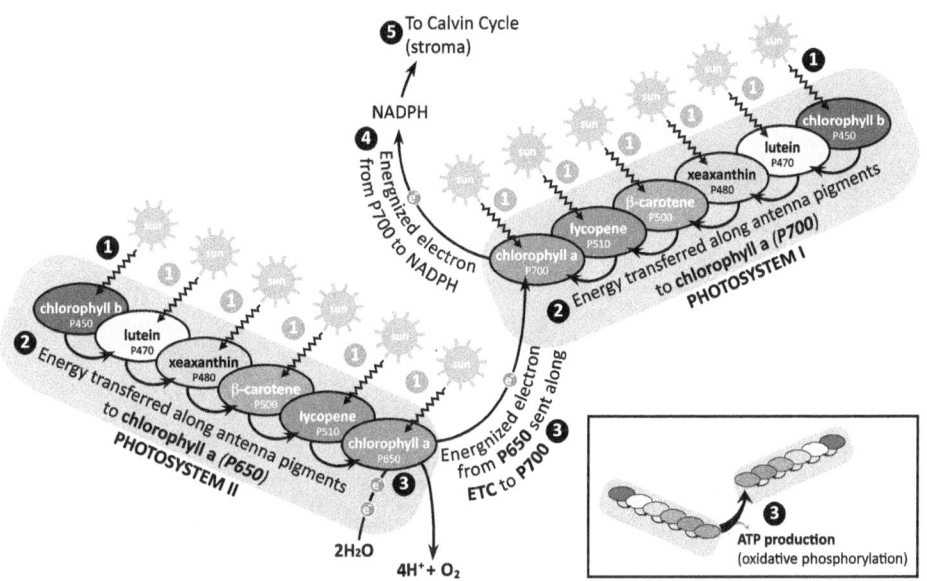

The steps of energy absorption and electron transport can be summarized as follows:
1. Simultaneous energy absorption by antenna pigment in both PSII and PSI
2. Simultaneous funneling of absorbed energy to chlorophyll a (P650 and P700) of PSII and PSI.
3. Excitement and loss of electrons from P650 to P700 (via the ETC) creates an electron vacuum on P650.
   - As water molecule is split to form proton (H^+) and oxygen (O_2), its electrons are used to refill the electron vacuum at P650.
   - As the P650 derived electrons travel along the ETC, some of their energy is released to oxidative phosphorylation for ATP production.
4. Due to the funneling of energy to P700, 2 electrons from P700 become excited and are accepted by NADP⁺, reducing it to NADPH.
   - Donation of electrons to NADP⁺ creates an electron vacuum at P700 that is refilled by the electrons received form P650
5. NADPH delivers the P700 derived electrons to the stroma where it is used in the Calvin cycle.

Photosynthesis action spectra

The figure below shows the action spectra for the photosynthesis reaction. Each of the photosynthetic pigments absorbs a unique range within the light spectrum. This collective effort at light absorption allows plants to harness a broader range of wavelength in the visible spectrum, and to increase the rate of photosynthesis (dotted lines)

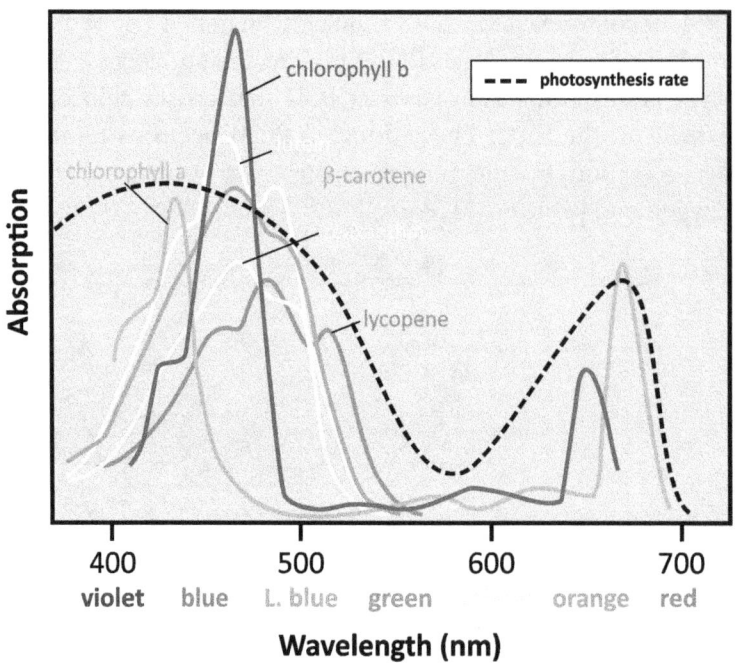

Link between energy, NADPH, ATP, and electrons

The absorbed energy from light is passed to the excited electrons that indirectly transfer some of this energy to ATP (oxidative phosphorylation) and the rest to NADPH for the Calvin cycle.

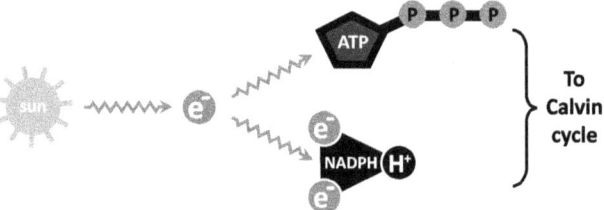

Although the energized electrons loose some of the absorbed energy in the ETC for ATP production, they are still sufficiently excited above their resting state. The remaining energy is sent with the electrons to the NADP$^+$/NADPH. Think of NADPH as a filled transport truck that has a capacity for 2 electrons, and NADP$^+$ as the same truck that has delivered its electron to the stroma and is now unfilled.

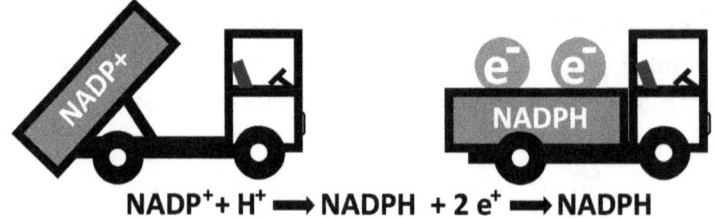

$$NADP^+ + H^+ \longrightarrow NADPH + 2\ e^+ \longrightarrow NADPH$$

Light reaction analogy

The two figures below summarize the key events in the capture and transfer of energy and electrons from water and light. Notice that oxidative phosphorylation involves the coupling of ATP synthase to the diffusion of protons (H^+). These protons concentrate in the lumen as a result of the splitting of water and ETC driven pumping.

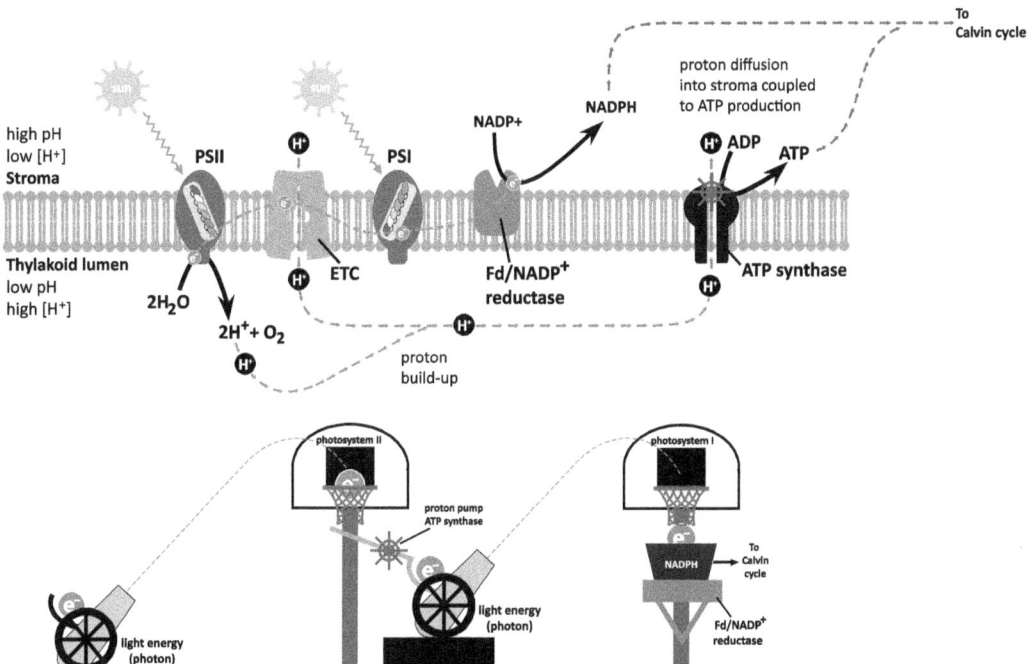

The Calvin cycle

The Calvin cycle uses 6 CO_2 to produce 1 glucose molecule. It is not necessary to recall each step of the cycle and intermediate products in the formation of glucose. However, be able to account for the carbon atoms that enter and leave the cycle, and the sources of these atoms. The cycle always has six resident cyclic 5-carbon compounds (6 × 5 = 30 C atoms) that combine with the C atoms from the six incoming CO_2 (30 + 6 = 36 C atoms). 12 ATP and 12 NADPH are used to produce 1 glucose molecule, and an additional 6 ATP are used to replenish the six cyclic 5-carbon compounds that start of the next round of the cycle.

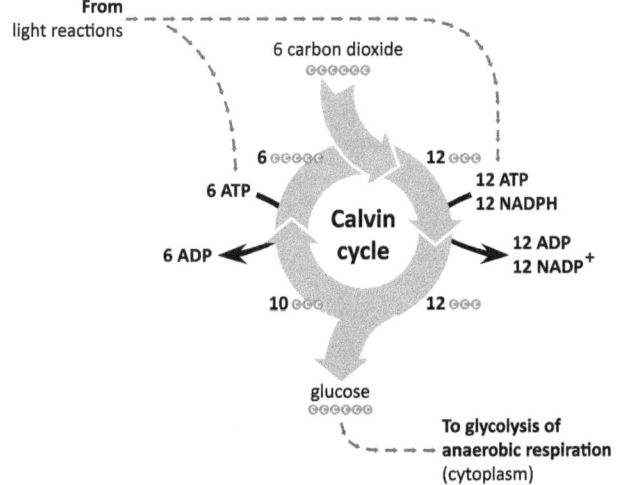

NAMING ACIDS
Lactate is the same as lactic acid, and pyruvate is pyruvic acid.

OXYGEN DEBT
Lactic acids can build-up in muscle tissue during strenuous exercise when the oxygen demand is greater than the oxygen supply. This build-up of lactate is called the "oxygen debt," and is due to lactic acids fermentation. The debt must be repaid when oxygen demand returns to normal by converting lactic acids back to pyruvate when oxygen levels return to normal.

Once produced, glucose is facilitated across the chloroplast membranes to the cytoplasm where one of the following can occur:

- Glucose undergoes anaerobic respiration via glycolysis and fermentation (prokaryotes and oxygen starved eukaryotic cells)
- Glucose undergoes anaerobic respiration via glycolysis to produce pyruvate. Pyruvate is then transported into the mitochondria for aerobic respiration (oxygenated eukaryotic cells).
- Glucose is transported to storage tissue such as the roots in potatoes, and linked up to form starch.
- Glucose is linked up to form cellulose that is used to build the cell walls and structural tissue such as the stem.

MITOCHONDRIA AND RESPIRATION

Mitochondria are organelles present in most eukaryotic cells. They amplify ATP production by further metabolizing pyruvate (pyruvic acid). Similar to the chloroplast, the mitochondrion has an outer and inner membrane.

Inner and outer membranes

The outer membrane separates the mitochondrion from the rest of the cell. It contains protein channels called **porins** that allow small molecules, ions, nutrients, ATP and other to enter the mitochondria.

Similar to the chloroplast, the inner mitochondrial membrane is separated from the outer membrane by the intermembrane space. It is also more selectively permeable, allowing unrestricted passage only to water, oxygen, and carbon dioxide. Other substance must use embedded transport proteins to cross. In addition to transport proteins, the inner membrane contains all of the protein complexes involved in the electron transport chain, and oxidative phosphorylation (ATP synthase). The entire inner membrane is folded into structures called cristae that greatly increase its surface area to improve functionality. Contained within the inner membrane is the matrix, which contains the mitochondrial DNA, ribosomes, and most of the other proteins.

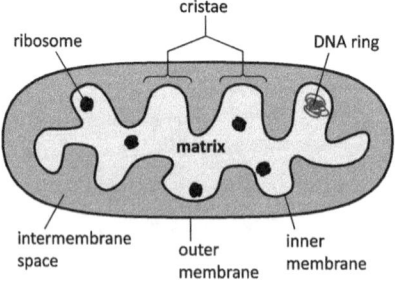

Respiration reactions

There are 2 types of respiration, oxygen-independent anaerobic respiration, and oxygen-dependent aerobic respiration. Anaerobic respiration is the most ancient form of respiration that occurs in the cytoplasm of both prokaryotes and eukaryotes. It involves 2 sets of reactions: glycolysis and fermentation. Aerobic respiration occurs in the eukaryotic mitochondria. It requires oxygen and involves the Krebs cycle (or citric acid cycle) and the mitochondrial ETC.

Glycolysis (anaerobic)

- Anaerobic respiration that occurs without oxygen.
- Evolved in common ancestral anaerobic bacteria.
- Produces a net of 2 ATP, 2 NADH (electron carrier equivalent to NADPH in photosynthesis), and 2 pyruvate.

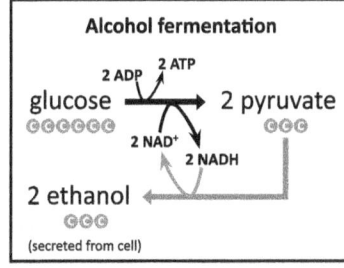

Fermentation (anaerobic)

- Anaerobic respiration that evolved in anaerobic bacteria to replenish 2 NAD+ for continued ATP production by glycolysis.
- Forms ethanol (yeast) or lactic acid (muscles) using the 4 electrons from 2 NADH molecules.

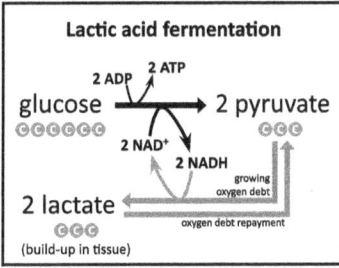

Krebs cycle (aerobic)

- Involves requires ETC.
- Releases energy from pyruvate to electrons.
- Delivers electrons to 2 cofactor electron carrying molecules (FAD and NAD$^+$)
- Cofactors reduction reactions: NAD$^+$ → NADH and FAD → FADH$_2$)
- NADH and FADH$_2$ can each carry 2 electrons to the ETC.

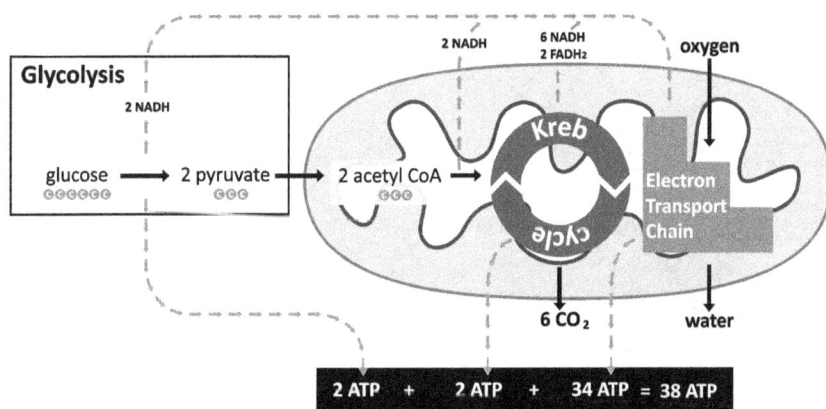

ATP ACCOUNTING
The net ATP produced is less than 38 because roughly 2 ATP are used for membrane transport.

NADH delivers its electrons at the first proton pump. These NADH electrons are responsible for driving the movement of 3 protons to produce 3 ATP per NADH. FADH$_2$ delivers its electrons after the first proton pump, but before the second. These FADH$_2$ electrons therefore pass through only 2 proton pumps to produce 2 ATPs per FADH$_2$.

Electron transport chain (aerobic)

- Requires oxygen as the final electron acceptor.
- Receives 10 NADH and 2 $FADH_2$ from glycolysis and Krebs cycle.
- Indirectly releases energy from the electrons for ATP synthase via the chemi-osmotic proton gradient (oxidative phosphorylation).
- Produces 3 ATP molecules per NADH, and 2 ATP molecules per $NADH_2$.

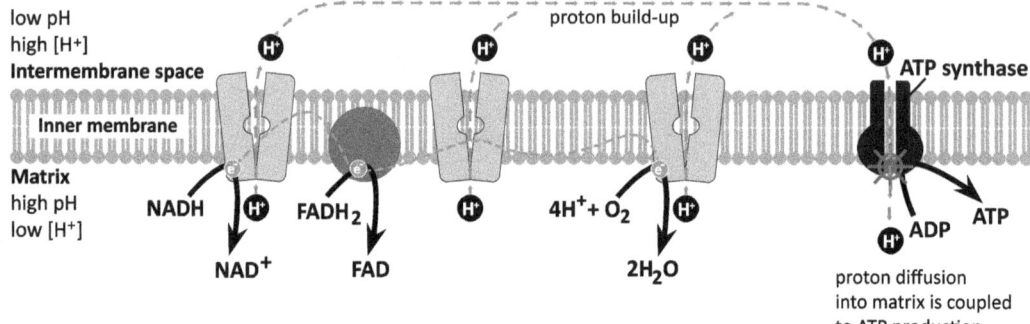

RESPIRATION AND METABOLISM

Metabolism includes all of the processes that allow organisms to exchange energy and matter with the environment. There are 2 interdependent phases of metabolism, the exergonic catabolic phase and the endergonic anabolic phase. Catabolic reactions result in the breakdown of organic matter, while anabolic reactions build new molecules. For example, glycolysis and the Krebs cycle are catabolic processes because they involve the breakdown of glucose and pyruvate to release energy. Oxidative phosphorylation, which captures this released energy and puts it into ATP, is an anabolic process. Subsequent catabolic hydrolysis of ATP to ADP is needed for make new macromolecules. The macromolecules (fats, carbohydrates, proteins, fatty acids, glycerol, sugars, and amino acids), store this energy and can also release the energy into the catabolic pathways of respiration at the points specified by the diagram on the right.

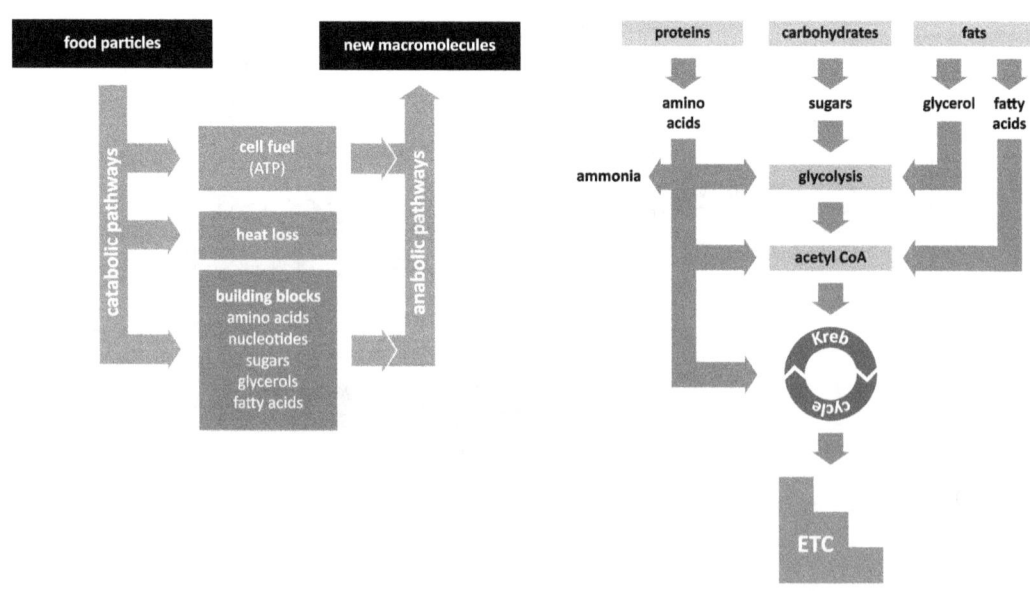

SUMMARY OF KEY CONCEPTS

- Autotrophs capture their energy from inorganic abiotic sources, with photosynthetic organisms harnessing energy from sunlight, and chemosynthetic organisms getting their energy from inorganic molecules.

- Heterotrophs get their energy from organic molecules that are produced by other organisms, such as carbohydrates, lipids, and proteins that can be hydrolyzed to release stored energy.

- Fermentation occurs in the absence of oxygen to produce either alcohol or lactic acid.

- NADP+ and oxygen are the final electron acceptors of the ETC in photosynthesis and respiration, respectively.

- The light-dependent reaction of photosynthesis involves several reaction pathways that capture light energy to produce NADPH and ATP that power the production of glucose in the Calvin cycle.

- PSI, PSII, and the ETC are located in the thylakoid membrane and are critical to the light reactions.

- The movement of electrons through the ETC establishes the proton gradient across the thylakoid membrane.

- Oxidative phosphorylation by ATP synthase is coupled to the ETC via the established proton gradient.

- Cellular respiration involves a series of reactions that release energy from fats proteins and carbohydrates and stores this energy in ATP.

- Glycolysis releases energy from glucose to ATP molecules and produces pyruvate.

- Pyruvate is transported into the mitochondria and undergoes oxidation by the Krebs cycle to release additional energy into ATP.

- The ETC of respiration and photosynthesis captures energy from electrons. Decoupling of ETC from oxidative phosphorylation is involved in thermoregulation.

CHECK YOUR UNDERSTANDING

1. Which of the following statements is true?

    (A) Anaerobic respiration in prokaryote can occur by glycolysis only (without fermentation) because, unlike multicellular eukaryotes, the bacteria can secrete any produced pyruvate to avoid the potential toxic acid build-up.

    (B) Glycolysis requires 2 NAD^+ to produce 2 ATPs.

    (C) Without bacterial fermentation, rapid depletion of NADH will cause glycolysis to stop and the bacteria will run out of ATP.

    (D) All of the above statements are true.

2. The electron donor of ETC in photosynthesis is

    (A) NADPH

    (B) Oxygen

    (C) Water

    (D) Glucose

3. The final electron acceptor of the ETC of respiration is

    (A) NADH

    (B) Oxygen

    (C) Water

    (D) Glucose

4. The light-dependent reaction occurs in the

    (A) lumen.

    (B) cytoplasm.

    (C) thylakoid membrane

    (D) stroma

5. Which of the following pigments can directly transfer electrons to the ETC

    (A) Chlorophyll a

    (B) Chlorophyll b

    (C) Carotene

    (D) lycopene

6. How many electrons does each of the electron carrying cofactors transport?

    (A) 1

    (B) 2

    (C) 3

    (D) NADH and NADPH carry 3 electrons and $FADH_2$ carries 2 electrons.

7. The enzyme that delivers the electrons to NADPH is called

   (A) Fd-NADP$^+$ reductase

   (B) Fd-NADPH oxidase

   (C) ATPase

   (D) None of the above.

8. If 10 NADH and 10 FADH$_2$ were supplied to the mitochondrial ETC, how many ATP will be produced?

   (A) 20

   (B) 30

   (C) 40

   (D) 50

9. Which of the following statements is true?

   (A) Proton build-up occurs in the stroma of the chloroplast.

   (B) Proton build-up occurs in the intermembrane space of the mitochondria.

   (C) Protons diffuses from the lumen to the stroma.

   (D) Aerobic respiration and glycolysis together produce a gross of 36 ATP.

ANSWERS AND EXPLANATIONS

1. (B) is correct. Options (A) and (C) are wrong because fermentation is needed to replenish NAD^+ (not NADH), which is used up by glycolysis. Without fermentation NAD^+ is rapidly depleted.

2. (C) is correct. Recall that water is split when to of its electrons are donated to chlorophyll a (P650) of photosystem II.

3. (B) is correct. Oxygen accepts the electrons and forms covalent bonds with protons to produce water.

4. (C) is correct.

5. (A) is correct. The other pigments are antenna pigments that funnel their absorbed energy to chlorophyll a.

6. (B) is correct.

7. (A) is correct.

8. (D) is correct because each NADH delivers enough energy for 3 ATP to be produced (or 10×3 = 30) , and each $FADH_2$ produces 2 ATP (or $10 \times 2 = 20$). $20 + 30 = 50$ ATPs.

9. (C) is correct. (D) is wrong because there is a gross of 38, but a net of about 36 ATPs because about 2 ATPs are used for membrane transport.

15 Dynamic homeostasis

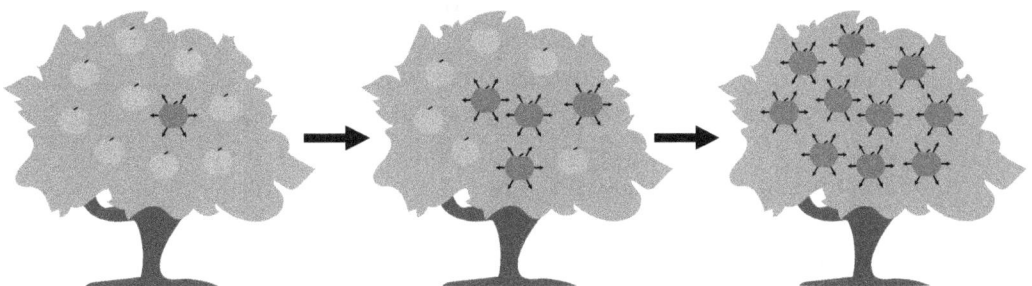

ALL STUDENTS MUST BE ABLE TO ANSWER THESE QUESTIONS

1. What is the difference between positive and negative feedback?
2. How is negative feedback used to regulate body temperature in animals?
3. What are the consequences of failure in feedback mechanisms? Reflect on diabetes, dehydration, Graves' disease, hemophilia.
4. What is the mechanism involving auxin, and outcome of phototropism? Photoperiodism?
5. What are biotic and abiotic factors, and how do they influence cell and organism activities?
6. How do biotic and abiotic factors affect the stability of populations, communities, and ecosystems?
7. What is the predator-prey relationship? How does negative feedback help keep it stable?
8. Why do algal blooms occur?
9. How are homeostatic mechanisms evidences of evolution? Reflect on plants response to drought, and homologous and analogous structures.
10. How can disruptions in homeostatic mechanisms at the molecular, cellular, and ecosystem levels affect the health of the biological system?
11. How does temporal regulation and coordination of development rely on homeostasis?
12. What is the role of hox genes, cytoplasmic determinants, and inducers in embryonic induction and development?
13. How is apoptosis important to development? Reflect on morphogenesis of fingers and toes.
14. How do plants, invertebrates, and vertebrates respond to pathogens?

ALL STUDENTS MUST BE ABLE TO COMPLETE THE FOLLOWING TASKS

1. Make predictions about the effects of positive feedback on a specific metabolic pathway.
2. Use a specific example to justify that positive feedback mechanisms amplify responses.
3. Make predictions about the outcome of a negatively regulated metabolic pathway.
4. Evaluate claims based on a describe scenario that involves a negatively regulated pathway.

 Big Idea 2C,2D,2E

Given that the boundaries of biological systems are porous, they alone are insufficient for resisting entropy driven diffusion. All biological systems require processes that can monitor and responds to disruptions in **homeostasis** and changes in external conditions that can predictably influence it. For example, a hungry predator that is low on energy and nutrients will naturally experience the sensation of hunger, stimulating its pursuit of a prey. When a prey is captured, the predator feeds until a satisfactory sensation signals a return to homeostasis (nutrient and energy balance). Likewise, the sight of the predator, stimulates prey's **fight-or-flight response**. Thus, while the predator is responding to existing disruption in homeostasis, the prey is innately responding to avoid future homeostatic disruptions.

FEEDBACK MECHANISMS AND HOMEOSTASIS

Homeostasis is used to maintain internal condition at a pre-programmed **set point**. For example, in humans the homeostatic body temperature is 98.6 °F, and pH is 7.4. A different homeostatic mechanism regulates blood glucose levels, electrolyte concentration and water balance, and calcium levels. **Negative feedback** control is used to maintain homeostasis within a narrow range by reversing a deviation above or below the set point. In a few rare instances, **positive feedback control** is used to amplify an effect that causes a deviation from the set point. Mammalian lactation is a good example of this, where the initial suckling effect to release breast milk, stimulates neurons in the mammary gland to stimulate increase milk production and secretion.

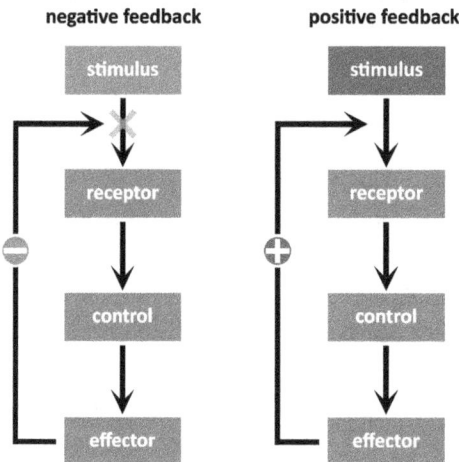

Positive feedback in fruit ripening

An apple starts to ripen when cells produce and release the plant hormone called **ethylene** (stimulus). Ethylene stimulates the physical changes (softening, pigmentation, sweetening) associated with fruit ripening. These changes amplify and accelerate the ripening process by promoting more ethylene biosynthesis. Besides accelerating it own ripening process, it also releases ethylene to the local environment to accelerate the ripening process in adjacent fruit. Through this positive control mechanism, all apples in the tree and in adjacent trees turn ripe in a coordinated, and relatively simultaneous manner.

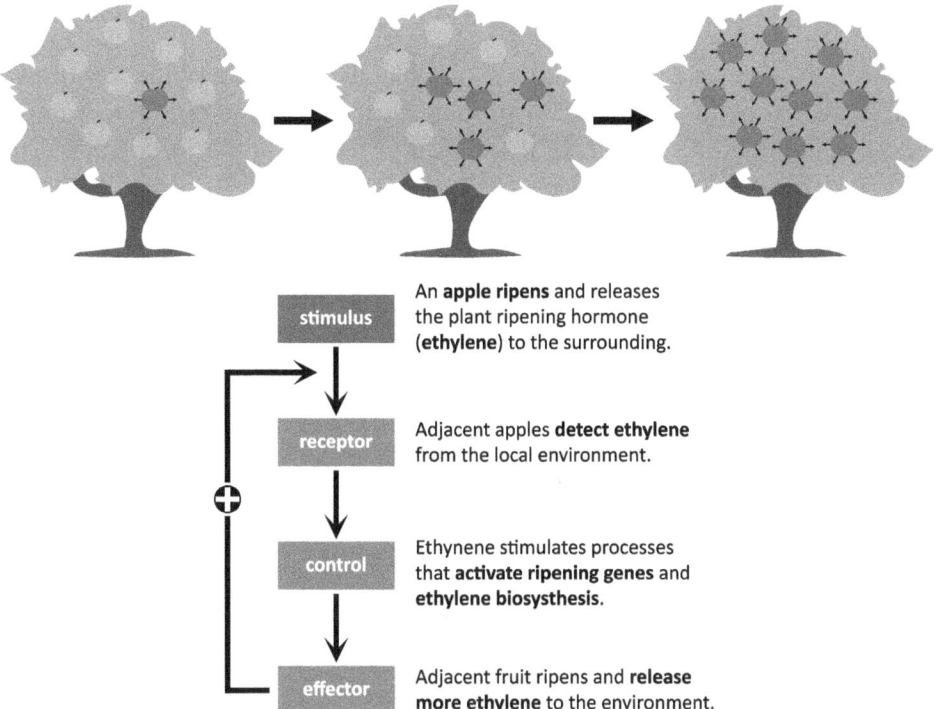

stimulus	An **apple ripens** and releases the plant ripening hormone (**ethylene**) to the surrounding.
receptor	Adjacent apples **detect ethylene** from the local environment.
control	Ethynene stimulates processes that **activate ripening genes** and **ethylene biosysthesis**.
effector	Adjacent fruit ripens and **release more ethylene** to the environment.

POSITIVE FEEDBACK
Other examples of positive feed back include: (1) Mammalian lactation, which is stimulated by the suckling effect and further amplified by the hormone called oxytocin. When the baby begin suckling, the hypothalamus releases oxytocin which stimulates the mammary glands to increasing milk production and release; and (2) Progression of labor in birthing is accelerated when the baby exerts pressure on the uterine wall, causing the release of the hormone called oxytocin , which further stimulates uterine contraction.

NEGATIVE FEEDBACK
Other examples of negative feedback include:

(1) Gene regulation by prokaryotic operons (see L22); (2) Plant response to drought by closing stomata or loosing leaves to reduce transpiration.

Negative Feedback in temperature regulation

The thermostat used to regulate home temperature during uses negative feedback control. If the selected set point is 60 °F, the heater remains on until the thermostat detects a temperature rise above this set point. This triggers a signal switch that turns off the heater. When the temperature falls below the set point, negative inhibition is deactivated and the heater turns on. A similar mechanism is used to regulate cooling by air conditioners in the summer months.

In mammals, when body temperature rises above the set point, vasodilation of the blood vessels allows more blood to reach the skin's surface where heat is released by increased sweating and air-cooling. When the body temperature drops below the set point, vasoconstriction (limits blood flow to the skin) and shivering (produces heat) are stimulated to restore the homeostatic temperature.

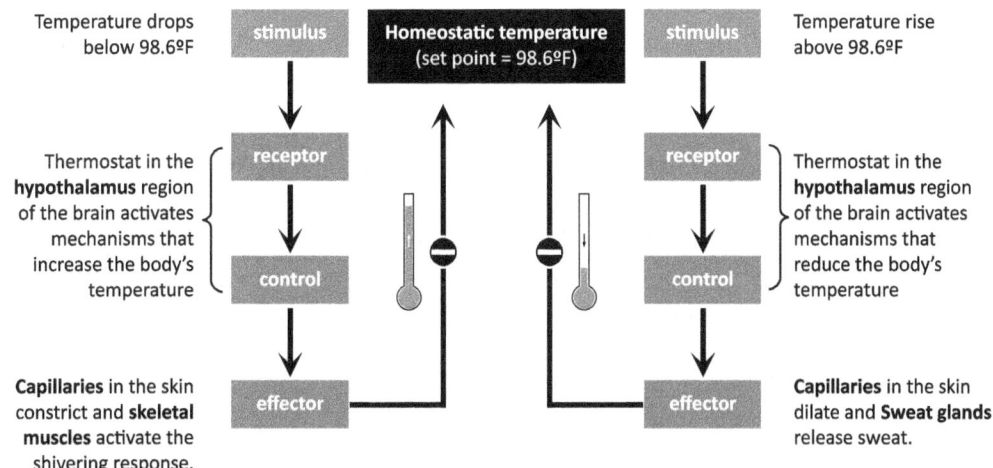

Failure in feedback mechanism

Because feedback mechanisms are critical to maintaining the set point, any disruption in their regulatory pathways can cause problems for the organism. Some examples of medical conditions that result from disruption in feedback control include diabetes, alcohol or caffeine induced dehydration, hyperthyroidism, and hemophilia.

Diabetes mellitus

Pancreatic hormones, insulin and glucagon, are responsible for regulating the blood glucose levels. The **pancreas** releases insulin when glucose levels rise after sugar consumption. The enzyme enters the blood and travels to target cells in the liver and adipose (fat) tissue. Liver cells are then stimulated to use glucose molecules for **glycogen** synthesis, {Glycogen is formed by linking together glucose monomers. It is the energy storage carbohydrate in animals. Recall that plants storage carbohydrate to starch.} while fat cells synthesize triglyceride. The outcome is that glucose is removed from the blood and stored in adipose and liver tissues.

When glucose levels are low, as a result of fasting or increased exercise, the pancreas instead releases **glucagon**. This enzyme stimulates the catabolism of glycogen (liver) and triglycerides (adipose tissue) to glucose. The outcome here is an increase in blood glucose levels. Diabetic patients who have a faulty form the insulin or glucagon genes are unable to properly regulate blood glucose levels, resulting in unusually high or low glucose levels. This can lead death or to various physiological complications including damage to eye, blood vessels, kidney, and nerves.

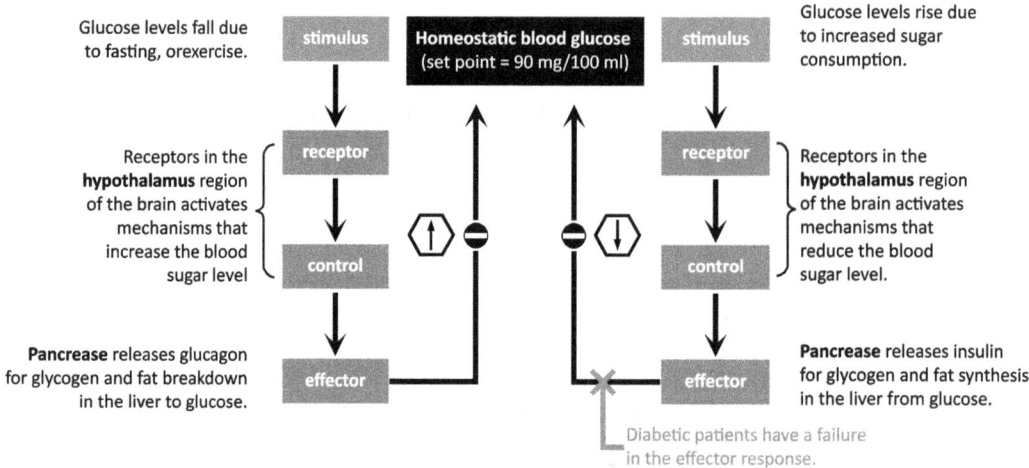

Hyperthyroidism

Hyperthyroidism is the over stimulation of the **thyroid gland,** causing the release of large quantities of the hormones triiodothyroxine (T3) and thyroxine (T4). These thyroid hormones act on nearly every cell type to regulate metabolisms by increasing the break down of energy storing compounds such as triglycerides and glycogen. By so doing, they directly raise body temperature (increased aerobic respiration), and breathing and heart rate (increased demand for oxygen). This increased body temperature stimulates negative feedback in thermal regulation to cause increased sweating.

As can be expected, the thyroid hormones play a critical role in the fight-or-flight response, which is a physiological response to an external threat. It prepares organisms to response appropriately by raising breathing and heart rates (increase blood flow and oxygen supply to muscles), stimulating the catabolism of glycogen and triglycerides (supplies glucose for respiration and ATP production), increasing muscle tension (increases strength and speed), and stimulating the release of clotting factors (limits blood loss).

Patients suffering from hyperthyroidism experience muscle fatigue, increased heart rates, heat intolerance, enlargement of the thyroid, and weight loss. Most incidences of hyperthyroidism are observed with Graves's disease patients. The cause of Graves disease is not fully understood, but it appears to be an autoimmune condition where self-produced antibodies bind to and stimulate receptors on thyroid cells, causing them to release T3 and T4 hormones.

Hemophilia

Blood clotting is regulated by a negative feedback loop involving clotting factors that act to stop blood loss. **Hemophiliacs** fail to produce a key clotting factors. Consequently, they are unable to form strong plugs over the wound and can experience life threatening blood loss from minor cuts.

RESPONSE TO ENVIRONMENTAL CHANGES

Organisms use a combination of behavioral and physiological mechanisms to respond to environmental changes that can affect their internal homeostasis.

Plant phototropism

Phototropism is the process that plants use to grow toward a light source. Phototropism is regulated by the plant hormone called **auxin**, which is produced in the cells of the elongating stem tip. Once produced, auxin diffuses to the shady side of the stem where it becomes concentrated. There, it stimulates cells to accelerate their growth and elongate. Consequently, cells on the shady side are longer relative to the light-facing side. This causes the stem to bend toward the light.

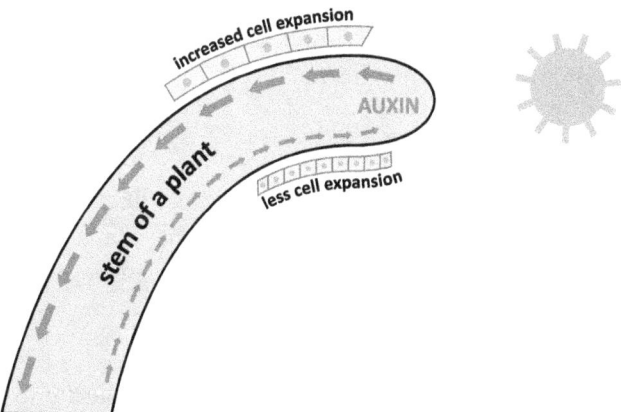

Plant photoperiodism

Photoperiodism is a physiological process that plants use to measure the length of the day in order to determine when during the year to flower and reproduce. Although the exact mechanism is not fully understood, it appears that a special photoreceptor protein detects the light and causes physiological changes equivalent to a kind of biological stopwatch for measuring the day length. Long-day plants produce flowers in the early summer when the nights are short and the days long. Short-day plants reproduce in the late fall when the days are short and the nights long. The exact number of hours of light required for flowering varies by plant species because of differences in the threshold number of light hours (or critical period).

Instead of photoperiodism, some day-neutral plants use vernalization (see L10) or developmental stage to regulating flowering.

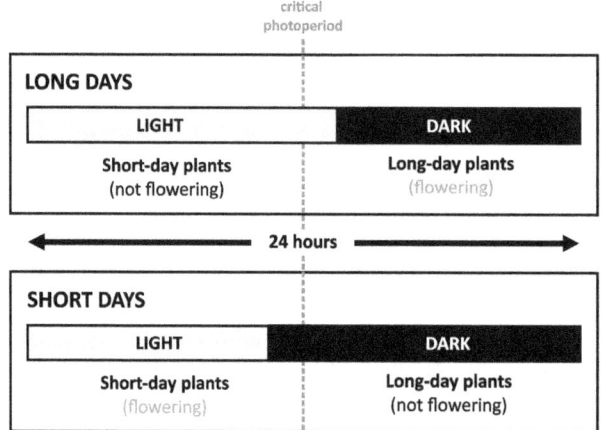

Other response mechanisms

Other examples of how organisms respond to environmental changes include:

- Seasonal hibernation and migration due to onset or winter or dry season when primary productivity dramatically decreases;
- Bacteria or animal taxis (guided movement toward or away from a stimulus) and kinesis (random motion in no specific direction in response to a stimulus);
- Bacteria or animal chemotaxis (movement toward or away from an increasing chemical gradient;
- Circadian rhythm in animals that influence nocturnal or diurnal activity;
- Initiation of mating process in fungal cells when stimulated by sex hormones released from compatible neighbors.

EFFECTS OF SYSTEMIC CHANGES

All levels of biological systems are affected by the availability of biotic and abiotic factors. Collectively, these factors influence the ecosystem biomass and, through natural selection, the genetic compositions of populations and species within.

Biotic and abiotic factors

The biotic factors of an ecosystem are the non-living physical and chemical components that organisms need and obtain from the land, atmosphere, or bodies of water. They include wind, humidity, weather, water, air, soil, minerals, and sunlight. Biotic factors are living organisms (and their organic remains) in the ecosystem that can directly or indirectly affect other living organisms.

Impact of drought on the ecosystem

Living organisms get material and energy from their environment. Amount of biotic and abiotic factors in the environment are directly linked to available energy and matter, and therefore affects individual and population fitness. For example, during drought, the primary production by plants becomes significantly reduced due to water shortage - recall that plants need water for photosynthesis and nutrient transport from roots by capillary action. Reduction at the first trophic level is felt in the higher trophic levels. The top predators at the top of the food chain have access to fewer preys, causing their population to decline. This bottom-up effect energy flow in the ecosystem can affect the entire food web and cause a loss of biodiversity, including extinction of especially vulnerable populations.

Temperature, water, salinity, and pH

Temperature decrease in an ecosystem can lower primary production, which reduces the available free energy for the higher trophic levels. Likewise, an increase in the temperature can cause dehydration, which also reduces primary production. In a similar manner, pH (acidity) and salinity (saltiness) levels in the soil and aquatic habitats can alter population density and community structures. For example, the current increase in global CO_2 levels is causing a decrease in ocean pH, which destroying coral reefs that are adapted to higher pH levels.

Predator-prey relationship

An increase in the number of predators can cause a decline in the prey population. In the near-term, this decline has a feedback effect on the predators by reducing the free that is available to them and causing their population to quickly drop after the prey population fall. In response to a predator decline, the feedback effect acts again to cause the prey population to rebound. This feedback loop of rise-and-fall is typical of an established predator-prey community.

Symbiosis

Organisms can affect each other by establishing symbiotic relationship, including mutualism (both organisms benefit), commensalism (one organism benefits, no affect to the other), parasitism (one organism benefits, the other harmed).

Algal blooms

Algal bloom is a rapid increase in algae population in aquatic ecosystem. It is often caused by agricultural run-off that carries fertilizer or animal waste to the waterways. The run-off is rich in key nutrients such as nitrogen and phosphorus that are limiting factors for primary plant productivity.

EVIDENCES OF EVOLUTION

Many organisms share mechanisms for maintaining internal homeostasis. Others acquired their mechanisms through independent adaptive evolution or by divergence from the common ancestor.

- Similarities in homeostatic mechanisms between organisms suggest a common ancestry. For example, all plants use specialized leaf structures called stoma that open and close during gas exchange. These structures also close to limit water loss (**osmoregulation**) by evapotranspiration caused by strong winds, low humidity, and high temperatures. Plants adapted to desert ecosystems have special water storing tissue and modified leaves (needle shaped, sunken stomata, and waxy leaf coating) that limit water loss.

■ Homologous differences in homeostatic mechanisms also suggest a common ancestry, where the observed variations are due to divergence from the common ancestor. For example, lungfish is a subclass of lobe-finned fish that retained the ability to breath air using its primitive lung that may have diverged from a lung shared by a common ancestor of all mammals.

■ Independently evolved mechanisms for obtaining nutrients and eliminating wastes are due to organisms adapting to different niches. For example, plant use stomata for gas exchange, while animals use a more complex respiratory system that includes lungs (mammals and birds) and gills (fish and amphibian).

DISRUPTION IN HOMEOSTATIC MECHANISMS

■ Disruptions in homeostasis at the molecular and cellular levels can affect the health or organisms. For example, snake venom often contains a neurotoxin that is used to paralyze a prey. These neurotoxins work by disrupting the ability of neurons to conduct a nerve impulse, resulting in the failure of various critical organs, and death.

■ Disruption in dynamic homeostasis at the ecosystem level can impacts the natural balance that exist between various tropic levels, by altering energy and nutrient flow in the food chain and the food web. For example, the North American gray wolf was nearly driven to extinction through systematic hunting to protect grazing domesticated cattle. As a keystone species, the removal of gray wolves from northern ecosystems had a rippling effect on other species and throughout the food web, including elk population that increased due to decreased wolf predation, and decreased scavenger population that no longer had the left-overs carcasses from the wolves meals.

Adaptive response to harmful agents

■ Plants use molecular detection systems to activate a non-specific (innate) immune response to pathogens. Infected cells secrete chemicals that kill surrounding cells and the infected cells, thereby isolating the infection.

■ Invertebrate also use non-specific immunity such as secreted hemolymph, a gel-like fluid that traps pathogens.

■ Vertebrate use both non-specific (skin, inflammation) and specific immunity (antibody mediated) to protect against pathogens.

TEMPORAL REGULATION AND COORDINATION

Internal homeostasis extends to the orderly growth, reproduction, and development of organisms, each regulated by multiple temporally coordinated pathways. **Embryonic induction** begins shortly after the egg is fertilized. It is the process that guides groups of cells along specific developmental pathways that determine the types of

NOTES

cells and tissue that their descendant cells will become. Embryonic induction is at work throughout the following stages of development:

- Zygote (embryo is a single cell)
- Morula (embryo is a ball of 8 cells or more without a cavity)
- Blastula (embryo is a ball of cells with a fluid-filled cavity)
- Gastrula (embryo is a mass of 3 distinct morphogenic germ cell layers that will develop into different body parts)

Cytoplasmic determinants

Cytoplasmic determinants are substances originating from the maternal gamete (unfertilized egg). During development, the fate of cells in the developing embryo is influenced by the unequal distribution of cytoplasmic determinants to daughter cells of the morula, blastula, and gastrula stages. The relative concentration of cytoplasmic determinant molecules in the early embryonic cells determines the fate of their descendants. Prior to this unequal distribution, cells are totipotent, meaning that they can become any cell type of the body. After unequal distribution, embryonic cells have began to differentiate and their developmental paths are now determined.

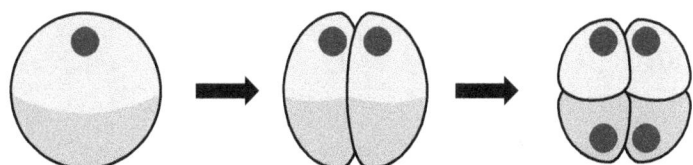

Unequal distribution of cytoplasmic determinants in the fertilized egg is retained as the cells divide, with higher concentrations accumulating in the lower cells.

Embryonic induction

At the gastrula stage, tissue cells begin secreting chemical signals molecules called inducers that act a cell-to-cell signaling molecules. They activate transcription factors that activate the genes needed to coordinate cell differentiation, and tissue and organ development.

Hox genes

Beside the inducers and cytoplasmic determinants, there are developmental genes, including a subset called the hox genes that regulate more specific aspects of development. The hox genes are the overseers of development, and are commonly referred to as the body plans genes because they instruct when during embryogenesis, and where in the body to assemble a specific body part. Because of the importance in embryonic development, mutations in hox genes can affect how the embryo's body parts are organized relative to each other.

Apoptosis in embryonic development

Some developmental cells are short lived, lasting only until they complete their temporally coordinated tasks. Once completed, they self-destruct through a programmed cell death pathway called apoptosis. For example, during human embryogenesis, apoptosis causes the death of the cells between the digits of fingers and toes to prevent the kind of webbing common to aquatic species such as ducks and seals.

SUMMARY OF KEY CONCEPTS

- Organisms use positive and negative feedback mechanisms to regulate homeostasis and to respond to changes in their environment.
- Positive feedback mechanisms are used to amplify response to stimulus.
- Negative feedback is used to downgrade response to stimulus.
- Disruption in feedback mechanisms can affect the health of an organism or ecosystem.
- Organisms respond to changes in their environment to ensure the maintenance of future homeostasis.
- The maintenance of homeostasis in all biological systems is influenced by the biotic and abiotic factors.
- Similarities and differences in homeostatic mechanisms are evidences of evolution.
- Plants and animals have mechanism to protect them against pathogens that can disrupt homeostasis.
- Homeostasis is involved in grow, reproduction, and development.
- Biological processes involved growth, reproduction, and development are often temporally regulated, and may involve apoptosis.

CHECK YOUR UNDERSTANDING

1. Which of the following is an example of negative feedback?

   (A) Since the start of the industrial revolution, the global human population has experienced exponential growth.

   (B) During a viral or bacterial infection, the body temperature rises as a mechanisms of the innate immune response to destroy the pathogen.

   (C) Hikers to the summit of Mount Everest must spend about 2 weeks at mid summit to acclimate to the lower oxygen levels. During this time, the body produces more red blood cells for increased oxygen transport.

   (D) Nursing mothers wear an absorbing pad over their breast nipple to absorb milk released due to oxytocin stimulation. Oxytocin can be released by the onset of suckling, or the sound of a crying hungry baby.

2. The body temperature is lowered to the set point by

   (A) capillary dilation.

   (B) increased respiration

   (C) releasing thyroid hormones.

   (D) decreased sweating.

3. Which of the following can result in unusually high blood glucose levels?

   (A) Over-stimulation of the pancreas to release insulin.
   (B) Autoimmune condition where self-produced antibodies bind the epinephrine (glucagon) receptors on liver and adipose cells.
   (C) Failure of liver cells to respond to glucagon.
   (D) None of the above.

4. Auxin is dissolved in the water supply for plants in a nursery. Which of the following is likely to occur?

   (A) The plants stem will grow relatively straight, irrespective of the light angle.
   (B) The plants will have deeper roots.
   (C) The plants will grow more toward the light.
   (D) The plants will grow less toward the light.

5. Which of the following is an expected organism response based on the provided information?

   (A) After several weeks of low precipitation, the trees began to loose their leaves.
   (B) A flock of migrating birds were seen flying north from Buffalo New York in late fall.
   (C) The number of squirrel newborns spike during the winter months
   (D) A flock of bees were seen flying out to sea.

6. Which of the following conditions will have a negative impact on population size at the second trophic level?

   (A) increased sunlight
   (B) decreased number of predators at the third tropic level
   (C) algal bloom
   (D) increased local temperature from set point

7. Which of the following statements is incorrect?

   (A) Cytoplasmic determinants are derived from the unfertilized egg.
   (B) Embryonic inducer activate genes needed for coordinated development of tissues and organs.
   (C) Inducers are produced by local tissue cells.
   (D) Cytoplasmic determinants and inducers are the same substances.

ANSWERS AND EXPLANATIONS

1. (C) is correct because at higher altitudes, the re is less oxygen available. The body compensates by supplying more RBC for binding and transporting the available oxygen. Remember that in negative feedback, the effect returns the system to set point. In this case, the production of more RBC return the body to it normal level of oxygen supply. (A) is positive feedback because as the population grows, the number of babies born increases - birth rate is tied to population size. If we assume that the set point is preindustrial revolution population, the population is moving away from set point. An affect that causes the system to drift away from set point is considered positive feedback. (D) is also positive feedback because, if we assume that the set point in no lactation, increasing lactation is drifting away from set point.

2. (A) is correct because the dilation of the capillaries will cause more blood to reach the skin surface where it is cooled by the air. Release of thyroid hormone (C) increases respiration rate and causes body temperature (B) to increase. (D) only happens as the body approaches set point temperature.

3. (B) is correct because the autoimmune condition will result in lots of glucagon be released. This hormone will than cause the liver and adipose cells to release lots of glucose into the blood by metabolizing glycogen and fats, respectively. (A) will cause increase glucagon and fats synthesis from blood glucose in liver and adipose tissue, respectively. This will lower the blood glucose level. (C) will block the liver from metabolizing glycogen and releasing glucose into blood.

4. (B) is correct because growth is stimulated in both the stem and roots, causing the plant to grow taller and deeper. However, there is no effect on the bend angle as suggested by (A), (C), and (D).

5. (A) is correct because plants tend to loose their leaves during drought in order to conserve water by reducing evapotranspiration. (B) is wrong because migrating birds will fly south as winter approaches. (C) is wrong because squirrel temporally time reproduction so that newborn arrive in the spring when excess free energy is available. (D) is wrong because bees should no fly out to see where they are sure to not find flowers.

6. (C) is correct because an algae bloom causes the oxygen depletion that suffocates heterotrophs in the higher trophic levels. (A) will increase the amount of free energy to the second trophic level. (B) results in a decreased predation on the second trophic level, causing their population to rise. A temperature variation from set point (D) is not ideal for growth since the species is well adapted to the set point temperature. Increasing or reducing set point temperature will likely cause a reduction in population size.

7. (D) is as incorrect statement. Cytoplasmic determinants are maternal factors derived from the unfertilized egg and lead to embryonic segmentation and determines the cell's fate. Determination does not physically change a cell. The determine cell will still look exactly the same as other cells, but it is restricted in what types of cells it can differentiate into. Embryonic inducers are produced and secreted by local embryonic cells and are therefore dependent on the embryos genotype. They induces cell differentiation, which leads to tissue and organ development. Unlike determinant, inducers cause a change in the cell's phenotype.

BIG IDEA 3

Information

- Information flow
- Cell reproduction
- Patterns of inheritance
- DNA replication
- Transcription and Translation
- Gene regulation
- Cell communication

16 Information flow

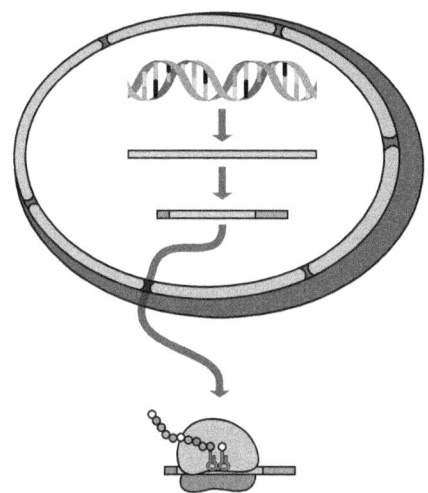

ALL STUDENTS MUST BE ABLE TO ANSWER THESE QUESTIONS

1. What are the differences between chromosomal and extra-chromosomal DNA?
2. Which scientist contributed to the discovery that DNA in the genetic material?
3. How is Chargaff's rule support the structure of DNA?
4. What is the significance of the (A+T)/(G+C) ratio to DNA denaturation?
5. What are the structural differences between DNA and protein that were exploited in the Hershey-Chase experiment?
6. What are the similarities and differences between the structure and function of RNA and DNA?
7. What does is meant by when DNA replication is described as being semiconservative?
8. How does semiconservative DNA replication ensure a continuity of genetic information?
9. What is the impact of mutations on the polypeptide sequence?
10. What is "triplet codon," and what causes codon degeneracy?
11. What is the central dogma of molecular biology? Reflect on the processes involved in information flow within cells?
12. What is transcription and translation, and how do they occur in eukaryotes? Prokaryotes?
13. What are the differences between a sense and antisense DNA sequence, and how do they compare to the mRNA?

ALL STUDENTS MUST BE ABLE TO COMPLETE THE FOLLOWING TASKS

14. Transcribe a provided DNA strand to mRNA and then translate the mRNA to polypeptide.
15. Determine which DNA strand is the sense strand based on provided information.
16. Distinguish between various types of DNA mutations.

 Big Idea 3A1a-d

DNA IS THE PRIMARY GENETIC MATERIAL

The organization of life depends on the storage of information, and the transfer of the information to progeny. Although DNA is the primary heredity molecule, some biological systems, such as the HIV retrovirus, use RNA instead of DNA.

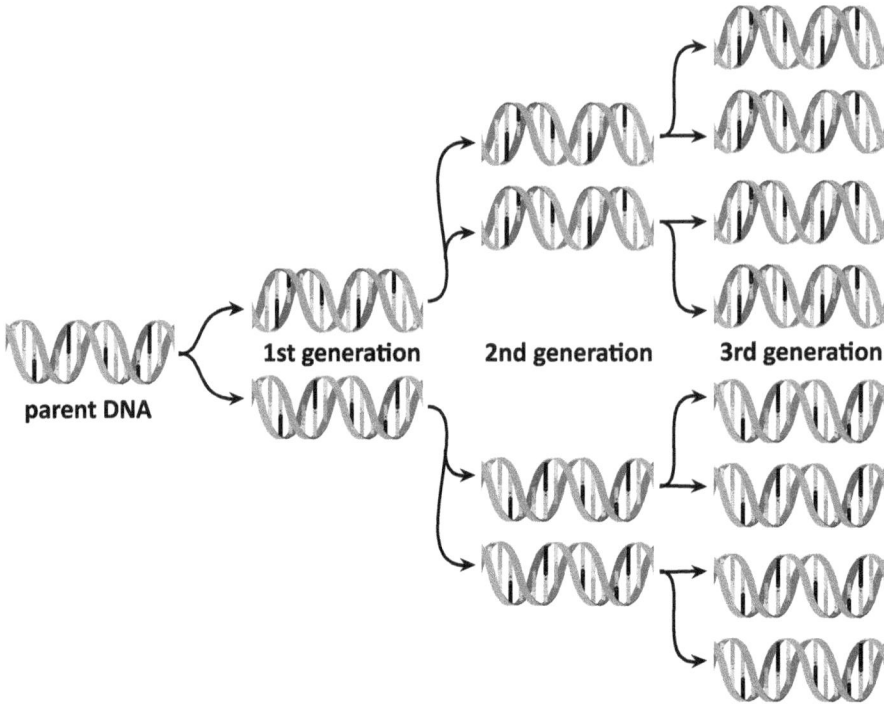

Chromosomal DNA

Cells tend to have multiple types of DNA. For example, in additional to chromosomal DNA, plant cells have chloroplast and mitochondria DNA, and bacteria cells have plasmid DNA. The chromosomal DNA is genomic, meaning that is store a majority of the organism's heritable information. In eukaryotes, the chromosomal DNA is divided into multiple linear chromosomes that are housed in the nucleus, while bacteria have a single circular chromosome that floats freely in the cytoplasm.

Plasmid DNA

Plasmid DNA is a small circular double-stranded DNA molecule that is of foreign origin (usually virus. A plasmid can be imported to the cell by diffusion through membrane pores, or virus mediated delivery. Plasmids can replicate independently from the chromosomal DNA, and can introduce beneficial genes such as antibiotic resistance. In biotechnology, artificially modified plasmids that include specific genes of interest are used to transform bacteria.

DISCOVERY OF DNA AS THE HERITABLE FACTOR

The discovery and proof that DNA was the primary heritable factor was based on the work of several individuals. The following table lists in chronological order, the most influential scientists and their specific contributions.

Year	Scientist	Contribution
1865	Gregor Mendel	He discovered that parents transmit a heritable factor to their progeny.
1902	Walter Sutton Theodor Boveri	They independently developed the chromosomal theory of inheritance, which states that the chromosomes contain the genetic information.
1908	Thomas Morgan	He expanded the chromosomal theory of inheritance by proving that genes are located on chromosomes.
1928	Frederick Griffith	He discovered that a transforming factor could cause a change in an organism's phenotype.
1944	Oswald Avery Colin MacLeod Maclyn McCarty	Their famous Avery-Macleod-McCarty experiment proved DNA was Griffith's transforming factor.
1947	Erwin Chargaff	He discovered the Chargaff's rule, which states that there is a 1:1 ratio of pyrimidine to purine with the amount of G=C and A=T in DNA.
1952	Alfred Hershey Martha Chase	Their famous Hershey-Chase experiment confirmed the Avery-Macleod-McCarty experiment that the DNA was the heritable factor.
1953	Rosalind Franklin James Watson Francis Crick	Franklin produced an X-ray diffraction image of DNA that Watson and Crick used to determine the double helical structure of DNA.
1958	Matthew Meselson Franklin W. Stahl	They discovered that DNA replication is semi-conservative.

Griffith's experiment

Griffith's experiment demonstrated that a transforming factor released from heat killed virulent smooth (S) bacteria could transform a harmless rough (R) bacteria into the virulent strain. Although Griffith experiment demonstrated the existence of a transforming factor, it did not address whether the transforming factor was DNA, RNA, or protein.

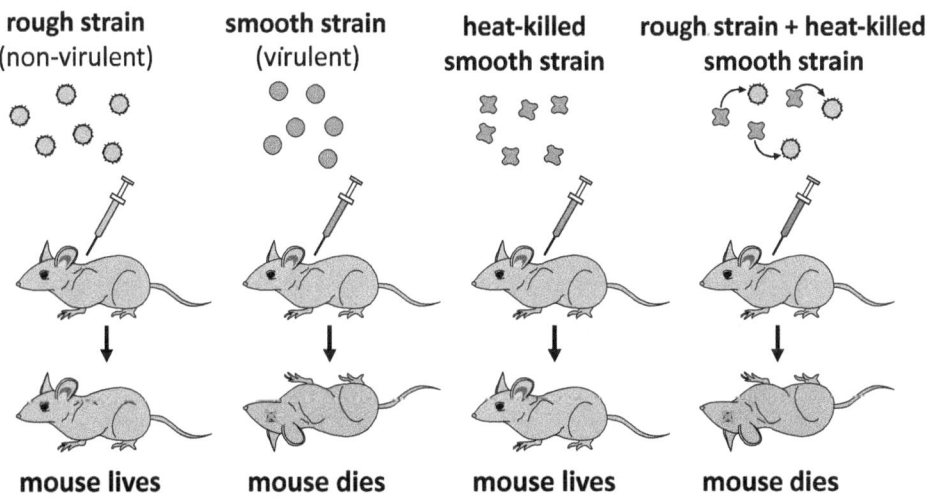

Avery-MacLeod-McCarty experiment

The S and R bacteria strains from Griffith's experiment were used to determine whether DNA, RNA, or protein was responsible for transformation. Heat-killed S bacteria were treated with DNase, RNase, or proteinase to destroy DNA, RNA, or protein, respectively. Each treatment was added to separate R bacteria cultures. The assumption was that transformation would not occur if the treatment destroyed the transforming factor.

For the RNase and proteinase treatments, transformation was still observed. However, transformation of the R bacteria to the virulent S strain failed when the DNase treatment was applied, suggesting that DNA was the transforming factor. Next, it had to be demonstrated that transformation required the insertion of DNA into the cell. The Hershey-Chase experiment addressed this problem.

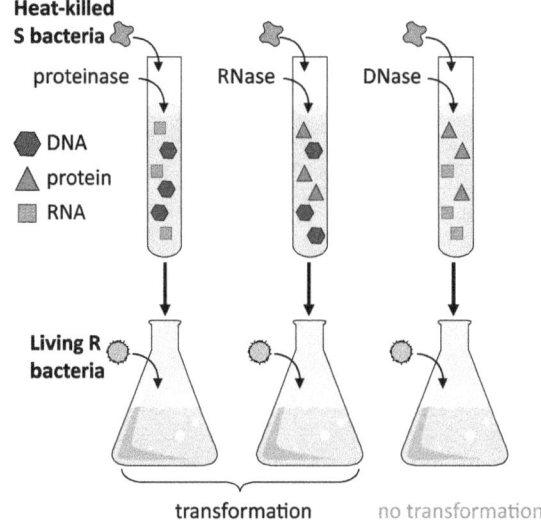

Hershey-Chase experiment

The Hershey-Chase experiment used structural differences between DNA and protein - proteins contain sulfur and DNA contains phosphorus - to show that bacteriophages transform bacteria cells by injecting their DNA into the cell's cytoplasm.

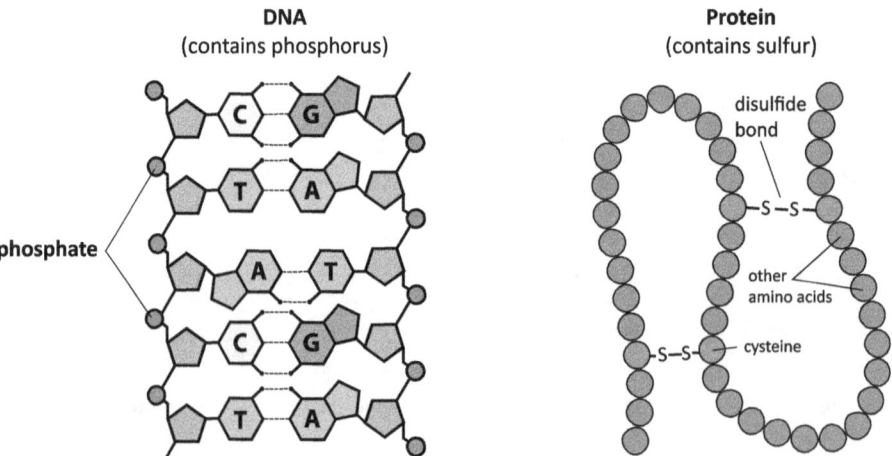

Two experiments were conducted. In the first, they inserted radioactive phosphorus-32 (32P) into bacteriophage and than used the radioactive bacteriophage to infect cells. They than traced the radioactivity to determine if it was present in the heritable factor that the bacteriophage injected into the bacteria cell. If the infected cells became radioactive, than the transforming material was DNA. In the second experiment, the scientists repeated the process, but replacing 32P with 35S so that proteins were instead detected.

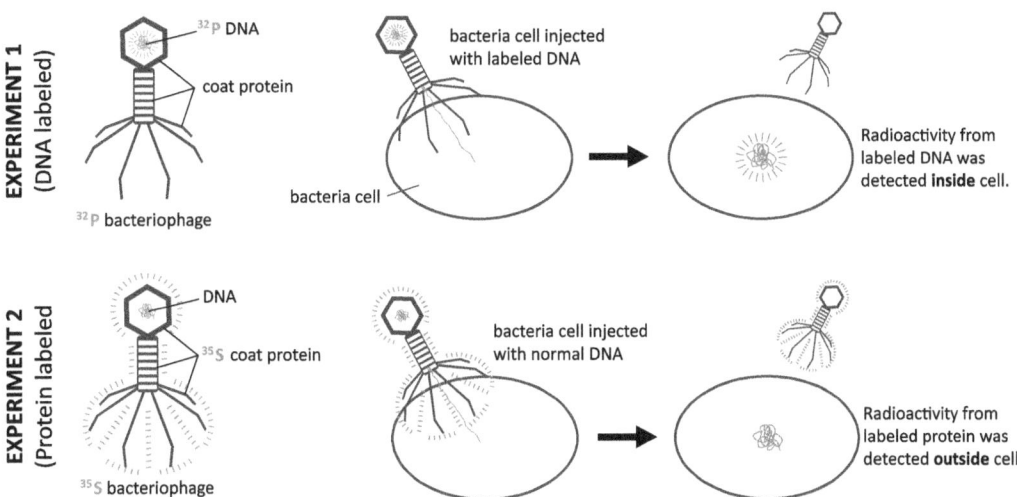

In the first experiment, the cells became radioactive, while in the second experiment only the external fluid of the cells were radioactive. This means that the viral DNA entered the cell and not it's proteins. This showed that the DNA and not protein, was the injected into the cell, proving that DNA was the heritable factor.

CHARGAFF'S RULE

Chargaff's rule states that in DNA, the amount of adenine is equal to thymine (A=T), and guanine is equal to cytosine (G=C). This rule is conserved across species and point to a common ancestry. This correlates with the structure of DNA where all G nucleotides of one strand form triple H-bonds with C of the complementary strand, while A and T form double H-bonds. The table below compares the relative percent of G, A, T, and C in five species. Notice that the A:T and G:C ratios are about 1, showing that the compositions of complementary nucleotides remain equal across species.

SPECIES	% A	%T	%G	%C	A:T ratio	G:C ratio	(A+T)/(G+C)
human	30.9	29.4	19.9	19.8	1.05	1.01	1.52
sea urchin	32.8	32.1	17.7	17.3	1.02	1.02	1.85
E. coli bacteria	24.7	23.6	26	25.7	1.05	1.01	0.93
S. lutea bacteria	13.4	12.4	37.1	37.1	1.08	1.00	0.35
T7 bacteriophage	26.0	26.0	24.0	24.0	1.00	1.00	1.08

Species that have more G and C relative to A and T have a higher ratio of triple to double H-bond between their complementary strands, making it more difficult to denature the strands during DNA replication. For example, because sea urchin has a high proportion of (A+T) relative to (G+C), denaturation of the sea urchin double stranded DNA occurs more readily than it does in S. lutea. For this reason, during DNA replication, the denaturation of DNA always begins at special A/T rich regions.

THE STRUCTURE OF DNA

DNA is made of 2 antiparallel sugar-phosphate backbones, with the nitrogenous bases paired in the interior of the molecule by H-bonds according to the Chargaff's rule.

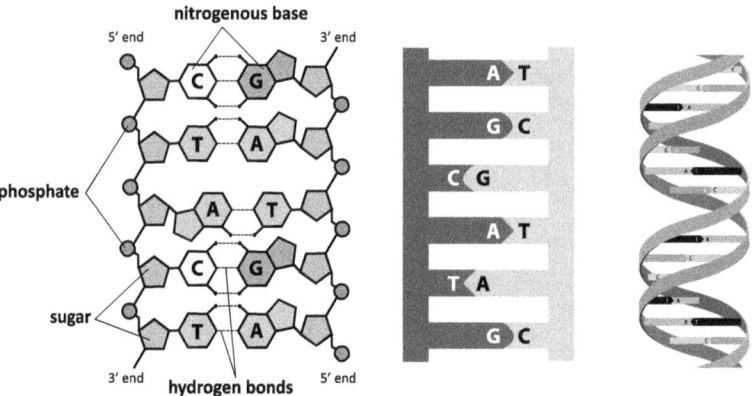

Three different ways of representing the DNA

Similarities between DNA and RNA

- DNA shares some structural features with RNA that give them their special information storage capabilities. The similarities include:
- Both are made of 3 components: sugar, phosphate, and nitrogenous base.
- Both are polymers formed by linking nucleotides together by covalent bonds.
- Both have directional orientation from the 5' to 3' end, with growth restricted to the 3' end.
- Both store genetic information in their nucleotide sequences.
- Both follow Chargaff's rule, except that Uracil (U) replaces T in RNA. Note that RNA is not double stranded, but when it is transcribed, Chargaff's rule applies for temporarily complementing the RNA nucleotides to the denatured single-stranded DNA. During transcription, the RNA transcripts as C (RNA) bonds to G (DNA), G (RNA) bonds to C (DNA), A (RNA) bonds to T (DNA), and U (RNA) bonds to T (DNA).

Pyrimidines and purines

The nitrogenous bases of DNA and RNA are classified according to the number of rings in their structure. Pyrimidines (C, T, and U) have a single ring, and can H-bond to doubled ringed purines, (G and A).

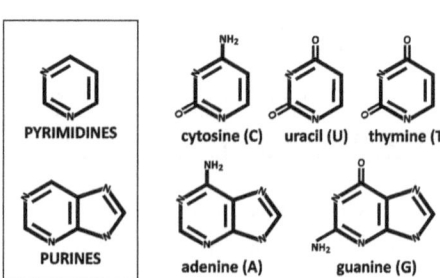

Differences between DNA and RNA

The main differences between DNA and RNA are summarized in the table below:

	DNA	RNA
sugar	deoxyribose	ribose
nucleotides	GATC	GAUC
strands	antiparallel double strand	single strand
stability	permanent	temporary
location	nucleus only	nucleus, cytoplasm
varieties	only 1 form	mRNA, tRNA, rRNA, RNAi

Semiconservative replication

DNA replication occurs prior to cell division so that there is enough DNA for each daughter cell to receive a complete set of genetic information. DNA replication is semiconservative, meaning that each of the two daughter DNA molecules is made of an old template strand derived from the parent molecule, and a newly synthesized complementary strand.

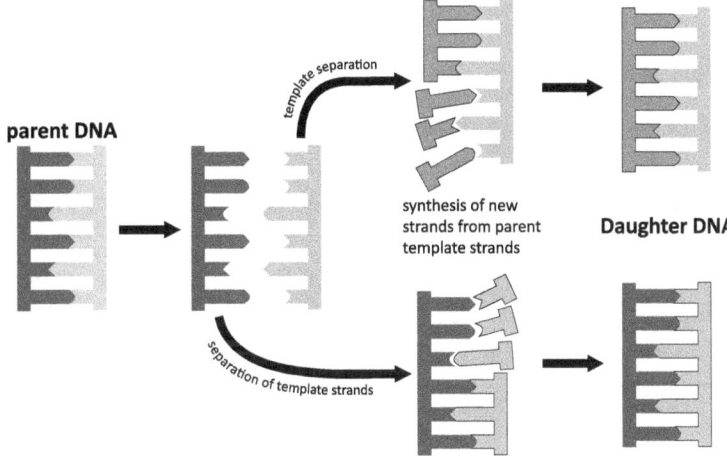

parent DNA

template separation

synthesis of new strands from parent template strands

Daughter DNA

separation of template strands

INFORMATION FLOW

The sequence of nucleotides in the DNA provides the instruction for protein synthesis. Proteins are the primary structural and functional units of biological system, performing a variety of functions that gives organisms their distinct phenotypes. Since the instructions for synthesizing proteins come from DNA, the DNA has an indirect and critical influence on the phenotype and fitness of organisms. Gene expression, the process that regulates protein synthesis, includes two important stages:

1. RNA transcription (DNA → mRNA)
2. Protein translation (mRNA → protein)

The central dogma of molecular biology

The central dogma is the idea that cells are governed by a cellular chain of command from DNA to RNA to protein through a series of evolutionarily conserved pathways. RNA is the bridge between segments of DNA called genes that code for specific proteins. Transcription is the synthesis of messenger RNA (mRNA) from the antisense template strand of the DNA molecule, and translation is the synthesis of a protein polypeptide based on the mRNA sequence.

Transcription, translation, and replication all require special proteins and enzymes that read the DNA or RNA in the 3' to 5' direction, and polymerize the new strand in the 5' to 3' direction (reading and polymerization direction explained later). Note that unlike RNA, DNA cannot self-replicate - DNA replication requires the enzyme called DNA polymerase. In some rare instances, such as with retroviruses, an enzyme called reverse transcriptase can synthesize DNA from a RNA template, a process called reverse transcription.

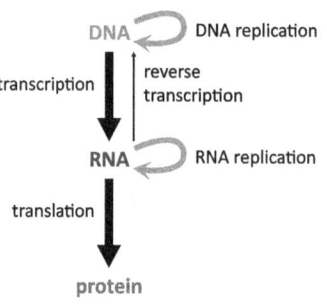

Prokaryotes versus eukaryotes

The central dogma governs how information flows in both prokaryotes and eukaryotes. In prokaryotes, the process is simpler, with both transcription and translation occurring simultaneously in the cytoplasm. In eukaryotes, transcription occurs in the nucleus to produce a pre-mRNA that undergoes further processing before leaving the nucleus for the cytoplasm. Translation of the mRNA occurs on ribosomes that are either free-floating (both prokaryotes and eukaryotes) or bound to the rough ER (eukaryotes only). Since in eukaryotes, transcription and translation are compartmentalized in the nucleus and cytoplasm, they do not occur simultaneously as in prokaryotes.

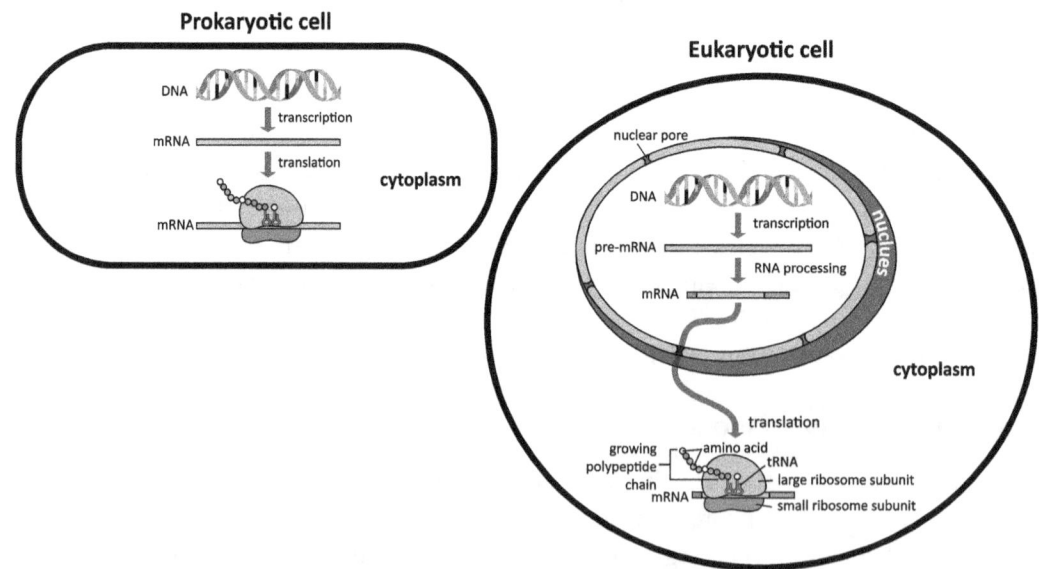

TRANSLATION AND THE TRIPLET CODON

The DNA is made of a sense and an antisense strand. Only the antisense strand is used for transcription. It is therefore complementary to both the sense strand and the mRNA. In other words, with the exception of U in mRNA and T in DNA, the sense strand holds the same nucleotide sequences as the mRNA.

Once in the cytoplasm, translation begins when the mRNA interacts with the rRNA in the ribosomes. The ribosome reads codon units (nucleotide triplets) in the 3' to 5' direction. There are 64 possible codons ($4^3 = 64$). With the exception of the stop codons (see codon chart on next page), each codon specifies one of 20 possible amino acids. Genes begin with a start codon (AUG) and end with one of the 3 stop codons (UAA, UAG, or UGA).

NOTES

PROTEIN & PHENOTYPES
Phenotypes are determined by protein activity, which include catalyzing reactions (enzymes), membrane transport (membrane-bound transport proteins), and metabolism of macromolecules (enzymes).

CODON CHART

second base of the codon

first base of the codon

	U	C	A	G	
U	UUU Phe UUC Phe UUA Leu UUG Leu	UCU Ser UCC Ser UCA Ser UCG Ser	UAU Tyr UAC Tyr UAA Stop UAG Stop	UGU Cys UGC Cys UGA Stop UGG Trp	U C A G
C	CUU Leu CUC Leu CUA Leu CUG Leu	CCU Pro CCC Pro CCA Pro CCG Pro	CAU His CAC His CAA Gln CAG Gln	CGU Arg CGC Arg CGA Arg CGG Arg	U C A G
A	AUU Ile AUC Ile AUA Ile AUG Met	ACU Thr ACC Thr ACA Thr ACG Thr	AAU Asn AAC Asn AAA Lys AAG Lys	AGU Ser AGC Ser AGA Arg AGG Arg	U C A G
G	GUU Val GUC Val GUA Val GUG Val	GCU Ala GCC Ala GCA Ala GCG Ala	GAU Asp GAC Asp GAA Glu GAG Glu	GGU Gly GGC Gly GGA Glu GGG Gly	U C A G

third base of the codon

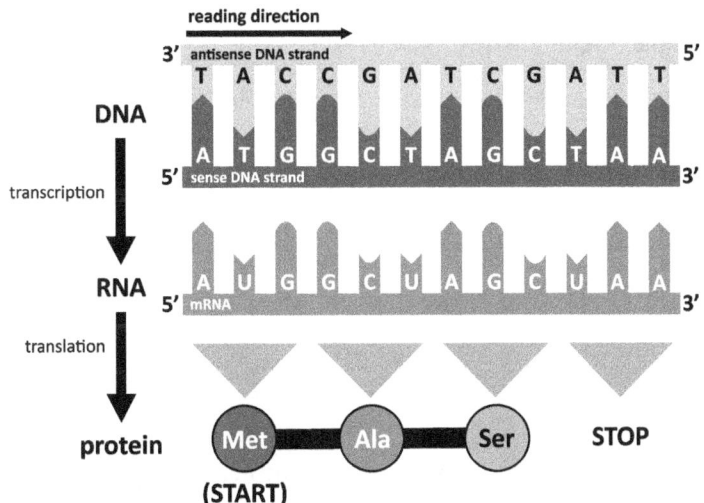

Four steps of translation

The entire process of translation can be summarized in the following 4 steps: 1) tRNA, charging, 2) initiation, 3) elongation, and 4) termination.

tRNA charging

The tRNA and amino acids float freely in the cytoplasm where they can come into contact. The tRNA is folded into a shape that gives it 2 functional domains: the amino acid and the mRNA codon binding domains. The codon-binding domain is called the anticodon and is complementary to the codon. The other domain binds to a specific amino acid specified by the complementing codon. tRNA charging occurs when the tRNA binds to its specific amino acid.

Initiation

When the start codon (AUG) of the mRNA interacts with the rRNA in the ribosome, the large and small ribosomal subunit comes together to sandwich the mRNA so that translation can begin.

Elongation

The tRNA brings its amino acid to the ribosomes in the order specified by the mRNA. Once the anticodon successfully binds the codon, the tRNA releases its amino acid to the ribosome. This step requires energy from ATP.

Termination

A release factor recognizes and binds a stop codon, causing the ribosomal subunits to

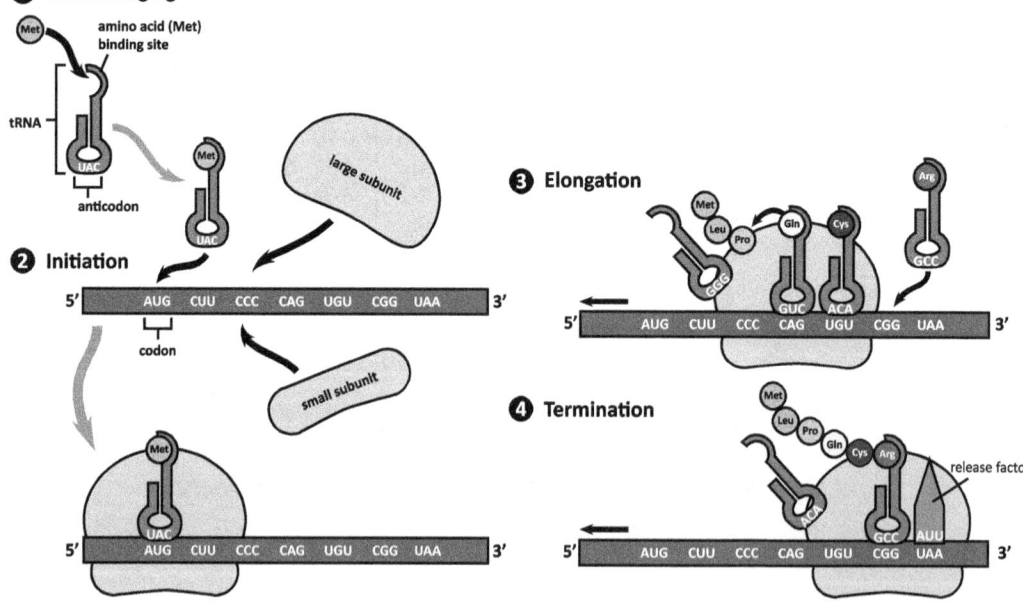

detach from the mRNA.

Codon Degeneracy

Although there are 64 different codons, each complementing the anticodon of a particular tRNA, there are only 20 different amino acids. Excluding the 3 stop codons, there are still 61 codons to 20 amino acids. This leads to codon redundancy (or degeneracy) where multiple codons are used for the same amino acid. For example, Leucine has 6 different codons (UUA, UUG, CUA, CUC, CUG, and CUU). This redundancy in the codon is caused by base pair wobble where Chargaff rules is violated and third nucleotide in the codon can match to more than one nucleotide from the anticodon. Also, some tRNAs have the nucleotide called inosine that complements U, C, and A.

MUTATIONS

Mutations are the ultimate source of all variation in the population. They can occur when the nucleotide sequence in the DNA changes. There are 2 broad classes of mutations: substitution mutations and frameshift mutations.

Substitution mutations

Base substitution mutation occurs when a nucleotide is switched out for another. Thanks to codon redundancy, many substitution mutations have no effect on the phenotype because they do not impact the polypeptide sequence. Mutations of this sort are called silent mutations. If a substitution mutation causes a change in the amino acid, it is called a missense mutation; if is causes an early stop codon, it is called a nonsense mutation.

Frameshift mutations

Frameshift mutation occur when a nucleotide is added (insertion mutation) or removed (deletion mutation), causing the entire reading frame to shift downstream of the mutation point by one nucleotide. Frameshift mutations have the greatest impact on the polypeptide sequence because they cause all amino acids from the mutation point in the sequence to change.

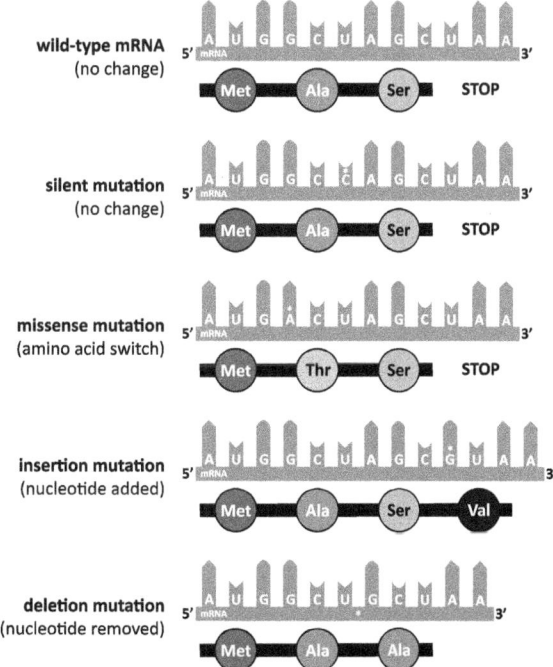

SUMMARY OF KEY CONCEPTS

- Genetic information is transmitted through the DNA to future generation.
- The central dogma states that genetic information flows from DNA to RNA to protein.
- Prokaryotes are less organized than eukaryotes.
- In addition to genomic DNA, the DNA of the mitochondria, chloroplasts, and viruses, and retrovirus RNA also transmits genetic information.
- Franklin, Wilkins, Watson and Crick helped discover the DNA structure.
- Avery-MacLeod-MacCarty & Hershey-Chase contributed to the discovery that DNA is the genetic material.
- Organisms are linked by a common ancestry. Evidence includes the shared processes across biological domains such as a universal genetic cod, and conserved metabolic pathways for DNA replication, transcription, and translation.
- Natural selection depends on genetic variation in the gene pool. Variation depends on the accumulation of mutations.
- DNA is the primary source of heritable information; RNA is a secondary source of genetic information.
- DNA replication is semi-conservative to ensure that the genetic information is continually passed down the generation with minimal opportunity for error.
- DNA and RNA have structural similarities and differences.

CHECK YOUR UNDERSTANDING

1. The organism's genome is stored in its

   (A) chromosomal DNA.
   (B) plasmid DNA.
   (C) both
   (D) neither

2. In the Griffith experiment

   (A) living S bacteria was used to transform dead R bacteria.
   (B) living R bacteria was used to transform dead S bacteria.
   (C) dead S bacteria was used to transform living R bacteria.
   (D) dead S bacteria was used to transform dead R bacteria.

3. DNA can be destroyed by treatment with

   (A) DNase
   (B) protease
   (C) RNase
   (D) none.

4. Which of the following statements best describes the differences between DNA and protein?

   (A) DNA can self-replicate, but protein cannot.
   (B) DNA contains phosphorus, while protein contains sulfur.
   (C) DNA can catalyze reactions, but proteins do not.
   (D) All of the statements are correct.

5. If an organism's genome contains 20% adenine, what % G does it have?

   (A) 20%
   (B) 30%
   (C) 40%
   (D) 80%

6. What is the polypeptide sequence for the following sense strand: ATGCCCAAATAA?

   (A) Met-Pro-Lys
   (B) Tyr-Gly-Phe-Ile
   (C) Met-Gly-Lys-Ile
   (D) None of the above.

ANSWERS AND EXPLANATIONS

1. (A) is correct.

2. (C) is correct.

3. (A) is correct.

4. (B) is correct.

5. (B) is correct because 20% A, 20% T, 30% G, and 30% C.

6. (A) is correct because the mRNA has the same sequence as the sense strand except that U is in place of T. The mRNA sequence is AUGCCCAAATAA.

17 Cell reproduction

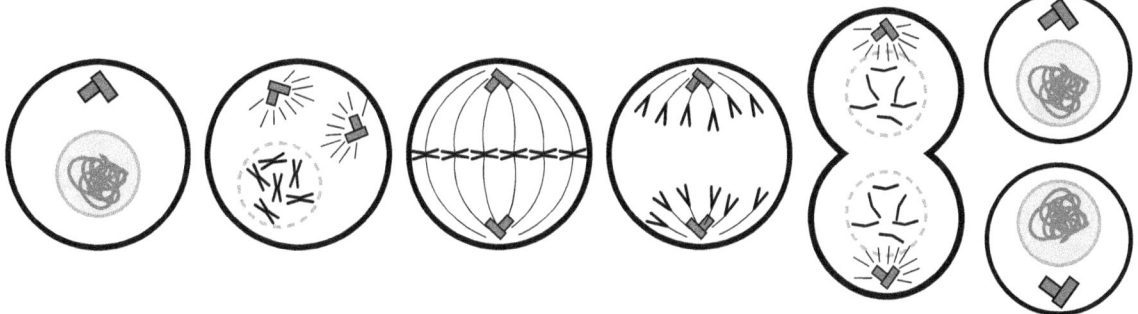

ALL STUDENTS MUST BE ABLE TO ANSWER THESE QUESTIONS

1. What is the difference between sexual and asexual reproduction?
2. What are the 3 different ways the cells can asexually reproduce?
3. What are the stages of interphase and mitosis, and how are they distinguished?
4. How does cytokinesis in plant cells differ from animal cells?
5. What are tumor suppressor genes? proto-oncogenes?
6. How and where in the cell cycle do p53 and MPF regulate cell division?
7. What is the role of CDK and cyclins in MPF mediated regulation of mitosis?
8. What are the 3 cell cycle checkpoint and what happens at each?
9. What are 5 different ways through which cells can sexually reproduce?
10. What is conjugation and how does it occur?
11. What are the 2 phases of the viral life cycle and how are they different?
12. What are the phases of meiosis and how is it similar to and different from mitosis?
13. Why does meiosis require fertilization in order to complete the process of sexual reproduction?

ALL STUDENTS MUST BE ABLE TO COMPLETE THE FOLLOWING TASKS

1. Make predictions about what is occurring during the cell cycle.
2. Describe the events of interphase, meiosis I/II, and mitosis.
3. Use self-produced models and illustrative analogies to explain the cell cycle.

 Big Idea 3A2

Cells use the cell cycle to reproduce either by sexual or asexual mechanisms. In sexual reproduction the daughter cell receives genetic information from 2 different individuals, while asexual reproduction is the cloning of the parent cell to produce 2 genetically identical daughter cells.

In both prokaryotes and eukaryotes, the cell goes through a 3-step process to divide into descendant (or daughter) cells. It begins with the replication of the chromosome so that each daughter cell receives a full set of the parent's genetic information. Once replicated chromosomes genetic information is separated and the cell begins to divide. Cell division concludes with cytokinesis, which is the actual process of splitting of the parent cell into two daughter cells.

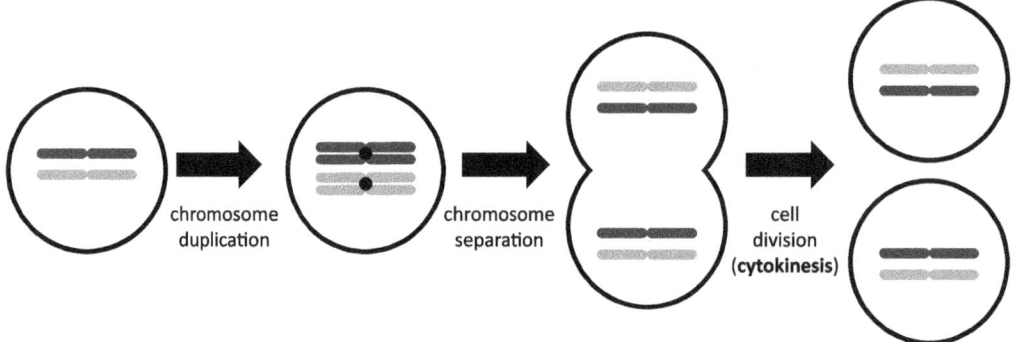

ASEXUAL REPRODUCTION

Asexual reproduction is the cloning of a cell to produce genetically identical daughter cells. In eukaryotes, asexual reproduction occurs in somatic cells by mitosis. Prokaryotes, mitochondria, chloroplasts, and some eukaryotic cells divide by binary fission. Some prokaryotes and eukaryotic cells also divide by budding.

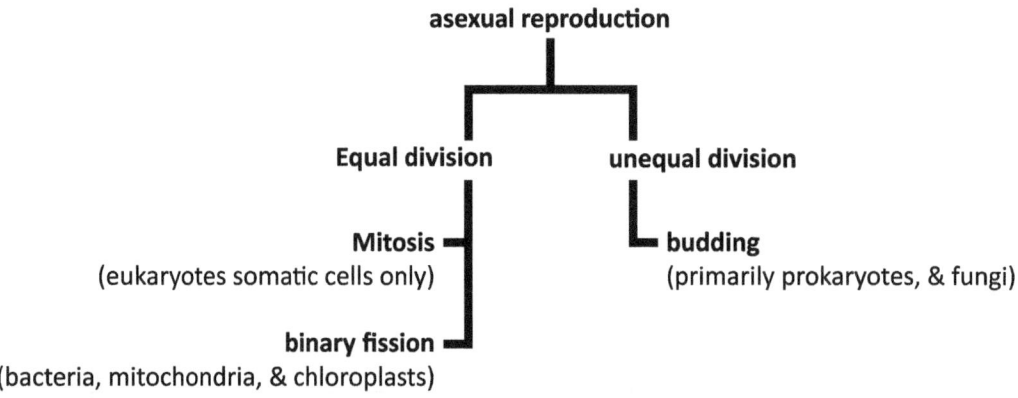

Binary fission and budding

In binary fission, the parent cell grows before splitting into two daughter cells of relatively equal sizes.

Budding is similar to binary fission except that a single daughter cell forms as an outgrowth from the parent cell and is initially much smaller. It remains attached to the parent until it grows to a mature size, before separating.

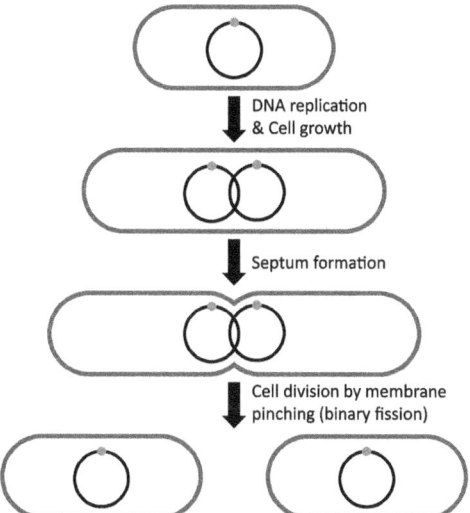

ASEXUAL REPRODUCTION IN EUKARYOTES

In eukaryotes, somatic cells go through the following 3 distinct phases of the cell cycle:

- Interphase – cell growth and DNA replication that duplicates the chromosome to form sister chromatids).
- Mitosis (or M phase) – nuclear division involves the separation of sister chromatids in preparation for cytokinesis.
- Cytokinesis – cell splitting into two daughter cells.

Interphase

The cell spends most of its time in interphase, which is divided into the G_0, G_1, S, and G_2 stages. Non-dividing cells enter and stay in the resting G_0 stage. Tissue damage or environmental factors can induce a cell to leave G_0 and prepare for cell division.

When a signal to divide is received, the cell enters the S stage of interphase, where DNA replication produces sister chromatids and forms the classic "X" shape of metaphase chromosomes. The centrioles are also duplicated in the S stage.

Following the S stage, the cell enters the G_2 growth stage, which is characterized by increases protein and RNA synthesis, and organelle production to prepare the cell to divide its cytoplasm and organelles between 2 daughter cells.

After G_2 the cell goes through mitosis and cytokinesis. Both daughter cells than return to interphase and enters the G_1 growth stage, where protein and RNA synthesis resumes, and the cell grows to its mature and functional size.

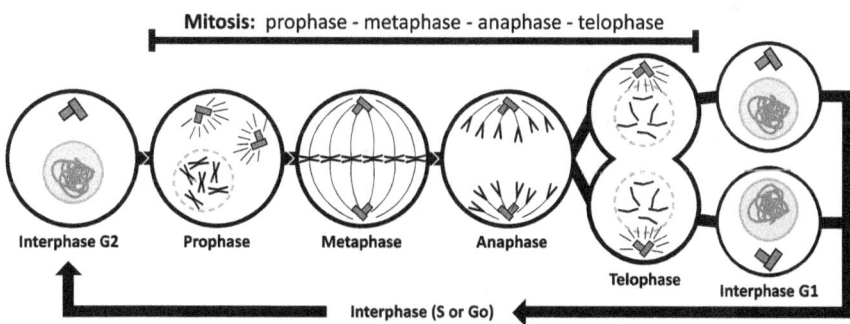

Mitosis and cytokinesis

Mitosis is divided prophase, metaphase, anaphase, and telophase, and is followed by cytokinesis.

- Prophase is the first phase of mitosis and is characterized by the movement of centrosomes (centriole pairs) to opposite poles of the cell, formation of spindle fibers from the polarized centrosomes, dissolution of nucleolus and nuclear envelope, and condensation of chromosomes so that they are visible.
- Metaphase is when the spindle fibers attach to the centromere of chromosomes and align them at mid-cell so that sister chromatids face opposite poles.
- Anaphase is when the centromere splits and sister chromatids are pulled by the attached spindle fibers to opposite poles
- Telophase differs in animal and plant cells. In animal cells, the membrane pinches at mid-cell and the nuclear envelop reforms. In plant cells, vesicles derived from the Golgi move to the center of the cell and fuse to form a cell plate that dissects the cell. The cell plant grows into the cell membrane and fuses with it to produce 2 daughter cells. The cell wall forms between them.
- Cytokinesis (cell division).

Cytokinesis in plant cells

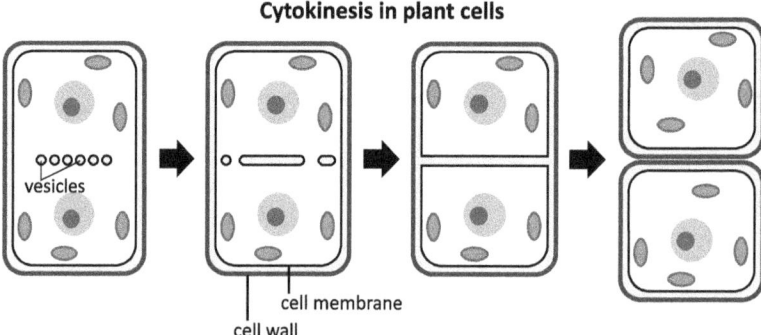

Cell cycle regulation

Each step in the cell cycle must be carefully monitors in order to ensure the health of the organism. A cell the contains DNA damage, or that has not completed DNA replication (S phase) will normally be prevented from progressing along the cell cycle by cell cycle regulators. MPF and p53 are the primary regulators that act at the G_1/S and G_2/M checkpoints. The metaphase checkpoint (or M checkpoint), which ensures the proper alignment of chromosomes along the mid-cell, is regulated by bipolar tension caused by the spindle fibers.

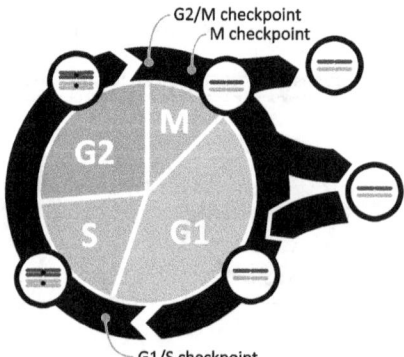

p53 and tumor suppression

p53 is a tumor suppressor gene that negatively regulates a cells ability to progress through the cell cycle by:

- Arresting non-dividing, or terminally differentiated cells in the resting G_0 phase. Tissue damage can stimulate certain cells to leave G_0 and begin dividing to repair the tissue.
- Blocking cell cycle progression when DNA damage is detected, so that repair can be attempted.
- Initiating apoptosis (automated cell death) when irreparable DNA or cell damage has occurred.

G_1/S checkpoint is also known as the restriction point because it is the last point of interphase when environmental and growth factors can influence the cell cycle. A cell that passes this point becomes committed to another round of the cell cycle and can no longer enter G_0. G2/M checkpoint is the last checkpoint before the cell leaves interphase for mitosis. p53 checks for DNA damage at both checkpoints.

Mitosis promoting factor (MPF)

MPF is a proto-oncogen that positively regulates a cell's progression through the cell cycle. It acts at the G_2/M checkpoint to promote a cell that is the right size and that is ready for mitosis to pass the G_2/M checkpoint.

MPF is a kind of cell cycle clock that keeps track of how much time was spent moving through interphase. It is made of two classes of proteins called cyclin and cyclin dependent kinase (CDK). While CDK levels in the cell remains relatively constant, cyclin levels gradually increases beginning in late G_1, and is rapidly degraded during mitosis. At the critical concentration of cyclin, sufficient MPF is produced to push the cell into mitosis.

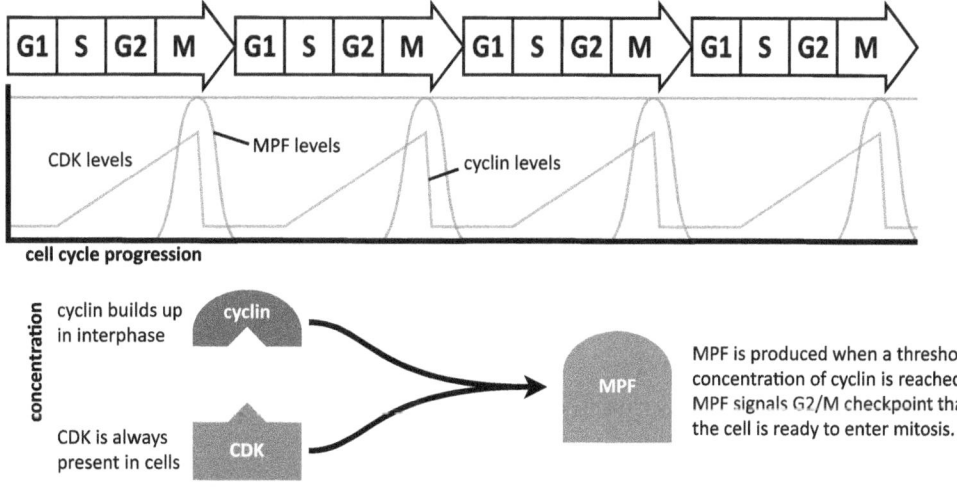

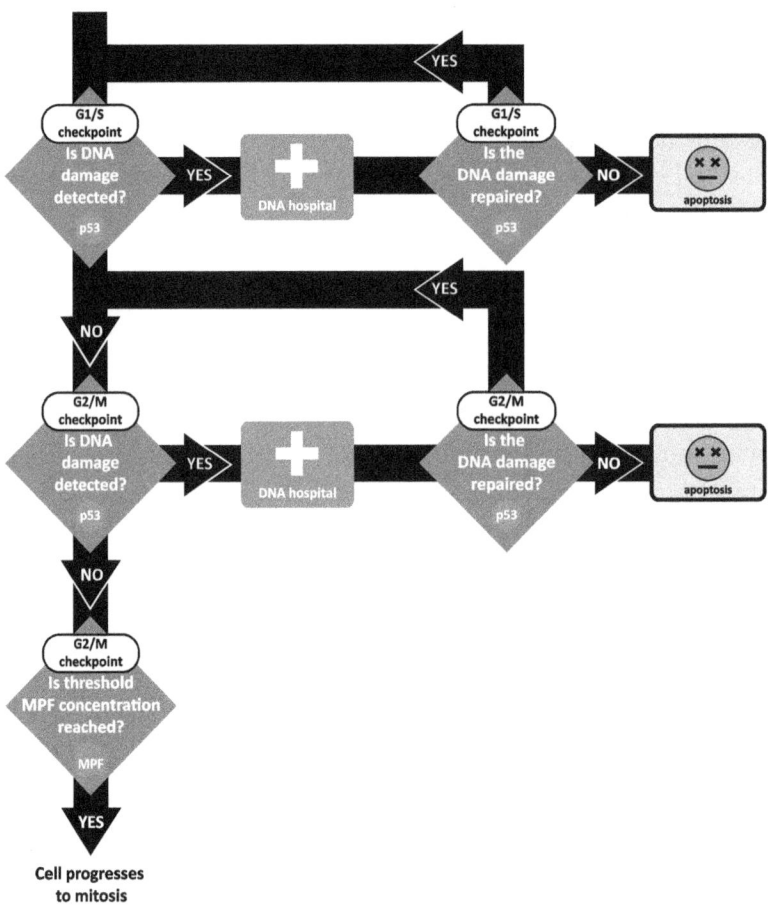

SEXUAL REPRODUCTION

"True" sexual reproduction requires 2 components:

- The formation of a recombinant chromosome from two parent cells.
- Cell division of a cell containing the recombinant chromosome.

Because of the above requirements, the only "true" sexual reproduction mechanism occurs in eukaryotic cells through meiosis and fertilization. However, the term is extended to other mechanisms that introduce foreign genetic DNA into cells. With this distinction made clear, we will henceforth use the term to include these other mechanisms.

Sexual reproduction through direct cell-to-cell contact occurs in eukaryotes by meiosis coupled with fertilization, and in prokaryotes by conjugation. Transduction is a viral mediated introduction of foreign genetic material into a cell. It is the primary mechanism of lateral gene transfer between different species. Transformation is the naturally occurring non-viral mediated mechanisms of introducing foreign genetic material into both eukaryotes and prokaryotes. However, the term is primarily reserved for prokaryotes, as the term is also used to refer to cells that have become cancerous. Transformation can also be induced artificially by altering the bacteria local environment in a manner that causes membrane pores to form. Foreign genetic material can then diffuse through these pores to enter the cell. Artificially induced transformation in eukaryotic cells is referred to as transfection.

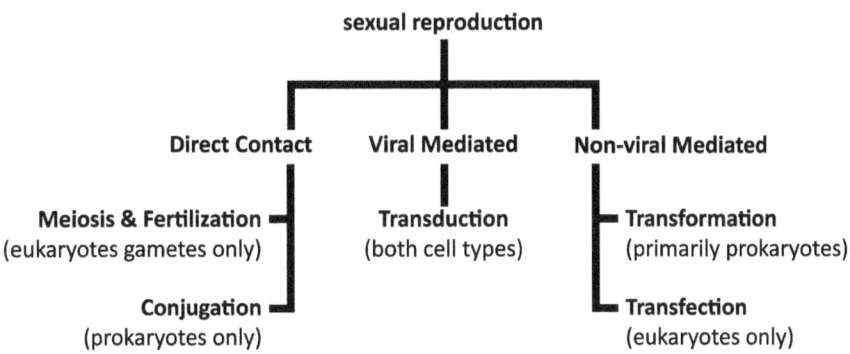

Conjugation in prokaryotes

During bacterial conjugation, the "male" F$^+$ cell that contains the F factor gene forms a cytoplasmic bridge called the conjugation pilus (or sex pilus) to a "female" F$^-$ cell. The F$^+$ cell than donates genetic material from its cytoplasm to the F$^-$ cell. If the F factor is on an F plasmid, rather than integrated in the bacterial chromosome, it is donated to the F$^-$ cells, which than becomes F$^+$.

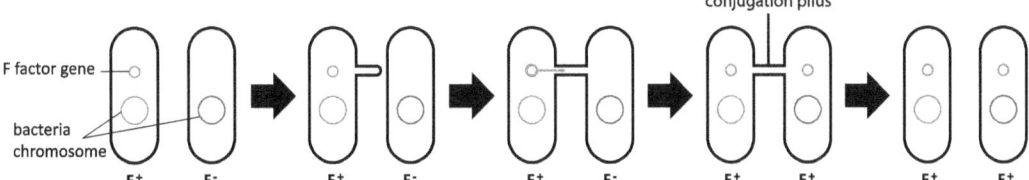

Transformation in prokaryotes

While bacteria conjugation requires specific genes, transformation involves the passive uptake of genetic material from the local environment by diffusion of DNA through membrane pores. In artificial transformation, pores are induced in the bacteria cell by exposure to calcium chloride followed by heat or electric shock.

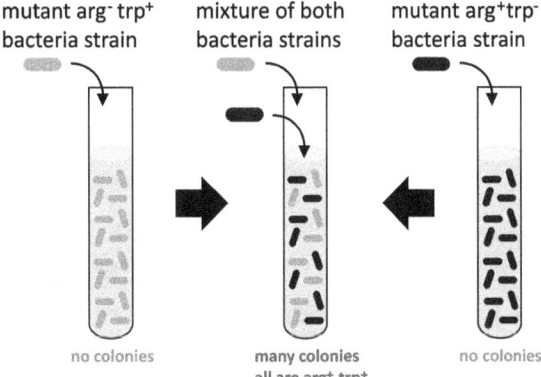

In the example, arg$^-$ trp$^+$ bacteria strain that produce the amino acid tryptophan, but not arginine are growth inhibited. Similarly, the arg$^+$ trp$^-$ mutant produces arginine, but not tryptophan, and thus are also growth inhibited. When both strains are cultured together, some cells acquire their specific missing gene from other cell and are transformed. These transformed parent cells give rise to a descendant population of arg$^+$ trp$^+$ bacteria that can make both amino acids.

Transduction in prokaryotes

In bacteria transduction, a bacteriophage injects its DNA into the bacteria cell. Once infection is completed, the virus life cycle can enter one of two possible reproduction pathways, the lysogenic or lytic phases. During the lysogenic phase, the viral DNA integration with the host cell DNA to produce a recombinant prophage DNA. The cell remains healthy, but whenever it replicates its DNA in preparation for cell division, the viral DNA is also replicated.

Environmental factors can cause the virus to exit the lysogenic phase and enter the lytic phase, where millions of viral particles are produced over a short time period, and the host cell is lysed to release new viral particles to infect healthy cells. Both mechanisms involve the production of recombinant DNA. During the lytic cycle, the virus will incorporate some of the host DNA into released viral particles that is subsequently introduced to other host cells. This is a major mechanism for lateral transfer of genetic material between unrelated species.

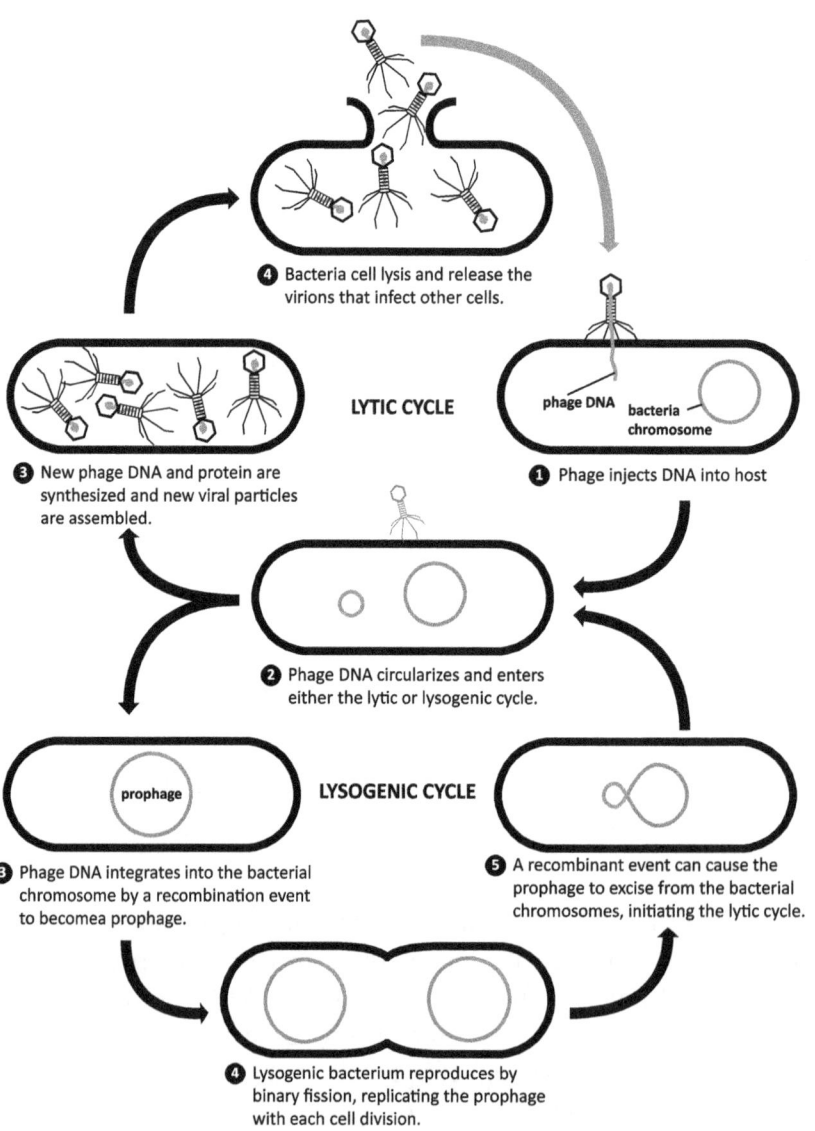

4 Bacteria cell lysis and release the virions that infect other cells.

LYTIC CYCLE

phage DNA   bacteria chromosome

3 New phage DNA and protein are synthesized and new viral particles are assembled.

1 Phage injects DNA into host

2 Phage DNA circularizes and enters either the lytic or lysogenic cycle.

prophage

LYSOGENIC CYCLE

3 Phage DNA integrates into the bacterial chromosome by a recombination event to becomea prophage.

5 A recombinant event can cause the prophage to excise from the bacterial chromosomes, initiating the lytic cycle.

4 Lysogenic bacterium reproduces by binary fission, replicating the prophage with each cell division.

Meiosis & fertilization in Eukaryotes

- Meiosis is reduction division that includes 2 rounds of cell division (meiosis I and meiosis II) to produce 4 haploid (n) gamete cells from a diploid (2n) parent cell.
- Meiosis I reduces the chromosome count from a diploid to a haploid state. This means that at the end of meiosis I, each daughter cell will have 1 copy of each chromosome, rather than a homologous pair of each.
- Genetic recombination via crossing over of chromosomes occurs during Prophase I of meiosis I.
- Meiosis II is exactly the same process as mitosis except that the cells are haploid rather than diploid.

During fertilization, 2 different gamete cells fuse to form a diploid zygote. In animals, the egg is fertilized by a sperm cell from different parents. In plants the ovum is fertilized by pollen. Self-pollinating plants can produce zygotes from a single parent.

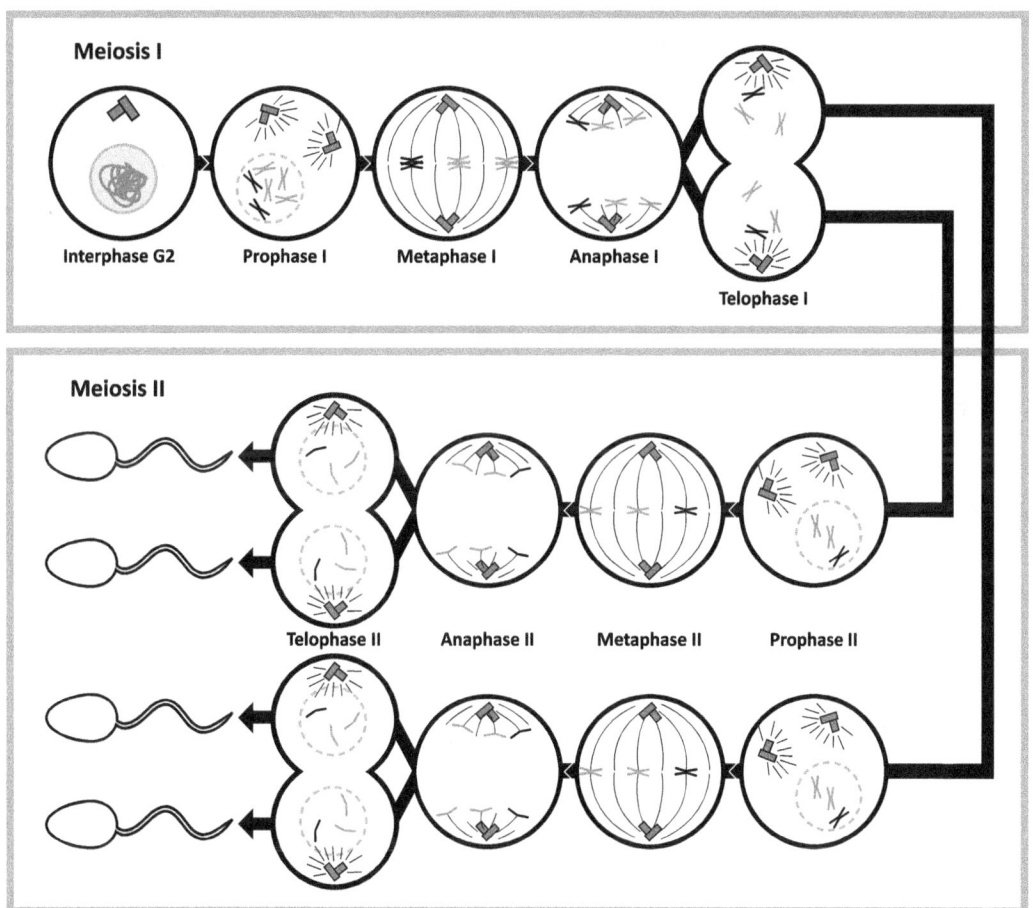

SUMMARY OF KEY CONCEPTS

- The transmission of genetic information involves the cell cycle (binary fission, mitosis, meiosis) and fertilization.
- Information is transmitted through sexual (meiosis and fertilization) or asexual reproductive processes (mitosis or binary fission)

- In eukaryotes, the cell cycle includes interphase, mitosis, meiosis, and cytokinesis.
- Regulation of the cell cycle require internal controls (checkpoints, mitotic machinery assembly) and external controls (MPF, GF, p53, cyclin, CDK)
- Breakdown in the regulation of the cell cycle lead to increased thermodynamic disorder and can cause uncontrolled cell growth and cancer.
- Some terminally differentiated cells no longer divide; they leave the cell cycle and enter G_0 state of mitotic inactivity.
- Some cells like skin cells can leave G_0 in response to tissue damage to undergo repair division
- Mitosis is cell cloning that produces cells that are identical in genetic composition to the parent cell.
- Mitosis is used for growth, repair, and asexual reproduction
- Meiosis is reduction division (2n to n) that produces 4 haploid gametes.
- Homologous chromosomes undergo genetic recombination by crossing over during prophase I to increased genetic variation in the population.
- Fertilization restores the diploid number

CHECK YOUR UNDERSTANDING

1. Cells spend most of their life cycle in

   (A) interphase
   (B) mitosis
   (C) cytokinesis
   (D) none of the above.

2. During cytokinesis in plant cells, the membrane between the two daughter cells form by

   (A) pinching.
   (B) invagination.
   (C) fusion of golgi derived vesicles.
   (D) inward growth of cell membrane from parent cell.

3. The three major cell cycle checkpoints are

   (A) G_1/S
   (B) G_2/M
   (C) M
   (D) All of the above.

4. p53 acts at which of the following checkpoints?

    (A) G_1/S
    (B) G_2/M
    (C) M
    (D) both A and B are correct.

5. MPF acts at which of the following checkpoints?

    (A) G_1/S
    (B) G_2/M
    (C) M
    (D) both A and B

6. Which of the following statements is incorrect?

    (A) Cyclin is more structurally stable than CDK.
    (B) CDK levels remains relatively constant throughout the cell cycle.
    (C) MPF requires both CDK and cyclin in order to form.
    (D) Cyclin is rapidly degraded by the end of mitosis.

7. Which of the following cell pairs can undergo conjugation?

    (A) F^-/F^-
    (B) F^+/F^-
    (C) F^+/F^{-+}
    (D) Both B and C.

8. During the lytic cycle

    (A) recombinant DNA is formed.
    (B) prophage forms.
    (C) both A and B are correct.
    (D) None of the above is correct.

9. Which of the following statements about meiosis is incorrect?

    (A) Crossing over occurs during prophase I.
    (B) meiosis I produces 2 haploid cells.
    (C) Meiosis II is most similar to mitosis.
    (D) During anaphase II, homologous chromosomes are separated.

ANSWERS AND EXPLANATIONS

1. (A) is correct.

2. (C) is correct.

3. (D) is correct.

4. (D) is correct.

5. (B) is correct

6. (A) is an incorrect statement because cyclin is rapidly degraded during mitosis, while CDK levels remain relatively constant.

7. (D) is correct because conjugation simply require the F factor, which is present in all F^+ cells.

8. (D) is correct.

9. (D) is an incorrect statement because anaphase II is similar to anaphase of mitosis where the sister chromatids (not homologous pairs) are pulled apart.

18 Patterns of inheritance

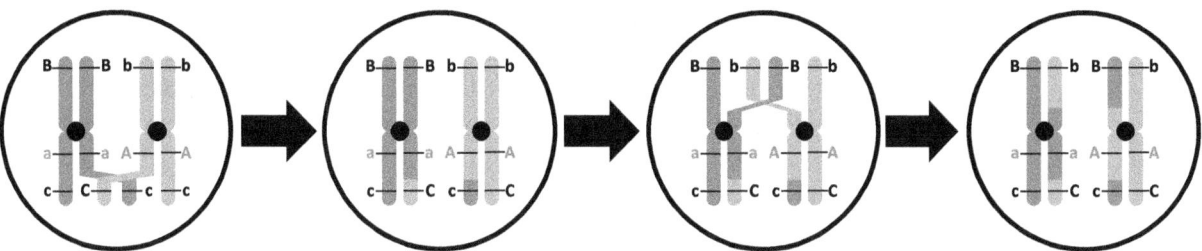

ALL STUDENTS MUST BE ABLE TO ANSWER THESE QUESTIONS

1. What is the primary difference between the 1st and 2nd law of inheritance?
2. How does a monohybrid cross compare to a dihybrid cross?
3. What is the difference between the P, F1, and F2 generations? P and F1 crosses?
4. What is a test cross and why is it performed?
5. What are the 7 conditions that can lead to a non-Mendelian pattern of inheritance? Why does each cause a deviation from Mendelian genetics?
6. What is the difference between incomplete dominance and codominance?
7. What are 3 types of non-disjunction and what are the effects on the cell?
8. What are mosiacism syndromes and why do they occur?
9. What are linked genes?
10. What is the relationship between linkage distance and recombination frequency?
11. How are parental and recombinant chromosomes distinguished?
12. What are sex-linked genes? Why are most sex-linked genes located on the X chromosome?
13. How does X chromosome inactivation influence the phenotype of heterozygous females?
14. Why are males more affected by sex-linked recessive disorders?

ALL STUDENTS MUST BE ABLE TO COMPLETE THE FOLLOWING TASKS

1. Perform P and F1 crosses (monohybrid and dihybrid), and determine the genotype and phenotype frequencies of offspring.
2. Use a test cross to determine the genotype of a dominant individual.
3. Use probability rules to determine the probability of an offspring from a cross having a particular genotype and/or phenotype.
4. Use provided information to calculate the map units between two genes.
5. Calculate the map units between three linked genes and determine their order on the chromosome.

 Big Idea 3A3-4

Mendelian genetics is used to make predictions about the traits of offspring from a cross between two individuals. However, there are different scenarios that produce a non-Mendelian pattern.

MENDELIAN GENETICS

Mendelian pattern of inheritance is based on the work of Gregor Mendel who studied pea plants to discover predictable patterns of inheritance and gene expression in progeny. Mendel's two laws of inheritance form the foundation of Mendelian genetics.

Mendel's 1st law of inheritance

The **law of segregation** states that during gamete formation, a paired set of alleles of a given gene is separated into two different gametes. In the illustration, the allelic pair A1 and A2 of gene A are separated during meiosis to form 2 possible sperm cells that each have only one copy of the gene. Notice that this law is concerned with the separation of the alleles of a single gene.

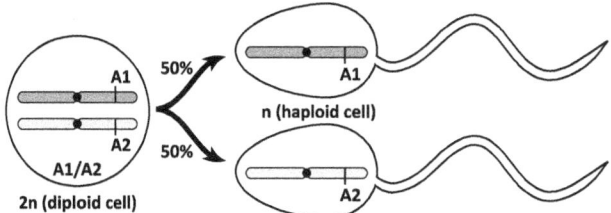

Mendel's 2nd law of inheritance

The **law of independent assortment** states that during gamete formation, alleles of different genes segregate independent of each other. In the illustration gene A and gene B sort independently from each other during meiosis to form 4 possible sperm cells (A1/B1, A1/B2, A2/B1, and A2/B2), each with 1 copy of gene A and 1 copy of gene B.

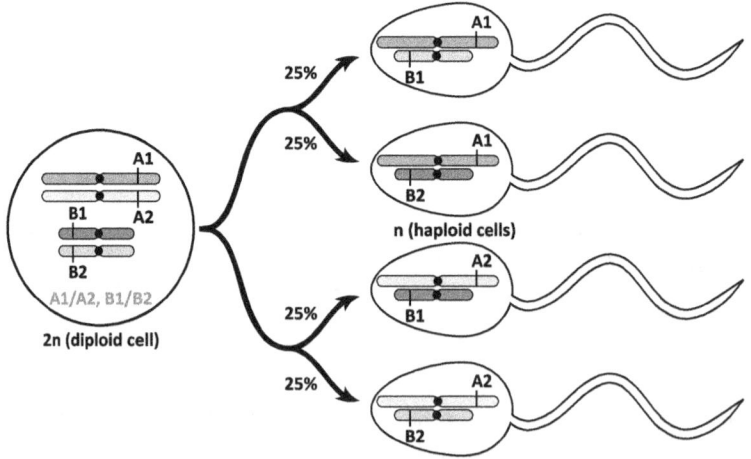

Exceptions to this law

While genes that are located on different chromosomes always sort independently, **linked genes** that are located on the same chromosome will only sort independently if crossing over occurs between them. The closer 2 genes are to each other on a chromosome, the less likely that they will experience independent assortment and the greater probability of a deviation from the Mendelian predictions.

Mendel's observation of pea plants

Mendel made his discovery by investigating 7 different characters in pea plants over several generation and keeping a record of the distribution of the characteristics in descendant generations.

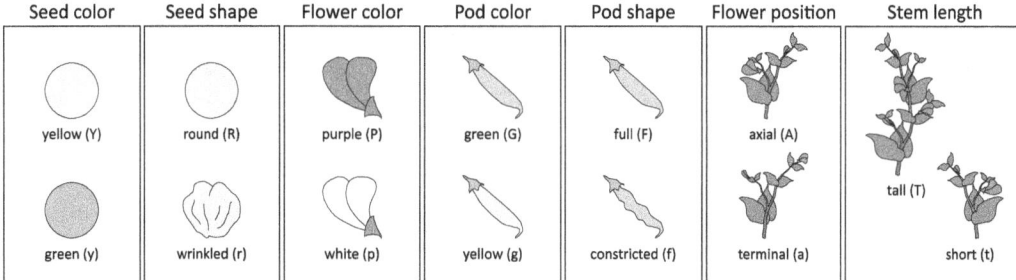

Mendel's P and F1 crosses

Mendel began his crosses with 2 individuals in the parent generation (or **P generation**) that were **true breeders** (**homozygous**) for different traits of the same character. For example, for the flower color character, he crossed homozygous **dominant** purple (PP) with homozygous **recessive** white (pp) plants. The seeds from the P cross were used to produce the hybrid plants of the first filial generation (or **F1 generation**), all of which expressed the dominant purple phenotype. The F1 individuals were then crossed to produce the **F2 generation**. The resulting phenotype ratio of purple to white was 3:1.

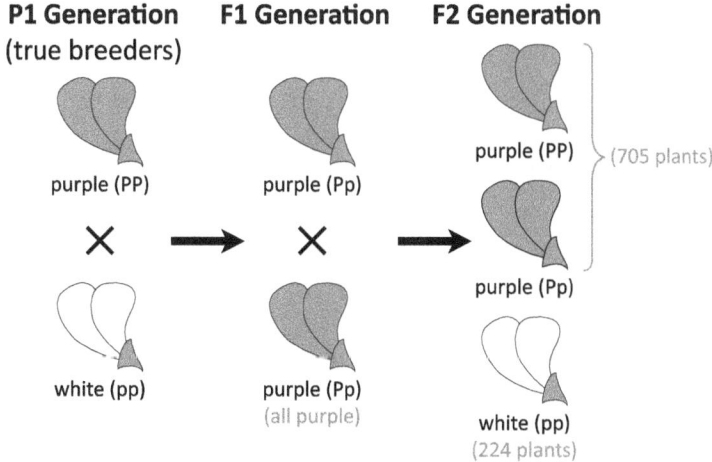

The results for crosses for the other 6 characters were consistent with those observed for flower color. All F1 individuals expressed the dominant phenotype, and F2 generation showed a roughly 3:1 phenotype ratio (dominant:recessive). The following table provides a full breakdown of Mendel's results.

Character	P1 cross	F1	F2 (quantity)	F2 ratio
Seed shape	round x wrinkle	all round	5475 round 1850 wrinkled	3.0 : 1
Seed color	yellow x green	all yellow	6022 yellow 2001 green	3.0 : 1
Flower color	purple x white	all purple	705 purple 224 white	3.2 : 1
Pod shape	full x constricted	all full	882 full 299 constricted	3.0 : 1
Pod color	green x yellow	all green	428 green 152 yellow	2.8 : 1
Flower posi-tion	axial x terminal	all axial	651 axial 207 terminal	3.1 : 1
Stem height	tall x short	all tall	787 tall 277 short	2.8 : 1

Monohybrid cross

Mendel's results fit the predictions of a **monohybrid** cross for true breeding parents (P generation). When the parents are true breeders for different phenotypes of the same character, all F1 offspring turn out as hybrids. An F1 cross than produces F2 offspring with a dominant to recessive phenotype of 3:1 - similar to Mendel's findings.

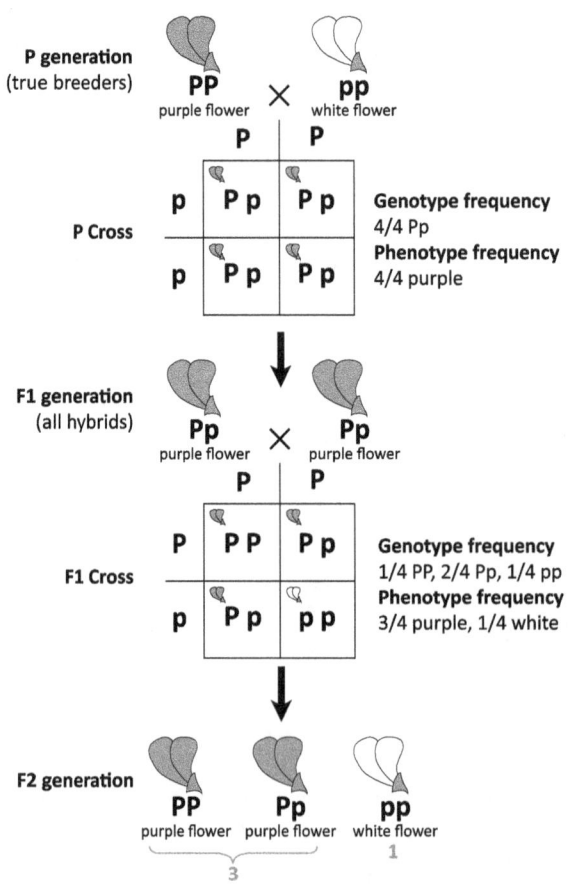

Dihybrid cross

While a monohybrid cross involves only 1 gene and uses a 2 × 2 punnet square matrix, a dihybrid cross involves 2 genes and uses a 4 × 4 matrix. In a monohybrid cross the phenotype ratio of the F2 generation is 3:1 (dominant: recessive), while in a dihybrid cross it is 9:3:3:1 (dominant/dominant: dominant/recessive: recessive/dominant: recessive/recessive). The following example shows a dihybrid cross for the seed shape and seed color characters.

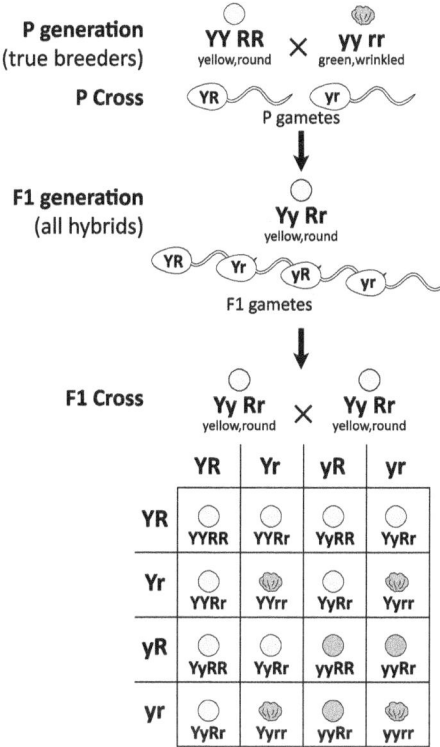

F2 generation phenotype frequency
9 yellow, round (dominant, dominant)
3 yellow, wrinkled (dominant, recessive)
3 green, round (recessive, dominant)
1 green, wrinkled (recessive, recessive)
(9:3:3:1)

Test cross

The genotypes of all individuals with the recessive phenotype is always homozygous recessive. For example, a wrinkled pea has a known genotype of rr. However, when an individual expresses the dominant phenotype, they can have either of two possible genotypes, such as round peas that is RR or Rr. For the P generation, only true breeding homozygous individuals must be used.

To determine whether the dominant individual is a true breeder, a test crosses is performed, where the dominant individual of unknown genotype (RR or Rr) is crossed with a recessive individual (rr). If any recessive offspring appear from this cross, the dominant individual is heterozygous, otherwise it is assumed to be homozygous. Note that test crosses are not performed for recessive individuals because the genotype of all homozygous recessive is known to be homozygous recessive.

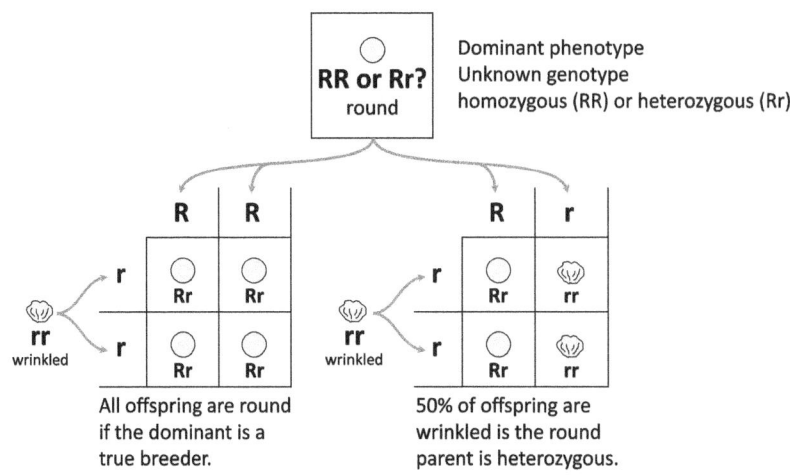

Probability rules

When more than 1 gene is involved in a cross, probability rules can be used predict the phenotype and genotype distribution of the offspring. The basic rules are as follows:

1. Perform a separate monohybrid cross for each gene.
2. For each monohybird cross, determine the genotype frequencies.
3. Multiply the predicted genotype frequencies for the genotype combination.

Example: Two pea plants with the genotypes Yy Rr Pp and Yy rr pp are crossed.

(A) What is the probability of producing offspring with the genotypes yy rr Pp, yy Rr pp, Yy rr pp, YY rr pp, or yy, rr, pp?

(B) What frequency of these offspring is expected to express at least 2 recessive or 2 dominant traits?

Solution: Multiple character cross is equivalent to multiple independent monohybrid crosses. To calculate the combined genotype probability of offspring, each character is considered separately, and then the individual probabilities are multiplied together.

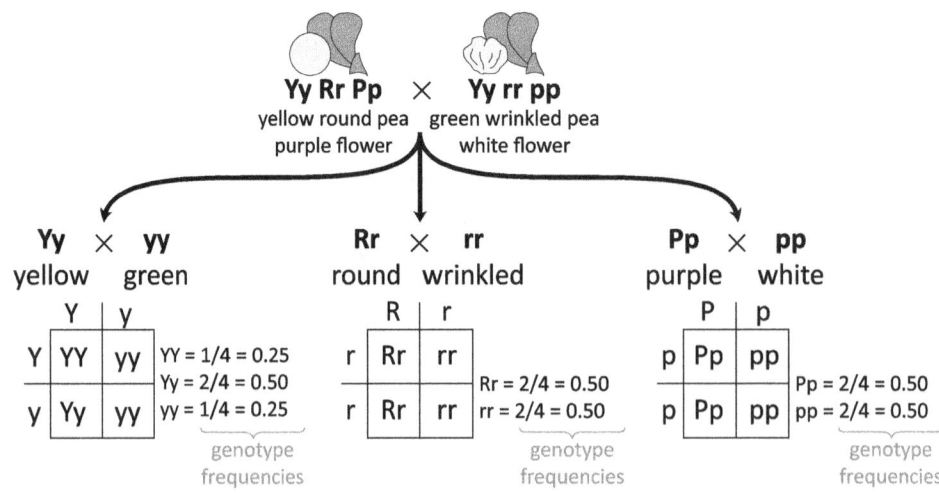

Offspring genotype	Genotype probability			Probability of offspring receiving the combined genotype
	Pea color	Pea shape	Flower color	
yy rr Pp	yy = 0.25	rr = 0.50	Pp = 0.50	= 0.25 x 0.50 x 0.50 = 0.063
yy Rr pp	yy = 0.25	Rr = 0.50	pp = 0.50	= 0.25 x 0.50 x 0.50 = 0.063
Yy rr pp	Yy = 0.50	rr = 0.50	pp = 0.50	= 0.50 x 0.50 x 0.50 = 0.125
YY rr pp	YY = 0.25	rr = 0.50	pp = 0.50	= 0.25 x 0.50 x 0.50 = 0.063
yy rr pp	yy = 0.25	rr = 0.50	pp = 0.50	= 0.25 x 0.50 x 0.50 = 0.063
Chance of offspring receiving at least 2 recessive traits				= 4(0.063) + 0.125 = 0.375
Chance of offspring receiving at least 2 recessive traits				= 3(0.063) + 0.125 = 0.313

NON-MENDALIAN GENETICS

1. The observed phenotype distribution of offspring will deviate from Mendelian genetics predictions under 7 conditions:
2. Variation in the degree of **dominance** can cause **codominant** or **incomplete dominant** expression in heterozygous individuals, rather than the classic dominant expression described by Mendelian genetics.
3. **Polygenic traits** have phenotypes that are influenced by many genes so that the phenotype distribution displays a bell curve distribution rather than distinct dominant or recessive traits.
4. **Nondisjunction of homologous chromosomes** during gamete formation can cause some gamete cells have additional or fewer chromosomes.
5. Males are **hemizygous** for **sex-linked genes** and need only one copy of the recessive allele to display the recessive phenotype.
6. Because **linked genes** are present on the same chromosome, they tend to violate Mendel's 2nd law of inheritance to sort together during gamete formation.
7. **Mitochondrial and chloroplast DNA** are maternally inherited from the egg and ovum so that the phenotype is influenced only be one parent - the mother.
8. **Maternal effect** is where the environment and genotype of the mother influences the phenotype of the developing embryo in a manner that is independent of the embryo's own genotype. The maternal effect occurs through the mother's secretion of cytoplasmic determinants, mRNA, and proteins to the embryo.

Incomplete dominance

If the hybrid (heterozygous) expresses both alleles in all cells and has a blended phenotype, the two alleles are said to be in an **incomplete dominant relationship**. For example, a heterozygous flower that has red (C^R) and white (C^W) alleles will appear pink.

C^WC^W (white)   × C^RC^R (red)

C^RC^W (pink)

POLYGENIC TRAITS
Polygenic traits can cause a deviation from Mendelian genetics by producing a broad spectrum of phenotypes that are not easily predictable. A good example of this is the variation in individual height. Because so many genes are involved, predicting an offspring's height is not always possible. Mendelian genetics addresses single gene traits with distinct predictable phenotype distribution that can be graphically represented by bar graphs. However, polygenic traits display many more phenotype categories with a continuous distribution. It is best represented by a bell curve.

MAMMALIAN SEX DETERMINATION & SRY
The Y-chromosomes determines sex in mammals. It contains a gene called the testis-determining factor that is coded in the sex-determining region (SRY). The expression on this gene induces the developing embryo to form male sex organs. Mutation in SRY can cause certain sex disorder, or can lead to and XY individual developing into a female instead of a male.

Codominance in blood type

Blood type is determined by two types of proteins displayed on the surface of blood cells. The gene for one of the proteins has 3 different alleles - I^A, I^B, and i^o. I^A and I^B are codominant and code the A and B antigens, respectively. The i^o allele is recessive and does not code for a protein. The other protein is coded by Rh factor gene which has 2 alleles. The dominant allele (Rh^+) produces the Rh antigen and the recessive allele (Rh^-) does not code for a protein.

Genotype	RBC	Blood type	Genotype	RBC	Blood type
$i\ i$ Rh -/-		O -	$i\ i$ Rh+/+ or +/-		O +
$I^A I^A$ or $I^A i$ Rh -/-		A -	$I^A I^A$ or $I^A i$ Rh+/+ or +/-		A +
$I^B I^B$ or $I^B i$ Rh -/-		B -	$I^B I^B$ or $I^B i$ Rh+/+ or +/-		B +
$I^A I^B$ Rh -/-		AB -	$I^A I^B$ Rh+/+ or +/-		AB +

Legend:
- I^A
- I^B
- i (NONE)
- Rh^+
- Rh^-
- ABO SYSTEM / RH SYSTEM

Codominance in chicken feathers

Two alleles of a gene display codominance with each other when some cells of the organism express one allele while others express the second allele. For example, the Black (B) allele that is responsible for the pigmentation in chicken feather is codominant with the white (W) allele so that heterozygous chickens have mostly black feathers that are speckled white.

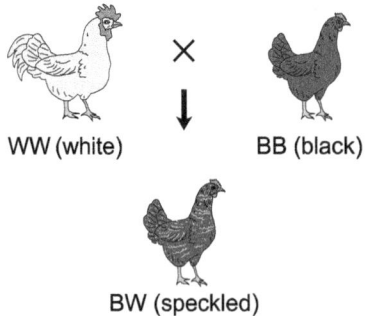

WW (white)   ×   BB (black)

BW (speckled)

Nondisjunction

Nondisjunction is the failure of homologous chromosomes or sister chromatids to properly separate during cell division. There are 3 types of nondisjuction:

- Failure of homologous chromosomes to separate during anaphase I (meiosis I) produces 2 gametes that have an extra chromosome (n+1) and 2 gametes that are missing a chromosome (n-1). If the N+1 gamete is fertilized, a trisomy embryo with an extra chromosome (2n+1) will form; and the fertilization of an n-1 gamete produces a monosomy embryo with an unpaired chromosome.
- Failure of sister chromatids to separate during anaphase II (meiosis II) produces two normal, one n+1, and one n-1 gametes.

■ Failure of sister chromatids to separate during anaphase of mitosis. Nondisjunction during mitosis effects only descendant cells.

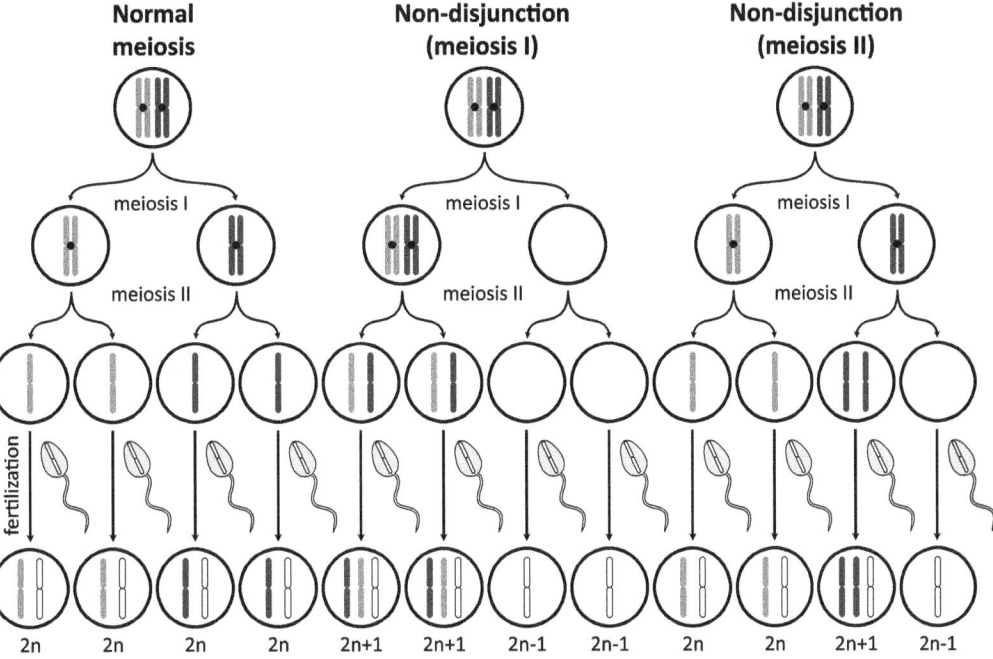

Normal somatic cells are diploid (2n=46) with homologous pairs of **autosomes** (chromosomes 1 thru 22) and 2 **allosomes** (XX females or XY males). A **somatic cell** can undergo normal meiosis to produce four **haploid** (n = 23) **gamete cells**. **Diploid** cells that are effected by non-disjunction are said to be **aneuploids**. Aneuploidy is responsible for a number of syndromes such as those listed in the table.

Syndrome	Nondisjunction	Karyotype notation
Normal	none	46, XY or 46, XX
Turner syndrome (monosomy X)	- X (meiosis)	45, XO
Down syndrome (trisomy 21)	+ 21 (meiosis)	47, XY,+21 or 47, XX,+21
Edwards syndrome (trisomy 18)	+18 (meiosis)	47, XY,+18 or 47, XX,+18
Patau syndrome (trisomy 13)	+13 (meiosis)	47, XY,+13 or 47, XX,+13
Klinefelter syndrome (multiple X)	+X or more (meiosis)	47, XXY, 48, XXXY, or more in males
XYY syndrome	+Y (meiosis)	47,XYY
Trisomy X	+X (meiosis)	47, XXX
Mosiacism syndromes	Varies (mitosis)	Varies

The occurrence of aneuploidy is rare due to spontaneous abortion from miscarriages. Down syndrome, which is caused by an extra chromosome 21, is the most common type of aneuploidy in humans.

When nondisjunction occurs during mitosis, only a subset of cells inherit the aneuploidy. Depending on how early during embryonic development the mitotic nondisjunction occurred, a mosiacism syndrome may develop where the individual displays an intermediate phenotype.

Aneuploidy is diagnosed by performing a chromosome spread and examining the chromosomes under a light microscope to determine how many of each chromosome are present. In order to distinguish the chromosomes from each other, they are stained with a dye that specifically binds to A/T rich regions, giving each chromosome its distinct dark and light bands.

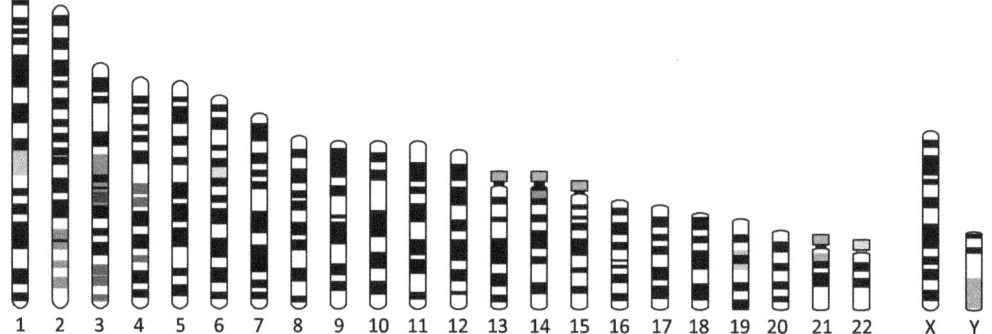

LINKED GENES

The 2nd law of inheritance states that during gamete formation, alleles of different genes sort independently of each other, meaning that 2 different genes are distributed to gamete as if they are on separate chromosomes. However, if 2 genes are located on the same chromosome, the genes are linked and will only sort independently when a crossover event occurs between them. Mendel was not aware of gene linkage because the traits that he examined were located on different chromosomes and always sorted independently. Thus, Mendel's 1st and 2nd laws do not account for gene linkage

Independent assortment of linked genes produces 4 possible gametes:
- 2 with parental chromosomes (A/B and a/b)
- 2 with recombinant chromosomes (A/b and a/B)

The closer that two genes are located to each other on a chromosome, the higher is the probability that they will violate the 2nd law and sort together during gamete formation. As a result, progeny in the next generation will display a phenotype distribution that deviates from those predicted by Mendelian genetics. If the two genes are directly adjacent to each other on the chromosome, crossover between them is improbable resulting in a very low probability of recombinant chromosomes forming and greater deviation from Mendelian predictions. However, if the two genes are located on opposite ends of the chromosomes (far apart), crossover between them is highly probable and they should almost always sort independently during gamete formation. For such distantly linked genes, phenotype distribution in the progeny generation should approach the Mendelian predictions.

Gene mapping

The relative position of linked genes is determined by calculating the recombinant percent between genes located on the same chromosome. The recombinant percent is calculated as the number of recombinant gametes divided by the total number gametes and multiplied by 100. Distantly linked genes will have a percent that approaches the Mendelian prediction of 50% each for the recombinant (A/b and a/B) and parental (A/B and a/b) chromosomes.

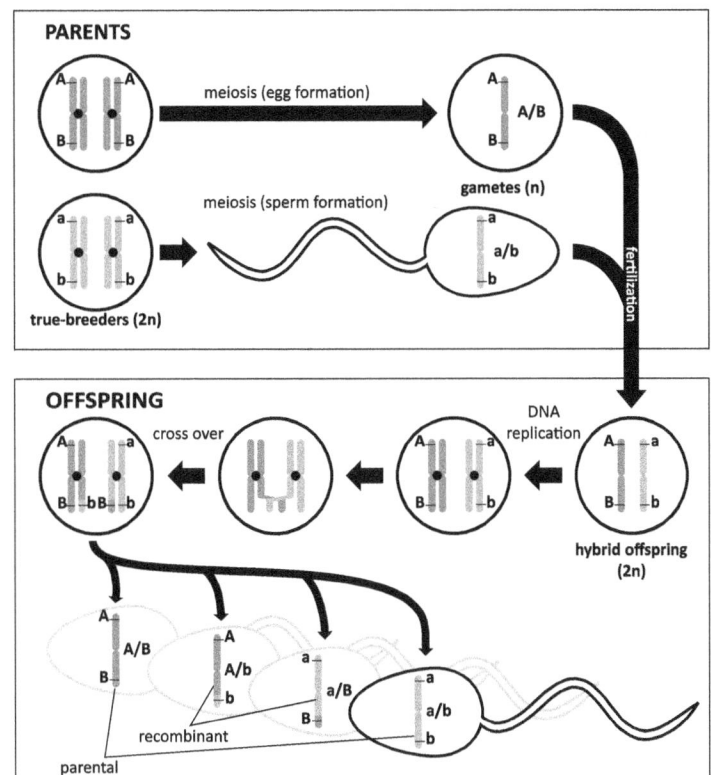

Gene mapping example 1: A cross of a hybrid individual (Aa/Bb) with a recessive true-breeder (aa/bb) produces the following offspring (letters indicate the observed phenotype): 150 AB, 50 Ab, 50 aB, and 150 ab. Calculate the map distance between genes A and B given that each has 2 alleles that display a complete dominance relationship.

Solution: According to Mendelian genetics Aa/Bb x aa/bb should produce 100 AB, 100 Ab, 100 aB, and 100 ab and a recombinant frequency of 0.5 or 50%.

1. **Identify the parental offspring.**
2. Parental offspring = AB + ab = 150 + 150 = 300
3. **Identify the recombinant offspring.**
4. Recombinant offspring = Ab + aB = 50 + 50 = 100
5. **Calculate the recombinant percent.**
6. Map distance = recombinant percent = 100 ÷ 400 × 100 = 25% = 25 map units.

Gene mapping example 2: A cross of a hybrid individual (Aa/Bb/Cc) with a recessive true-breeder (aa/bb/cc) produces the following offspring (letters indicate the observed phenotype): 190 ABC, 10 ABc, 1400 AbC, 100 Abc, 100 aBC, 1400 aBc, 10 abC, and 190 abc. Determine the order of genes A, B, and C and their relative map distances on the chromosome.

Solution: According to Mendelian genetics all phenotypes should have equal distribution in the progeny.

1. **Identify the parental offspring.**
   The parental offspring are the two most numerous classes that are reciprocal

phenotypes of each other. The parental offspring are the AbC and aBc.

Parental offspring = AbC + aBc = 1400 + 1400 = 2800

2. **Identify the double crossover class**
   The double crossover are those that result from a cross over between both pairs of genes. The double crossover offspring will be the two smallest classes.

   Double crossover = Abc + abC = 10 + 10 = 20

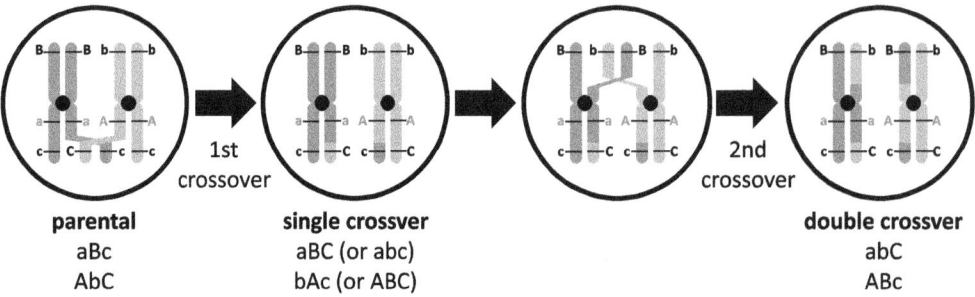

parental
aBc
AbC

single crossver
aBC (or abc)
bAc (or ABC)

double crossver
abC
ABc

3. **Determine the middle gene**
   Compare the parental and the double crossover. Only the middle genes recombines in the double crossover because the first and third genes are regrouped by the second crossover. Make a best match between each double crossover and a parental. The genes that is unmatched is the middle gene.

Matching the parental and double crossover

parental	a	B	c		parental	A	b	C
	×	|	|			×	|	|
double cross over	A	B	c		double cross over	a	b	C

4. **Identify the single crossover classes**
   These are all of the remaining classes bac, BAC, bAc, and BaC

5. **Calculate the recombinant percents between genes A and B**
   If you focus only on genes A and B, the parental classes are aB and Ab, therefore the AB and ab single crossover classes are recombinants. All double crossover classes are also recombinants. The 4 classes that had recombination between genes A and B are 190 BAC, 190 bac and both double crossover classes, 10 BAc and 10 Bac.

   AB recombinants = 190 + 190 + 10 + 10 = 400

   Total number of offspring = 3400

   AB map distance = 400 ÷ 3400 × 100 = 11.8 % = 11.8 map units

6. **Calculate the recombinant percents between genes A and C**
   If you focus only on genes A and C, the parental classes are AC and ac, therefore the Ac and aC single crossover classes are recombinants. All double crossover classes are also recombinants. The 4 classes that had recombination between genes A and C are 100 bAc, 100 BaC and both double crossover classes, 10 BAc and 10 Bac.

AC recombinants = 100 + 100 + 10 + 10 = 220

Total number of offspring = 3400

AC map distance = 220 ÷ 3400 × 100 = 6.5 % = 6.5 map units

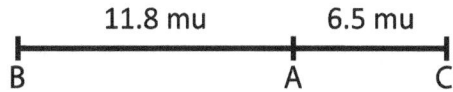

Sex-linked genes

Sex-linked genes are those that are located on either of the sex chromosomes. Since the Y chromosomes contains very few genes, most sex-linked genes are located on the X chromosome.

The pattern of expression of sex-linked recessive traits in males differ from females in that, females have two copies of all sex-linked genes while males have only one copy. Consequently, a males that has a single copy of a recessive allele will display the recessive phenotype. This explains why sex-linked recessive disorders such as color blindness is more common in males than females.

In females, one X chromosome is inactivated and forms a chromosome clump in the cell called a **barr body**. Since inactivation occurs in each cell separately and randomly, a woman that is heterozygous for color blindness will have some cells expressing the recessive allele if the normal dominant allele is on the inactive chromosome. This can potentially result in a mixed phenotype.

A good example of how **X-chromosome inactivation** can lead to mixed phenotype in females is in the coat color of cats. One of several genes responsible for coat color in cats is an X-linked with a recessive black (b) and dominant orange (B) allele. Male cats are either black with genotype $X^b Y$, or orange with genotype $X^B Y$. Females have 3 possible phenotypes, orange ($X^B X^B$), black ($X^b X^b$), or a mixed ($X^B X^b$).

In both males and females, the phenotype distribution of sex-linked traits deviate from Mendelian predictions.

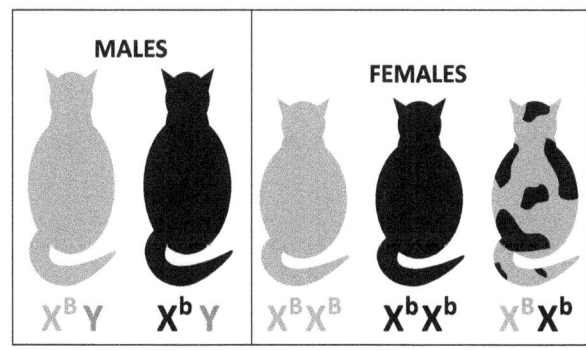

PEDIGREE CHARTS

A pedigree chart is a family tree diagram that shows how genetic traits have passed through the family tree from the ancestral generation to the present. Genetic traits are classified based on whether they are transmitted by gene that are autosomal or sex-linked, and dominant, or recessive. Autosomal genes are located on the autosomal chromosomes (chromosome 1 – 22 in humans), while sex-linked gene are located on the X or Y chromosomes (most sex-linked genes are on the X chromosome).

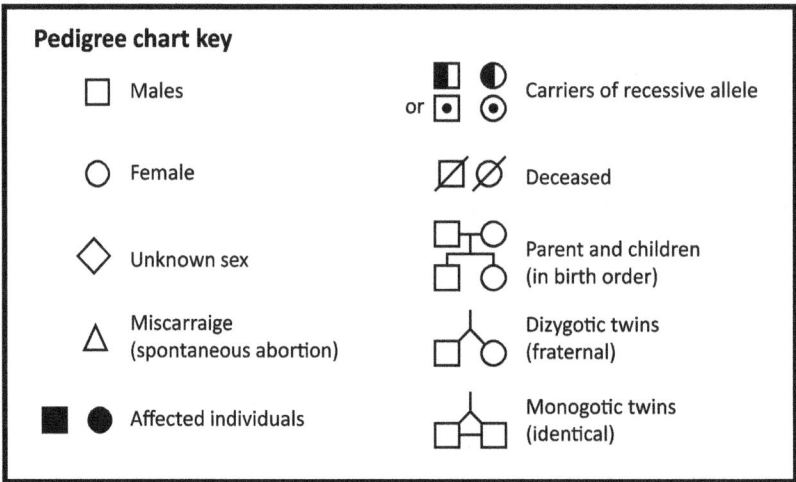

Pedigree chart key

□ Males	■ ● or ⊡ ⊙ Carriers of recessive allele
○ Female	⊘ ⊘ Deceased
◇ Unknown sex	Parent and children (in birth order)
△ Miscarraige (spontaneous abortion)	Dizygotic twins (fraternal)
■ ● Affected individuals	Monogotic twins (identical)

Autosomal dominant traits

Traits that are autosomal dominant are those coded by genes that are located on an autosomal chromosome. The specific allele for the trait is dominant over the normal allele. The following are key features that distinguish autosomal dominant expression from other traits:

- Each individual has two copies of the related gene
- Affected individuals have at least one copy of the allele for the trait
- Heterozygous individuals have one copy of the allele and are affected
- Homozygous dominant individuals have two copies of the allele and are affected
- Homozygous recessive individuals have two normal alleles and are not affected
- The trait does not skip generations because the heterozygous individuals show the affected phenotype.
- Males are females are equally affected

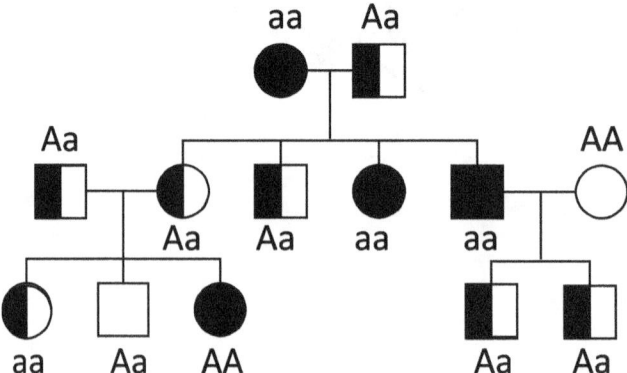

NOTES

Autosomal recessive traits

Traits that are autosomal recessive are also coded by genes located on an autosomal chromosome. However, the specific allele for the trait is recessive over the normal allele. The following are key features that distinguish autosomal recessive expression from other traits:

- The related gene is located on an autosomal chromosome
- Each individual has two copies of the related gene
- Affected individuals must have two copies of the allele
- Heterozygous individuals are carrier with one copy of the allele; they are not affected
- Homozygous recessive individuals have two copies of the allele and are affected
- Homozygous dominant individuals have two normal alleles and are not affected
- The trait can skip generations because the heterozygous individuals are carriers who do not show the affected phenotype

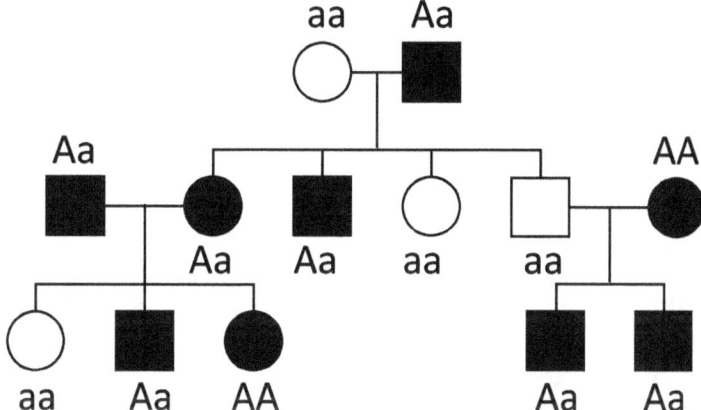

Sex-linked recessive traits

Traits that are sex-linked recessive are coded by genes that are typically located on the X chromosome and sometimes on the Y chromosomes. The specific allele for the trait is recessive over the normal allele. The following are key features that distinguish sex-linked recessive expression from other traits:

- The related gene is located on either the X or Y chromosome
- Females have two copies of a related X-linked gene, while males have only one copy
- Female have no copy of a related Y-linked gene, while males have one copy
- Affected females must have two copies of the allele
- Affected males have one copy of the allele (males are hemizygous)
- Heterozygous females are carriers and are not affected
- The trait can skip generations in heterozygous (carriers) female only who do not show the affected phenotype
- Males and females are unequally affected because while females require two copies of the allele to become affected, males with one copy are affected.

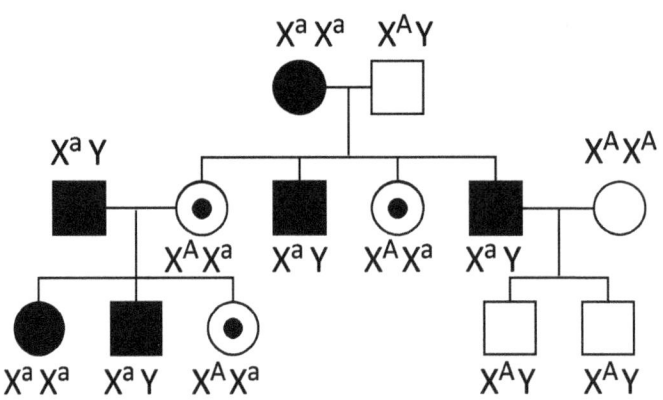

Sex-linked dominant traits

Traits that are sex-linked dominant are coded by genes that are typically located on the X chromosome and sometimes on the Y chromosomes. The specific allele for the trait is dominant over the normal allele. The following are key features that distinguish sex-linked recessive expression from other traits:

■ The related gene is located on either the X or Y chromosome
■ Females have two copies of a related X-linked gene, while males have only one copy
■ Female have no copy of a related Y-linked gene, while males have one copy
■ Both males and females are affected when they have a single copy of the allele (homozygous dominant or heterozygous females, and hemizygous males)
■ Heterozygous females are carriers and are not affected
■ The trait can skip generations in heterozygous (carriers) female only who do not show the affected phenotype
■ Males and females are equally affected

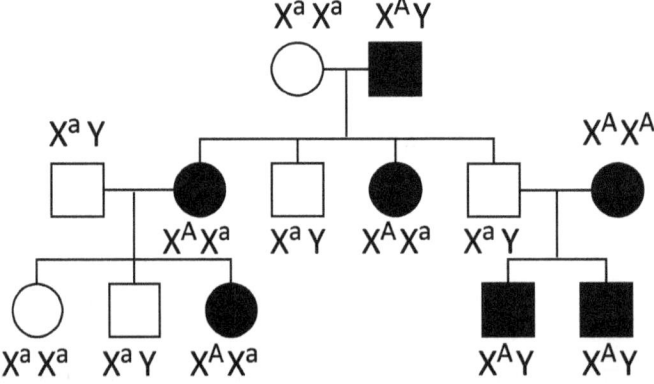

SUMMARY OF KEY CONCEPTS

■ Mendel's laws of inheritance are used in punnet square analysis to predict the phenotype and genotypes of progeny.
■ Probability rules are used to determine multiple traits expression in progeny.
■ Linked genes tend to violate Mendel's 2nd law because they are located on the same chromosome and tend to segregate together as if they were a single gene.
■ Codominance and incomplete dominance, sex-linked traits, polygenic traits,

mitochondria genes all display a non-Mendelian pattern of inheritance.

- Maternal effect can cause the offspring's phenotype to deviate from the Mendelian predictions.
- True breeders are homozygous and hybrids are heterozygous for a particular trait or set of traits, and produce offspring that are phenotypically similar.
- The dominant allele are always expressed when present while the recessive allele are expressed only in the homozygous recessive individuals.
- Test cross analysis is used to determine the genotype of dominant individuals
- Recombinant frequency can be used to map out genes on a chromosome.

CHECK YOUR UNDERSTANDING

1. A plant with the genotype Yy Gg Ff tt can produced how many different gametes?

   (A) 2
   (B) 4
   (C) 6
   (D) 8

2. If 23 out of 137 individuals in a population show the recessive phenotype, what percept of the population will express the dominant phenotype?

   (A) 17%
   (B) 35%
   (C) 59%
   (D) 83%

3. In the scarlet tiger moth population the wing coloration gene is a two-allele system occurring at a single locus and show an incomplete dominance pattern of expression. In a given population there are 242 white spotted (AA), 232 intermediate spotted (Aa), and 53 small spotted (aa). What is the recessive allele frequency?

   (A) 0.57
   (B) 0.32
   (C) 0.54
   (D) 0.68

4. A rat breeder has 723 rats, including 32 individuals that express the recessive albino allele (a). All of the albinos were removed from the population and only the remaining normal rats were allowed to breed. In the next generation, there are 8,273 baby rats. How many of the babies are albino? Assume the population if in HWE.

   (A) 0
   (B) 249
   (C) 366
   (D) 438

5. A pea plants with the genotype GG Ff tt a crossed with a plant that is hybrid for all three genes. What is the probability of producing offspring that are true breeders for all three genes?

(A) 0.0625

(B) 0.125

(C) 0.250

(D) 0.500

6. A cross of a hybrid individual (Aa/Bb/Cc) with a recessive true-breeder (aa/bb/cc) produces the following offspring (letters indicate the observed phenotype): 2281 ABC, 1000 Abc, 712 aBc, 1 ABc, 1 abC, 712 AbC, 1000 aBC, and 2281 abc. What is the gene order and map distance between the three genes?

(A) A-23.1-B-16.4-C

(B) B-25.1-A-42.9-C

(C) B-17.8-C-25.1-A

(D) A-17.8-C-25.1-B

7. A black male cat crosses with an orange female cate. What is the probability of producing an offspring that is mixed black/orange?

(A) 0.00

(B) 0.25

(C) 0.50

(D) 1.00

ANSWERS AND EXPLANATIONS

1. (D) is correct
    $$\# \text{ gametes} = 2^{\# \text{ heterozygous genes}} = 2^3 = 8$$

2. (D) is correct
    Recessive frequency $= q^2 = 23/137 = 0.17$
    $p^2 + 2pq + q^2 = 1$
    Dominant frequency $= p^2 + 2pq = 1 - q^2$
    $1 - q^2 = 1 - 0.17 = 0.83$ or 83%

3. (B) is correct.
    pop. size: $n = 242 + 232 + 54 = 527$
    $q^2 = 54/527 = 0.102$
    $q = (0.102)^{1/2} = 0.32$

    <u>Alternative method:</u>
    Total allele $= 527 \times 2 = 1054$
    \# recessive a alleles $= 2(54) + 232 = 340$
    $q = 340/1054 = 0.32$

4. (B) is correct.
    Initial population: $n_1 = 723$
    Initial \# recessive individuals $= 32$
    $q_1^2 = 32/723 = 0.044$
    $q_1 = (0.044)^{1/2} = 0.21$
    Initial total alleles $= 2 \times 723 = 1446$
    Initial total a alleles $= 0.21 \times 1446 = 304$
    Removed a alleles $= 2(32) = 64$
    Remaining a alleles $= 304 - 64 = 240$
    Remaining alleles $(A + a) = 1446 - 64 = 1382$
    $q_2 = 240/1382 = 0.17$
    $q_2^2 = (0.17)^2 = 0.03$
    \# albinos $= 8273 \times 0.03 = 249$

5. (B) is correct.
    $GG \times Gg \rightarrow$ 0.5 GG, 0.5 Gg
    $Ff \times Ff \rightarrow$ 0.25 FF, 0.5 Ff, 0.5 ff
    $Tt \times tt \rightarrow$ 0.5 Tt, 0.5 tt
    True breeders have the following genotypes:
    GG FF tt ($0.5 \times 0.25 \times 0.5 = 0.0625$) and
    FF ff tt ($0.5 \times 0.25 \times 0.5 = 0.0625$)
    Probability $= 0.0625 + 0.0625 = 0.125$

6. (C) is correct
   Parental chromosomes:
   ABC, abc ($2 \times 2281 = 4562$)
   Double crossovers chromosomes:

   abC, ABc ($2 \times 1 = 2$)
   Single cross overs

   AbC, aBc, Abc, aBC
   ($2 \times 1000 = 2000$; $2 \times 712 = 1424$)

   Gene order:
   B-C-A or A-C-B; The alignment between the double cross overs and parental chromosomes (abC aligned to abc or ABc to ABC) always leaves C unmatched, therefore C is the middle gene.

   Calculate recombinant frequencies:
   Calculate AB recombinant percent using the chromosomes that have cross over between A and B (Abc, aBC, aBc, AbC and the doubles)
   AB recombinant % = $3426/7988 = 42.9\%$

   Do the same for AC recombinant using Abc, aBC, and both doubles.
   AC recombinant % = $2002/7988 = 25.1\%$

   BC recombinant % = $42.9 - 25.1 = 17.8\%$
   You can also calculate the BC recombinant % using aBc, AbC, and the doubles as follows:
   AC recombinant % = $1426/7988 = 17.8\%$

7. (C) is correct. This is a X-linked (or sex-linked) trait. Perform the following cross using a punnet square: $X^bY \times X^BX^B$. The outcome is **0.5** X^BY and 0.5 X^BX^b. Therefore all males are black and all females are mixed. Since half of the offspring are predicted to be females, the frequency for mixed is 0.5.

19 DNA replication

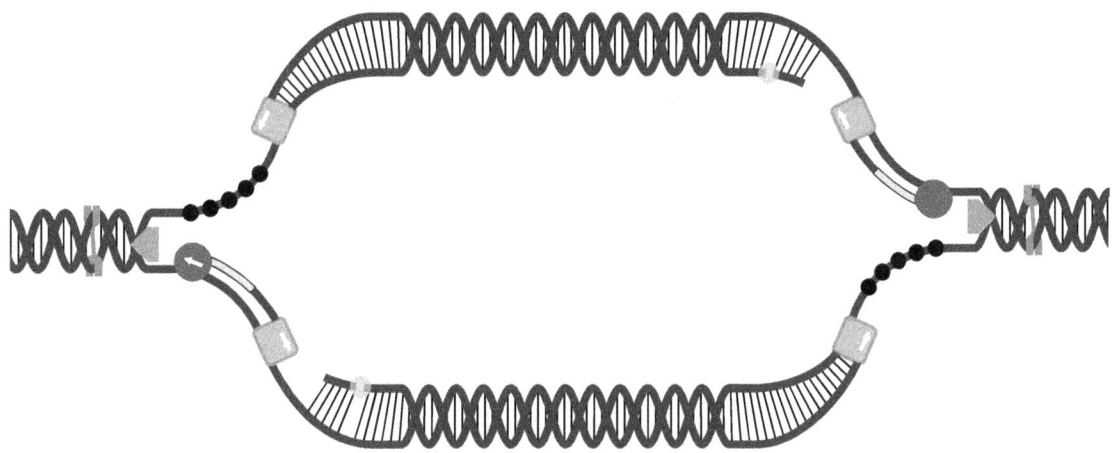

ALL STUDENTS MUST BE ABLE TO ANSWER THESE QUESTIONS

1. What are the three phases of DNA replication?
2. How does DNA replication in prokaryotes compare to eukaryotes?
3. At which stage of the cell cycle does DNA replication occur?
4. In which direction (5' or 3') does DNA replication occur? Is the template DNA read by DNA pol?
5. Why is DNA replication said to be bidirectional?
6. Why is the origin of replication A/T rich?
7. What are 3 limitations of DNA polymerase and how are they resolved?
8. What is the difference between the leading and lagging strand in terms of direction of growth, number of polymerase enzymes involved in producing the strand?
9. What are Okazaki fragments and how are the relevant for the lagging strand?
10. What are telomeres? Onto which strand (leading or lagging) are they added? To which end (5' or 3') are they added?
11. What is the role of the following enzymes and accessory proteins in DNA replication: helicase, Ssb protein, topoisomerase, primase, DNA pol I, II, & III, ligase, telomerase?

ALL STUDENTS MUST BE ABLE TO COMPLETE THE FOLLOWING TASKS

12. Make predictions about what is occurring during the stages of DNA replication.
13. Describe how the lagging strand is produced from Okazaki fragments.
14. Explain why Okazaki fragments are necessary, and why the lagging strand grow in the 5' direction.

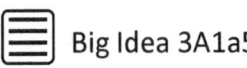 Big Idea 3A1a5

Before the cell can divide, its chromosomes are duplicated so that there is enough DNA for each daughter cell to receive a full set of genetic material. DNA replication in prokaryote, particularly in the E. Coli bacteria, has been extensively studied and is better understood than replication in eukaryotes. However, there is some evidence that the replication machinery of eukaryotes is similar to that of E. Coli.

PHASES OF DNA REPLICATION

DNA replication in prokaryotes and eukaryotes can be divided into 3 phases:
1. **Initiation** occurs when initiator proteins bind to the origin of replication, (ori)
2. **Elongation** follows initiation, and is the actual process of growth of the daughter strands using parent strands as templates.
3. **Termination** is the end of the replication, when DNA polymerase and accessory proteins leave the DNA.

Replication in human versus E. Coli bacteria

	E. Coli bacteria	Human
Chromosomes per cell	1	46
Base pairs per genome	4.6 million	6,000 million
Replication error	1 per 10,000,000,000	1 per 10,000,000,000
Replication time	~ 20 minutes	~ 2 hours
Number of Ori	1	Multiple

DNA REPLICATION IS SEMICONSERVATIVE

A human somatic cell in G_1 or G_0 phase has 46 chromosomes, each made of a linear double-stranded DNA (dsDNA) molecule that forms a single chromatid. When a cell receives the appropriate signal to begin cell division, it crosses the G_1/S checkpoint to enter the S phase of interphase. The dsDNA then denatures bidirectionally from the centromere toward the both ends of the molecule. Meanwhile, new complementary daughter strand (red) are made using the parent strands as templates. The resulting products are 2 sister chromatids that are attached to each other at their centromere to form the classic "X" shape of the metaphase chromosome.

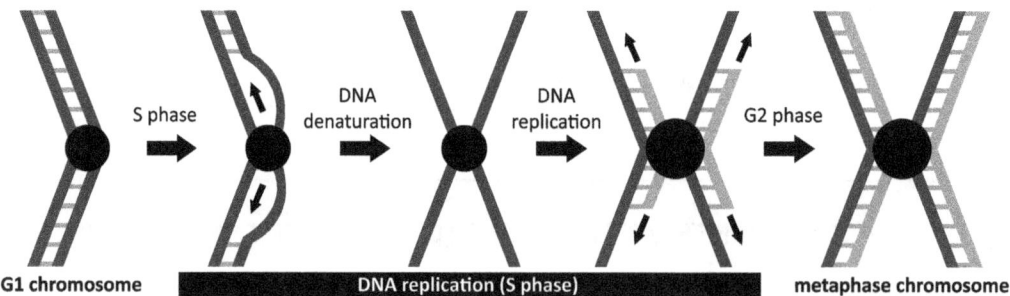

DNA REPLICATION IS BIDIRECTIONAL

During the S-phase, DNA replication occurs by bidirectional growth of both new strands in the 5' to 3' direction. This means that the DNA molecule is simultaneously unzipped toward both ends of the linear molecule as the new daughter strand elongates. Accessory proteins and enzyme that break the hydrogen bonds between complementary DNA strands facilitate unzipping at the growing fork.

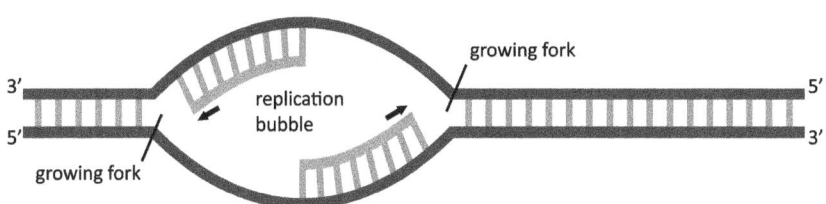

Bidirectional replication of viral DNA

Bidirectional growth has been observed with plasmid DNA from infected cells. Over-time, the replication bubble expands as the growing fork unzips the DNA and new strands are copied in the 5' to 3' direction. The parent strands serve as templates for making the complementary daughter strands.

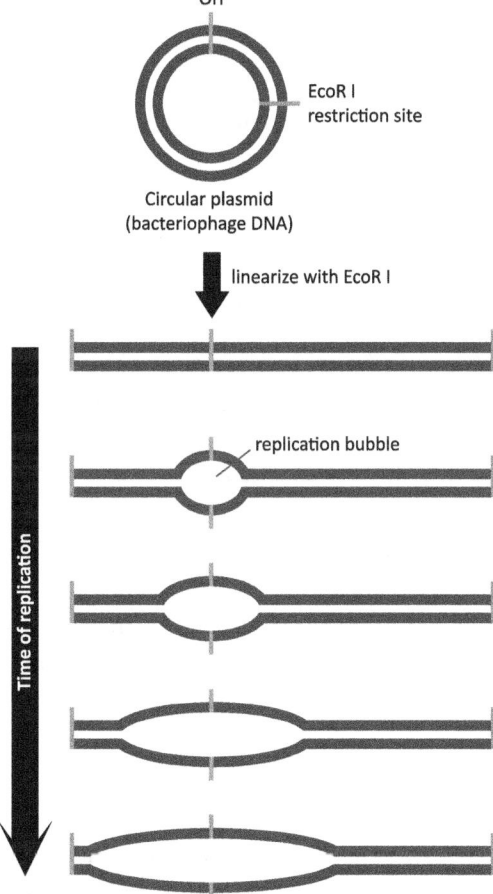

THE ORIGIN OF REPLICATION

DNA replication always begins at a specific site called the origin of replication or *ori*. The E.Coli ori, called oriC, includes multiple short-repeat sequences (shown in yellow and green) that are recognized by ori-binding proteins. These proteins help assemble the accessory proteins and enzymes that are needed for initiating replication. There are 2 types of repetitive sequences called 13-mer and 9-mer, because they are made of a 13 and 9 nucleotide sequences, respectively. These sequences are highly conserved across several bacterial species, suggesting a common ancestry. In addition, there is an adenine (A) and thymine (T) rich region (shown in red) located adjacent to the ori. This A/T rich region facilitates the unwinding of the double helix in preparation for unzipping.

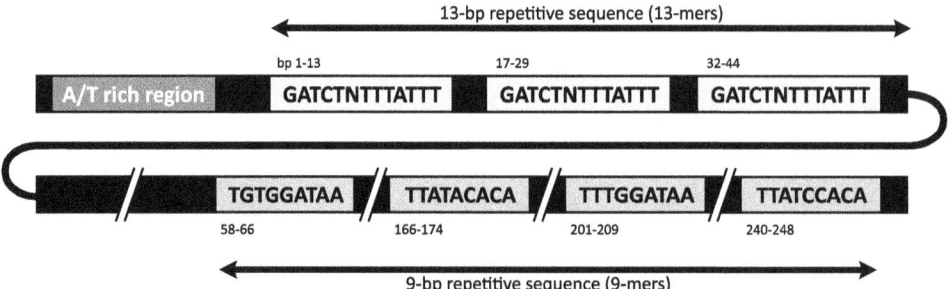

DNA POLYMERASE

DNA polymerase (DNA pol) is the enzyme responsible for polymerizing the new daughter strands. There are 3 different forms of the enzymes, DNA pol I, II, and III.

- DNA pol III is the form that is primarily responsible for growing the majority of the DNA strand.
- DNA pol I replaces RNA primers (discussed later) on daughter strands with DNA.
- DNA pol II corrects DNA replication errors.

Due to certain limitations of the polymerization process and restrictions in DNA polymerase, the three forms need help from other proteins. The challenges that DNA polymerase faces include:

1. DNA pol III is unable to unwind the DNA double helix and denature the complementary template strands.
2. DNA pol III cannot initiate daughter strand elongation. It can only grow a DNA strand from preexisting RNA primers or DNA.
3. DNA pol III grows the new strands only in the 5' to 3' direction. As the strand denatures in the opposite direction, new strands need to be made in the 3' to 5' direction.

Solutions to DNA polymerase challenges

1. **Helicase** unwinds and unzips with DNA double-helix and the **single-stranded-binding (Ssb) protein** binds to the single strands to keep them from reannealing. The unwinding of the double helix adds sufficient torque in the molecule so cause the DNA to supercoil. Supercoiling is similar to what happens to a spring that is unwound from its natural state - it deforms by supercoiling. The enzyme **gyrase** (also called **topoisomerase**) stabilizes the uncoiled DNA to prevent supercoiling.
2. The enzyme called **primase**, adds RNA primer to the parent template strand so that **DNA Pol III** can polymerize the complementary daughter strand in the 3' direction.

3. During replication there is a continuous **leading strand** that grows in the 5' to 3' direction, and a **lagging strand** composed of **Okazaki fragments** that grow in the 3' to 5' direction. DNA Pol I removes ribonucleotides of the RNA primer between adjacent Okazaki fragments on the lagging strand and fills the gap with deoxynucleotides. **DNA ligase** joins the adjacent fragments.

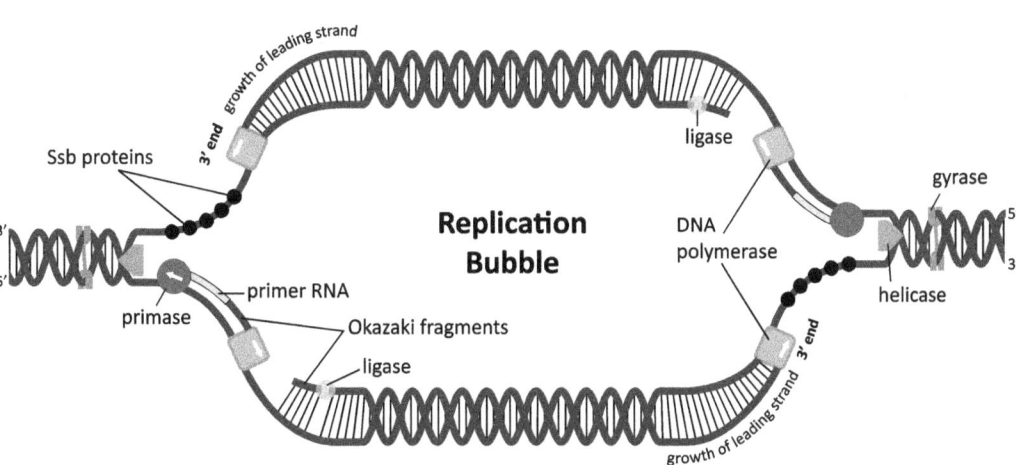

LEADING & LAGGING STRANDS

■ The leading strand begins at the ori with the RNA primer and elongates without breaking in the 3' direction (downstream of the ori) until it reaches the end of the DNA molecule.

■ Okazaki fragments are used for replicating the rest of the DNA that is upstream of the ori. Like the leading strand, Okazaki fragments are elongated from RNA primers in the 5' to 3' direction, but until a downstream Okazaki fragment or the leading strand it reached (recall that the leading strand elongates to the 3' end of the daughter DNA).

■ The lagging strand is made of all the Okazaki fragments of a single daughter strand that are fused together by Polymerase I and ligase. It begins at the ori and is assembled with successive Okazaki fragments, growing in the 3' to 5' direction rather than the polymerization direction of 5' to 3'.

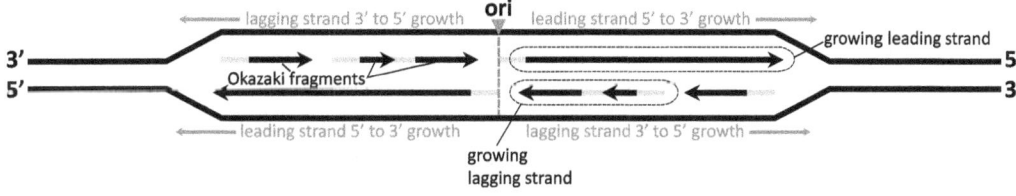

OVERVIEW OF THE ACCESSORY PROTEINS

Replication proteins	Role in DNA replication
Initiators DnaA, DnaB (or helicase), DnaC	Binds DNA at ori; Unwinds and unzips the double-stranded DNA at the replication fork.
Single-stranded binding proteins (or Ssb protein)	Binds separated single DNA strands and prevents them from re-annealing
Gyrase (or topoisomerase)	Moves ahead of the replication fork to release torque that builds up as a result of DNA unwinding.
Primase	Adds the RNA primer from which the DNA polymerase elongates the daughter strand.
DNA pol I	Remove the RNA primers and replace them with DNA.
DNA pol II	Correct DNA replication errors.
DNA pol III	The primary polymerase responsible for DNA replication.
Ligase	Joins together adjacent Okazaki fragments after DNA Pol I has removed the RNA primer.

TELOMERES

Telomeres are located at the non-coding ends of eukaryotic DNA molecules. They protect the coding regions from telomere DNA deletion that occurs during replication. Because DNA polymerase require a primer RNA to add new nucleotides, when the RNA primers on the 5' ends of the lagging strands are removed, DNA pol I cannot replace it with DNA. This means that whenever the DNA replicates, a section of DNA is deleted. After a few rounds of replication, genes and, eventually, the entire genome may be deleted.

Nature has devised a cleaver solution to this problem. Most important is the high activity of the enzyme called telomerase in embryonic stem cells. This enzyme increase the length of telomeres at the ends of DNA so that cells have additional non-coding DNA that can be deleted without affecting gene coding regions. Some differentiated somatic cells that have increased mitotic rates, such at blood cells, also have relatively high telomerase activity. However, activity of telomerase decreases with the age of cells, leading some scientist to hypothesize a role of telomere length in the aging process.

REPLICATION & MUTATIONS

While DNA pol III is elongating, DNA pol II is simultaneously proofreading and correcting replication mistakes. Proofreading lowers the error rate to 1 in 10 million nucleotides. Mismatched errors that escape causes a deformity in the secondary helical structure of DNA that can be detected and corrected by DNA pol II. In order to do this, the enzyme must be able to distinguish between the old and new strands (old strands are methylated). Fortunately, the old strand is methylated and new strands are only methylated after proofreading and DNA repair. Pol II uses methylation to determine which strand is the template and has the correct sequence. The enzyme also has both exonuclease and primase activity - the ability to cleave out the mismatched nucleotide, replace it with the correct nucleotide.

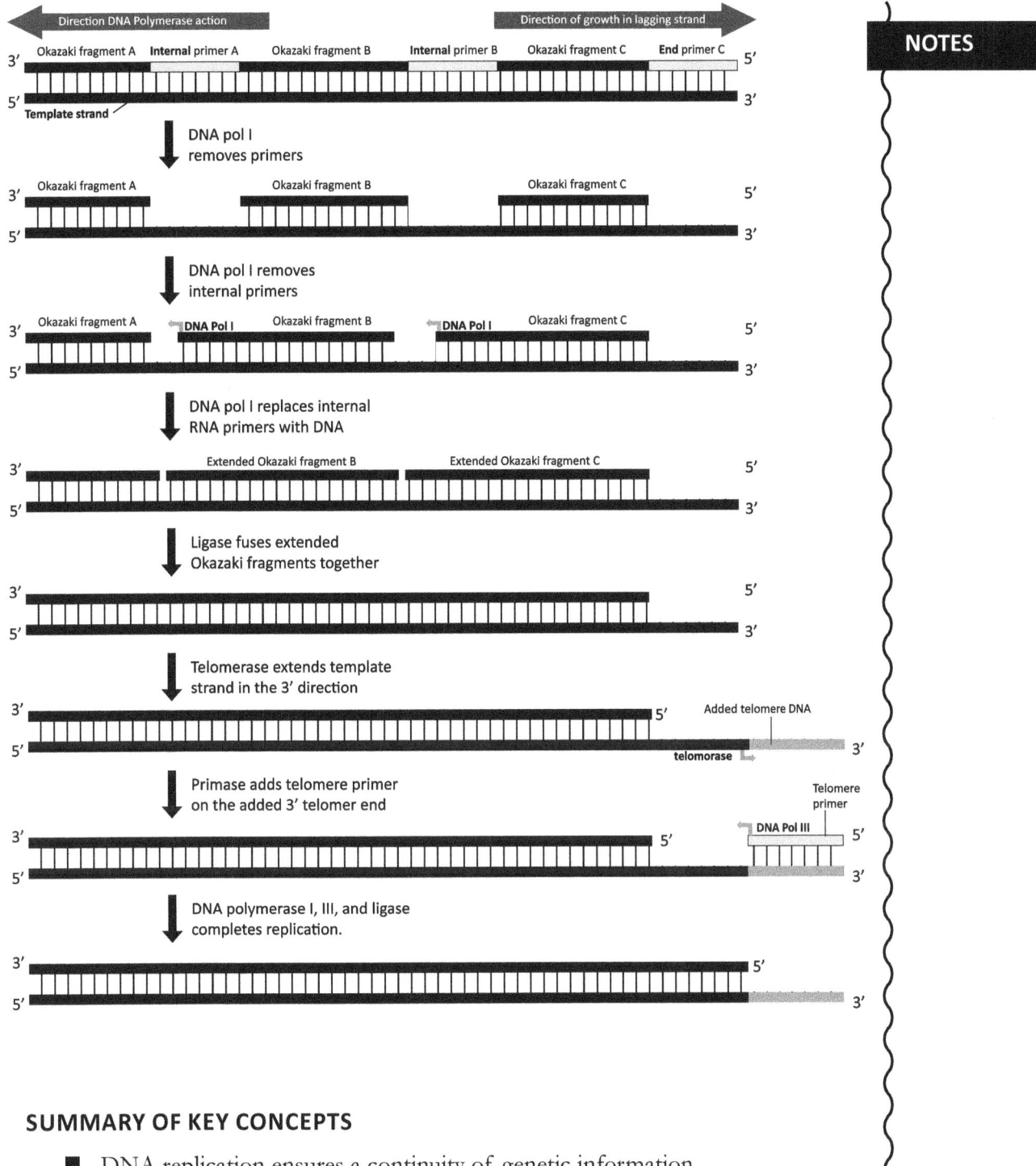

SUMMARY OF KEY CONCEPTS

- DNA replication ensures a continuity of genetic information.
- DNA replication is semiconservative and bidirectional.
- DNA replication requires DNA polymerase, and other accessory proteins .
- DNA replication produces a leading strand, and a lagging strand that is made of Okazaki fragments.
- The leading strand and Okazaki fragments grow in the 3' direction.
- The lagging strand grows from existing Okzaki fragment in the 5' direction.
- Telomeres are necessary for preserving the coding regions of eukaryotic DNA.

CHECK YOUR UNDERSTANDING

1. A chromosomal deletion resulted in a loss of portions of a bacteria cell's 13-mer and 9-mer repetitive sequences. What is the expected consequence of this deletion?

   (A) Gene regulation is repressed.

   (B) Genes are continuously expressed.

   (C) Negative regulation of DNA replication is prevented.

   (D) The cell fails to complete cell division.

2. Why has the primase genes been evolutionarily conserved across biological domains?

   (A) The gene is not actually conserved. Instead, each domain independently derived the gene through gene duplication, and natural selection and modification of a different ancestral gene. It is through convergence that the various primase genes adapted to the same function.

   (B) DNA polymerase is unable to initiate the elongation of a new DNA strand without the RNA primer.

   (C) Without primase, RNA transcription fails because RNA polymerase needs an RNA primer that primase adds to the DNA.

   (D) All of the above statements are correct.

3. Which DNA polymerase is responsible for correcting DNA replication error?

   (A) DNA Pol I

   (B) DNA Pol II

   (C) DNA Pol III

   (D) All of the above are equally capable of correcting replication error.

4. Which of the following statements is correct?

   (A) The leading strand grows in the 5' direction.

   (B) The lagging strand grows toward the 5' end of the template strand.

   (C) Okazaki fragments grow toward the 5' end of the template strand.

   (D) DNA I fuses adjacent Okazaki fragments together.

5. Which of the following statements is correct?

   (A) DNA replication in eukaryotes result in the shortening of the daughter strand at its 3' end.

   (B) Telomerase adds new telomeres to the 5' end of the new strand.

   (C) Helicase releases the torque that builds-up ahead of the replication fork.

   (D) Telomeres are added to the 3' end of the template strand.

ANSWERS AND EXPLANATIONS

1. (D) is correct because the repetitive sequences are part of the bacteria cell's origin of replication. These are highly conserved sequences that are used by ori-binding proteins to identify the ori. Because these proteins are necessary for the assembly of DNA polymerase and other accessory proteins, any alteration will prevent DNA replication, thereby preventing the cell from duplicating its DNA for the purpose of cell division.

2. (B) is correct because primase adds the RNA primer that DNA polymerase uses as a starting point for its continued elongation of the new daughter strand.

3. (B) is correct.

4. (C) is correct. Recall that all polymerization occurs in the 3' direction of the new strand, and the 5' direction of the template strand.

5. (D) is correct. Recall that the new strand is shortened at it 3' end. To fix this, the telomer is added to the 5' end of the template strand. Primase than adds a primer on this telomer, and DNA Pol III can complete the replication without loss of sequence.

20 Transcription & translation

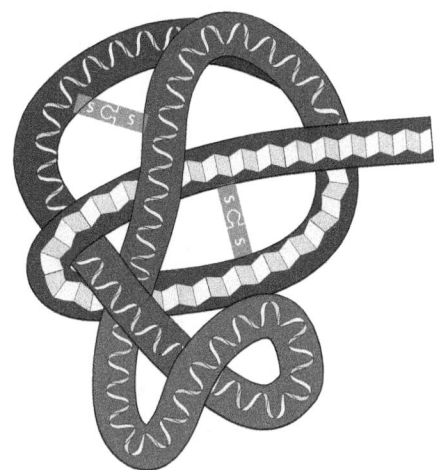

 Big Idea 3A1b4,c-d, 3B1a,c1,c3

THE ANTISENSE STRAND

DNA is made of 2 complementary strands that, for simplicity, will be referred to as Strand A and Strand B. For a particular gene, only one of the two strands is used for RNA transcription. For example, in the figure that follows, strand A is the template for RNA transcription of genes B and D. This means that RNA polymerase reads strand A (not strand B) to produce the mRNA transcripts of these two genes. The same can be said about strand B with respect to genes A and C.

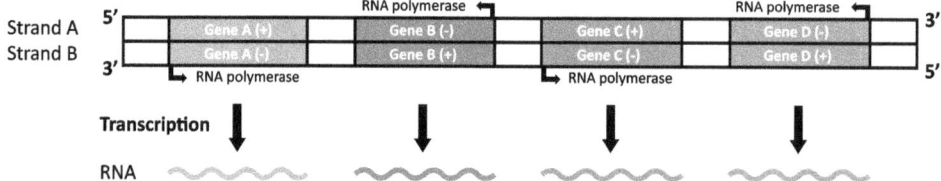

The DNA strand that complements the mRNA also called the sense strand (or + strand), while its complementary strand is the antisense (or - strand). Therefore, based on the above figure, the following assumptions can be made:

- The antisense (-) strand is used for transcription.
- RNA polymerase reads in the 3' to 5' direction.
- Different genes can be transcribed from different strands.
- Strand A is the antisense (-) strand for genes B and D
- Strand A is the sense (+) strand for genes A and C
- Strand B is the antisense (-) strand for genes A and C
- Strand B is the sense (+) strand for genes B and D

TRANSCRIPTIONAL UNITS

A **transcription unit** is the part of the DNA that is transcribed into RNA. Each transcriptional unit contains an **RNA-coding region** located downstream (5' direction) from a gene regulatory site called the **promoter**. The RNA coding region includes a specific **gene**, a **terminator sequence**, and the **termination site**. RNA polymerase binds at the promoter and transcribes the mRNA until it detects the transcription terminator and termination site.

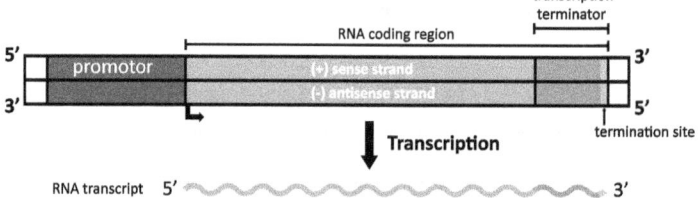

TRANSCRIPTION MACHINERY

The ATP consuming process of RNA transcription is similar to DNA replication except that RNA polymerase does not require accessory proteins in order to bind to the DNA (recall that DNA polymerase requires initiator proteins). However, accessory proteins called **transcription factors** can interact with RNA polymerase to greatly increase its affinity for the promoter, thereby accelerating the rate of transcription.

Prokaryotes have a simple transcription mechanism that involve only one type of RNA polymerase for transcribing all three classes of bacterial RNA (mRNA, tRNA, rRNA). Eukaryotic cells use three types of RNA polymerase to transcribe various classes of RNA:

- RNA pol I transcribes rRNA
- RNA pol II transcribes pre-mRNA, snRNA, snoRNA, and miRNA
- RNA pol III transcribes tRNA, small rRNA, snRNA, and miRNA

Location and function of RNA classes

RNA class	Cell type	Cell location	Function
Ribosomal (rRNA)	Both	Cytoplasm	Structural and function component of ribosomes.
Messenger (mRNA)	Both	Nucleus & cytoplasm	Carries the genetic code to the ribosome
Transfer (tRNA)	Both	Cytoplasm	Transfers the amino acids to the ribosome
Small nuclear (snRNA)	Eukaryotes only	Nucleus	Pre-mRNA processing
Small nucleolar (snoRNA)	Eukaryotes only	Nucleus	rRNA processing and assembly
micro (miRNA)	Eukaryotes only	Cytoplasm	Negative regular of mRNA translation (blocks translation)
Small interfering (siRNA)	Eukaryotes only	Cytoplasm	Triggers the breakdown of other RNA molecules

STAGES OF TRANSCRIPTION

Similar to DNA replication, transcription is divided into 3 stages: initiation, elongation, and termination.

Initiation involves the attachment of transcription apparatus, which includes RNA polymerase and its accessory proteins, to the promoter. This requires that RNA polymerase recognize the promoter and begins denaturing the dsDNA to form the transcription bubble (analogous to the replication bubble). After the first two nucleotides are linked and RNA polymerase leaves the promoter to enter the RNA-coding region, the second stage of elongation takes over.

Transcriptional elongation is similar to elongation in DNA replication. After RNA polymerase leaves the promoter, it undergoes a conformation change and is no longer able to bind the promoter. RNA polymerase continued to unwind and denature the DNA in the 5' direction to create a small transcription bubble. Addition of nucleotide in the 3' direction of the mRNA.

The third and final stage is **transcription termination**, which involves RNA polymerase's recognition of transcription terminator sequence, and the separation of the transcription apparatus and RNA from the DNA at the termination site.

NOTES

PROKARYOTIC CONSENSUS SEQUENCES VERSUS EUKARYOTIC TATA BOX

The promoter regions of many prokaryotes tend to share a short highly conserved stretches of nucleotides called **consensus sequences** (evidence of common ancestry and evolution). Eukaryotes have a similar region called the **TATA box**. In both case, these are AT-rich sequences with weaker hydrogen bonding between adenine and thymine, relative to guanine and cytosine in GC-rich regions. These weak bonds make it easier to denature the dsDNA after the promoter has been recognized by RNA polymerase and transcription factors. The two most common consensus and TATA box sequences are:

- **-10 consensus sequence** (5'-TATAAT-3'), located 10 base pairs upstream from the prokaryotic transcription start site.
- **-35 consensus sequence** (5'-TTGACA-3'), located 35 base pairs upstream from the prokaryotic transcription start site.
- **TATA box sequence** (5'-TATAAA-3'), located 25 base pairs upstream from the eukaryotic transcription start site.

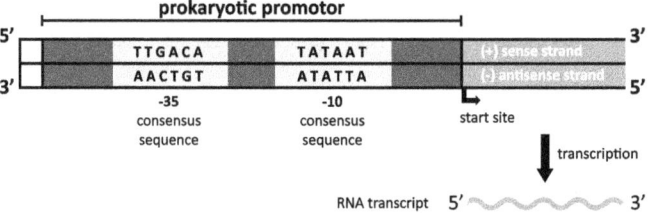

SPECIAL TRANSCRIPTION CHALLENGES IN EUKARYOTES

The DNA is packed into chromosome by proteins called histones. There are two types of histones called the core and H1 histones. Core histones normally have positively charged tails that have a high affinity for the negatively charge DNA. DNA wraps around sets of 8 core histones to beads units called nucleosomes that are connected by linker DNA. H1 histones interacts with the linker DNA to allow for an orderly, tight packing of the chromatin. Tight packing negatively impact transcription by making the DNA highly inaccessible to RNA polymerase. In order for transcription to take place, a special collection of proteins and enzymes, including acetyltransferase and chromatin-remodeling proteins (CRP) work to loosen-up the chromatin structure by removing nucleosomes from promoter and RNA-coding regions..

Acetyltransferase adds acetyl groups to the histone tails, converting them from a normally positive to negative charge. The newly negatively charged tails tend to have a reduced attraction for the negatively charged DNA, causing the chromatin to loosen. Histone acetylation is therefore associated with increased transcription and gene activity in the acetylated region, and vice versa for deacetylation. Similar mechanism by other enzymes include transcriptional repression or activation by histone methylation, depending on the particular amino acids that is methylated.

Chromosome packing

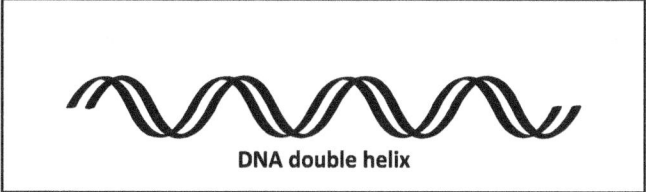

DNA double helix

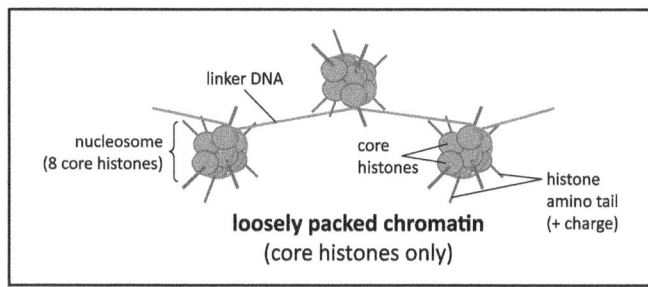

linker DNA

nucleosome
(8 core histones)

core
histones

histone
amino tail
(+ charge)

loosely packed chromatin
(core histones only)

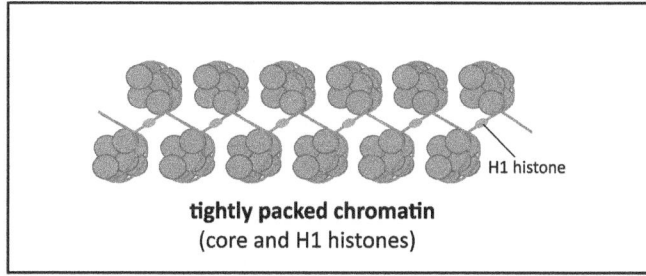

H1 histone

tightly packed chromatin
(core and H1 histones)

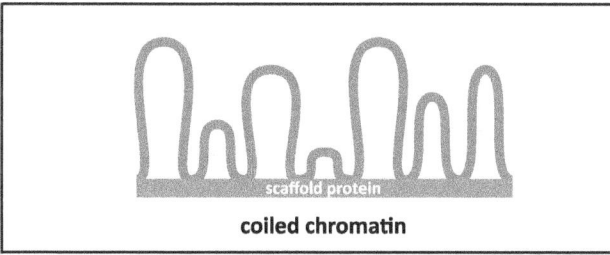

scaffold protein

coiled chromatin

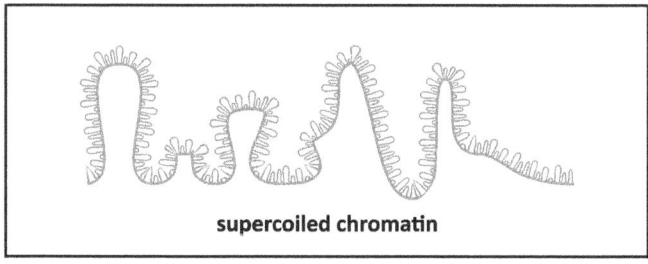

supercoiled chromatin

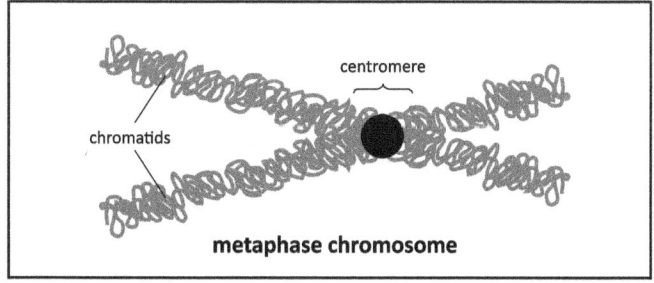

centromere

chromatids

metaphase chromosome

Transcription factors

Each type of eukaryotic RNA polymerase (RNA pol I, II, or III) needs to be able to recognize and bind to different type of RNA-coding region based on the type of RNA being transcribes. Promoter binding proteins resolve this problem by recruiting a specific type of RNA polymerase. The most important of these proteins are the previously mentioned general transcription factors, and the transcription activators that will be discussed in more details in L21 and L22. The general **transcription factors** interact with RNA polymerase to form the basal transcription apparatus that can independently sustain a low level of transcription. **Transcription activators** enhance transcription by stimulating assembly of basal transcription apparatus.

Pre-mRNA processing

In eukaryotic cells, there are 3 major types of post-transcriptional processing of the pre-mRNA.

pre-mRNA modification	Function
RNA splicing	Removes non-coding regions called **introns** from pre-mRNA Allows for the mRNA export from nucleus to cytoplasm Through alternative splicing, it can produce multiple polypeptides from a single coding region.
Cleavage of 3' end and addition of 3' poly A tail	Increases the mRNA stability by protecting it from cytoplasmic RNase. Helps in the export of mRNA from the nucleus. Alternative cleavage position can result in multiple mRNA and proteins being produced from a single pre-mRNA.
Addition of 5' GTP-methyl cap	mRNA from cytoplasmic RNase Facilitates the binding of ribosome to the 5' end of mRNA Supports mRNA splicing

pre-mRNA processing

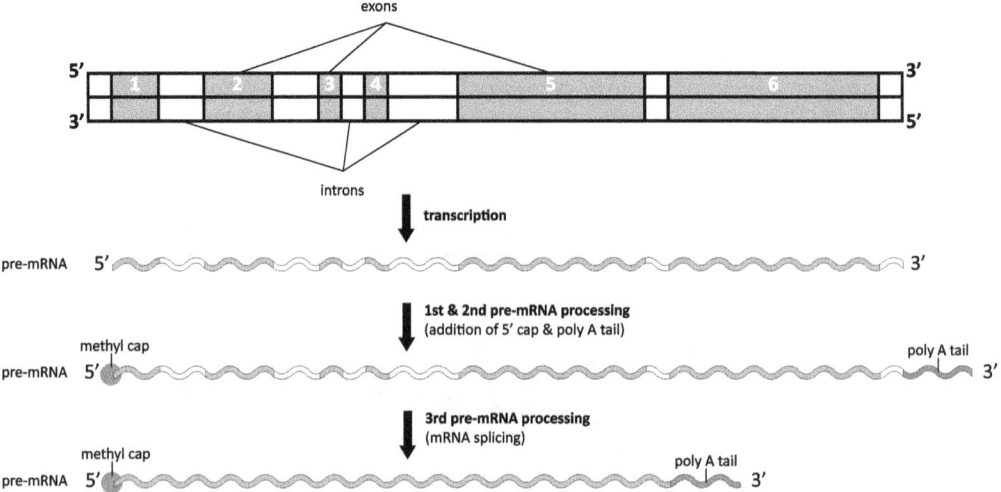

Overview of alternative splicing

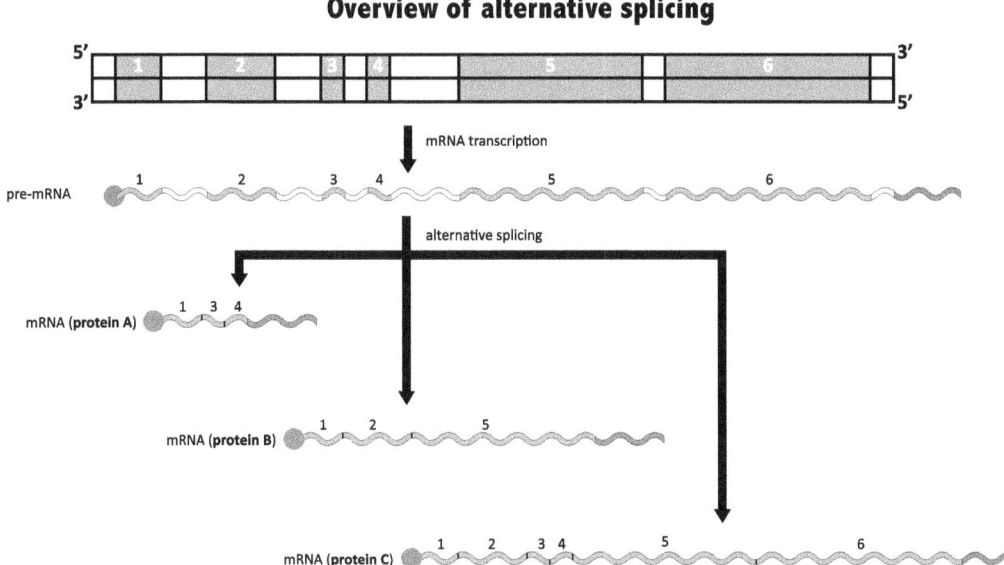

Translation

The completed mRNA exits the nucleus through the nuclear pores and enters the cytoplasm where it comes in contact with the ribosomes. Translation proceeds via the 3 steps illustrated below and discussed in L16.

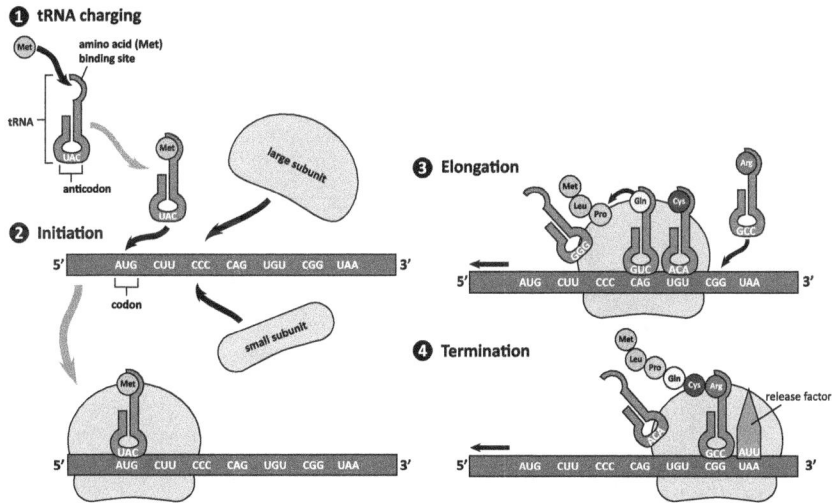

Translation, like transcription, is an endergonic process that requires the input of energy from an ATP-like molecule called GTP (not shown in the illustration). 3 GTPs are used for each amino acid in the polypeptide sequence. Several enzymes called initiation factors (not shown) are involved in the initiation step.

The protein structure

The protein has up to four structural levels: **primary**, **secondary**, **tertiary**, and **quaternary** structures. The primary structure is determined by its amino acid sequences that is coded by the mRNA. **Hydrogen bonding** between amino acids of the polypeptide orients that protein into its secondary structure that forms either a spiraling α-helix or a β-pleated sheet.

The secondary structure undergoes further folding to form the final functional tertiary structure of the protein (or protein subunit). The tertiary structure is held in place by **disulfide bonds** that form between amino acids.

Some large proteins are composed of two or more polypeptides. Proteins containing multiple **subunits** form a quaternary structure in which the various subunits combine to form the final protein product.

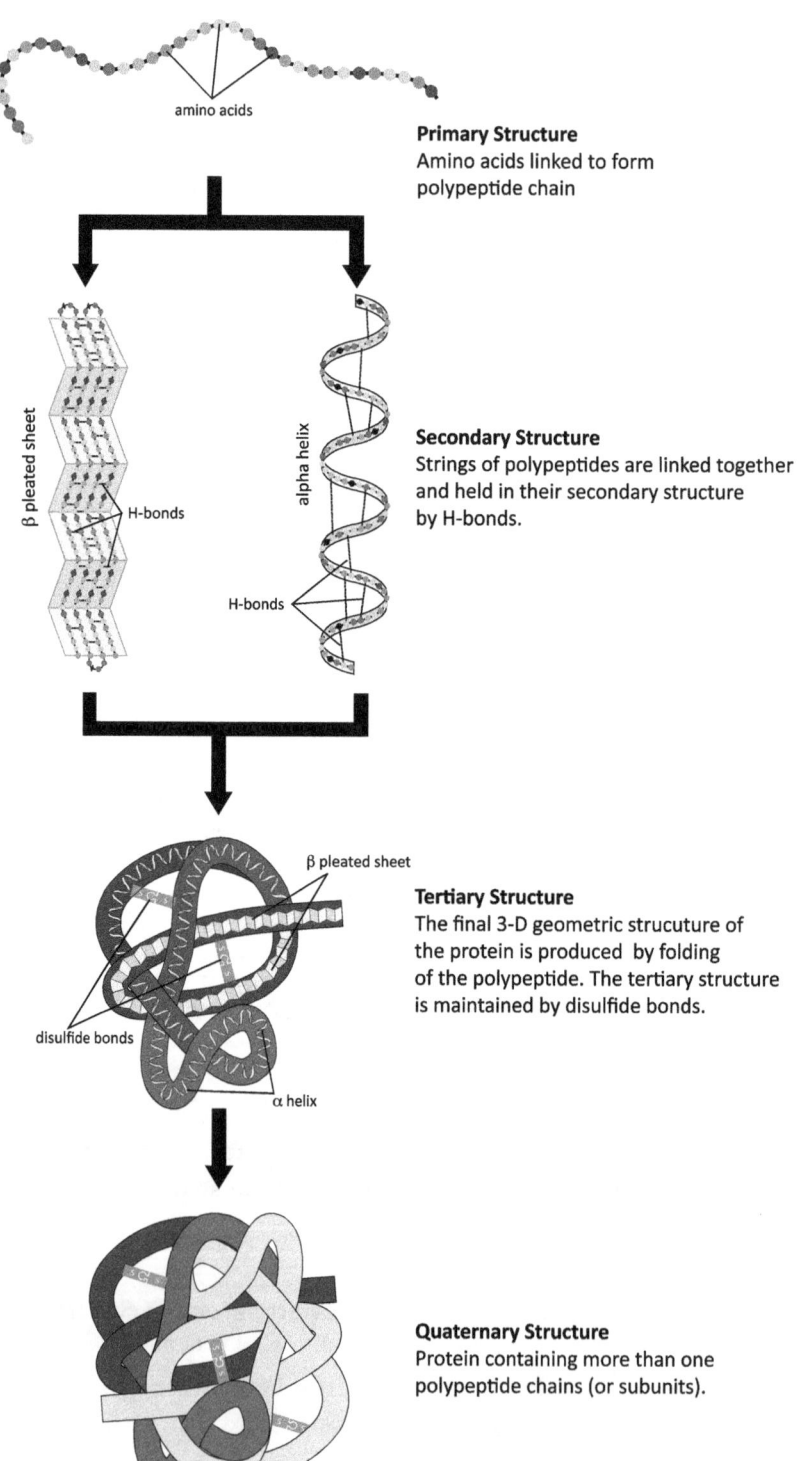

amino acids

Primary Structure
Amino acids linked to form polypeptide chain

β pleated sheet

alpha helix

H-bonds

H-bonds

Secondary Structure
Strings of polypeptides are linked together and held in their secondary structure by H-bonds.

β pleated sheet

disulfide bonds

α helix

Tertiary Structure
The final 3-D geometric strucuture of the protein is produced by folding of the polypeptide. The tertiary structure is maintained by disulfide bonds.

Quaternary Structure
Protein containing more than one polypeptide chains (or subunits).

Protein function

The function of proteins influences the organisms phenotype and is determined by the expressed allele of the encoding gene, and the roper folding the translated polypeptide chain. The major functions that is performed by proteins include:

- **Defense** through special proteins called antibodies that help protect the body from toxins and pathogens by binding to specific foreign particles like viruses and bacteria.
- **Catalyzing reactions** via proteins called enzymes that act to accelerate most metabolic reactions that are involved in the synthesis and degradation of macromolecules, and the biosynthesis of RNA through transcription and DNA through replication.
- **Structural support** functions at the cellular level, such as proteins like microfilaments that play a critical role in the cytoskeleton and extracellular support apparatus, and at the organismal level where proteins are involved in the movement of organisms in their local environment.
- **Communication function** through special proteins like hormone that serve as messengers that transmit chemical signals between cells, tissues, and organs in order to coordinate biological processes.
- **Transport & storage function** through proteins such as ferritin that can bind and store substances, while others like protein ion channels that can transport ions across the cell membrane.

SUMMARY OF KEY CONCEPTS

- Transcription is the process of generating an RNA sequence based on the genetic information stored in the DNA molecule; it is the critical intermediate step in the flow of genetic information from the DNA to the protein;
- There are different classes of RNA molecules, each with a structure that supports its specific function; the linear mRNA molecule is the form that carries the genetic code to the ribosome for protein translation.
- Transcription produces a pre-mRNA that is further post-transcriptionally modified (splicing, 3' poly (A) tail, and 5' GTP-methyl cap) for purposed discussed earlier.
- Transcription involves initiation, elongation, and termination.
- The mRNA is transcribed by the RNA polymerase from only one of the two DNA strand – the antisense strand.
- A transcription unit includes a promoter, and the RNA-coding region.
- The promoter contains consensus sequences that are recognized by the RNA polymerase and accessory proteins.
- Translation of mRNA into polypeptide occurs in the cytoplasm on ribosomes.
- The 3-D geometric shape of a proteins determine its function (defense, enzymatic, messenger, structural support, transport, or storage).
- The are four levels of organization of protein structure: primary, secondary, tertiary, and quaternary.

CHECK YOUR UNDERSTANDING

1. Which of the following statements is true?

    (A) If strand A of a dsDNA is the antisense strand for gene A, it can also be the sense strand for other genes on that chromosome.

    (B) In a transcriptional unit, the promotor is always located at the 5' end of the (+) strand.

    (C) If an RNA-coding region lacks a transcription terminator, the mRNA may be longer than normal.

    (D) All of the above are correct statements.

2. Which of the following RNA molecules is involved in degrading native mRNAs?

    (A) snRNA

    (B) snoRNA

    (C) miRNA

    (D) siRNA

3. If gene A is upstream from gene B, then gene A is located at the

    (A) 3' direction of gene B's (-) strand.

    (B) 5' direction of gene B's antisense strand.

    (C) 3' direction of gen B's (+) strand.

    (D) direction from gene B towards which RNA polymerase reads.

4. Which of the following statements is correct?

    (A) 9-mer and 13-mer consensus sequences are needed for DNA replication.

    (B) -10/-35 and TATA box consensus sequences are needed for RNA transcription.

    (C) Transcription consensus sequences are located upstream from the transcriptional start site.

    (D) All of the above are correct.

5. The function of transacetylase is to

    (A) make the histone tails negatively charged so that they loose their affinity for the negatively charged DNA.

    (B) loosen up the chromatin structure around the promotor and RNA-coding regions.

    (C) positively regulate transcription.

    (D) All of the above.

6. An RNA-coding region has three exons located in the following order on the chromosome: exon A, exon B, exon B (A-B-C). What is the maximum number of proteins that can be produced from this RNA-coding region via alternative splicing.

   (A) 1
   (B) 3
   (C) 7
   (D) 27

ANSWERS AND EXPLANATIONS

1. (D) is correct.

2. (C) is correct.

3. (A) is correct.

4. (D) is correct.

5. (D) is correct.

6. (C) is correct. The 7 possible mRNAs are A, B, C, A-B, A-C, B-C, A-B-C

21 Gene regulation

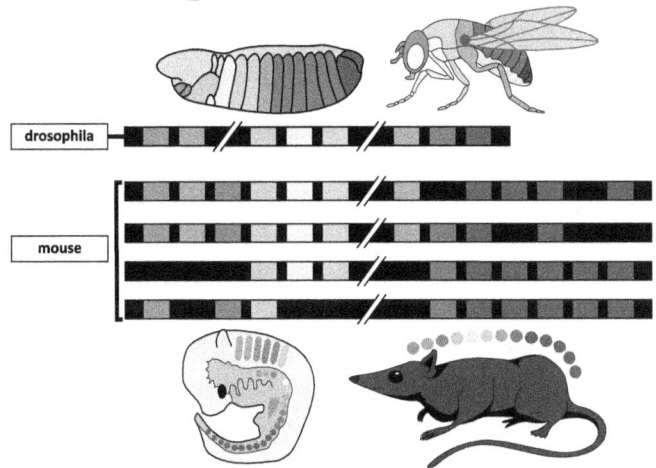

 Big Idea 3B1

A cell is analogous to a city, a tissue to a county, an organ to a state, and a multicellular organism to a country. These levels of organization did not abruptly appear, they grew out of the need to efficiently govern and manage a complex system that has millions of different parts (i.e., natural resources, people, land, taxes, businesses). A disruption at any level in the hierarchy can affect the health of the whole country.

If we take a closer look at cities, we find that they too have levels of organization to ensure internal order. There are rules and laws that must be followed, and there are enforcers of these rules and laws. Imagine the chaos that will ensue if all the traffic signs were removed, or garbage collection was halted, or teachers, police and fire fighters refused to work. Likewise, all of the cell's components have a part to play and rules for when and how they should function. Collectively, they keep the cell healthy and efficient, and the organism alive. For example, the rule for insulin production specifies that its transcription and translation must be directly tied to rising blood sugar levels. By regulating the enzyme in this manner, the cell increases its energy efficiency by only producing insulin when it is needed to lower glucose levels.

LEVELS OF GENE REGULATION

Genes are the rules that the cell uses to maintain internal order. **Gene regulation** is the cell's control of how a gene is expressed (transcribed, translated, and post-translational modification) into is specific protein product. Proteins are the workhorses of the metabolic pathways where it operates. By regulating genes, the organism is able to:

- Increase **energy efficiency** by only turning on genes only when their protein products are needed, and turning them off when not needed.
- **Respond to the environment** based receiving external signals and activating processes that cause an appropriate response. For example, plants respond to light by activating processes involving the plant hormone auxin that cause phototrophic growth toward the light.
- Orchestrate differential cell expression of gene that leads to **cell specialization** (multicellular organisms only). For example, kidney cells express a different combination of proteins than red blood cells, allowing each to perform its specific set of tasks.
- Develop **unique phenotype** based on the expressed genes and alleles, and the influence of environment factors on gene expression. This explains the phenotypic variation observed between identical twins, and interspecies variation in phenotypes for identical or similar genes

As shown in the figure below, gene regulation can take place at any of the following points along the pathway of information flow from the DNA to mRNA to protein:

1. **Chromosomal level** (eukaryotes only) regulation involves histone methylation and acetylation processes that cause the chromatin to become relaxed. This allows RNA polymerase and transcription factors to access the transcriptional unit and turn on (induce) the gene.
2. **Transcriptional level** regulation involves regulatory proteins called repressors and interfering RNA (RNAi) that interact directly with the DNA to repress genes. Repressor and RNAi bind to the DNA so as to prevent RNA polymerase from reading the gene. Other regulatory proteins called activators bind the DNA to enhance (induce) RNA transcription.

3. **mRNA processing** (eukaryotes only) involves exon splicing to remove introns, and addition of the 5' cap and 3' poly A tails that facilitate the export of mRNA to cytoplasm, binding of mRNA to ribosomes, and protection from cytoplasmic lytic enzymes.

4. **mRNA translation** is the actual process of synthesizing the polypeptide by the ribosomes. The stability of the mRNA is critical at this point because of processes that can degrade an unstable mRNA.

5. **Post-translational modification** involves outing on the final touches on the translated polypeptide so that it becomes functional. This regulatory level require additional peptide cleavage, and it's folding into the native 3D shape. Protein folding may involved special protein structures called chaperonin, which are protein-folding compartments that isolate the polypeptide from the rest of the cytoplasm.

NOTES

RNA STABILITY
The mRNA stability is influenced by polyadenylation (addition of poly A tail) during mRNA processing which protects the mRNA from cytoplasmic degradation by endonuclease enzymes.

MicroRNA (miRNA) removes poly A tails from mRNA after it has been translated. This reduces the mRNA stability and tags it for endonuclease destruction.

Small interfering RNA (siRNA) works in a similar manner to miRNA. However, siRNA marks foreign RNA for destruction by exonucleases.

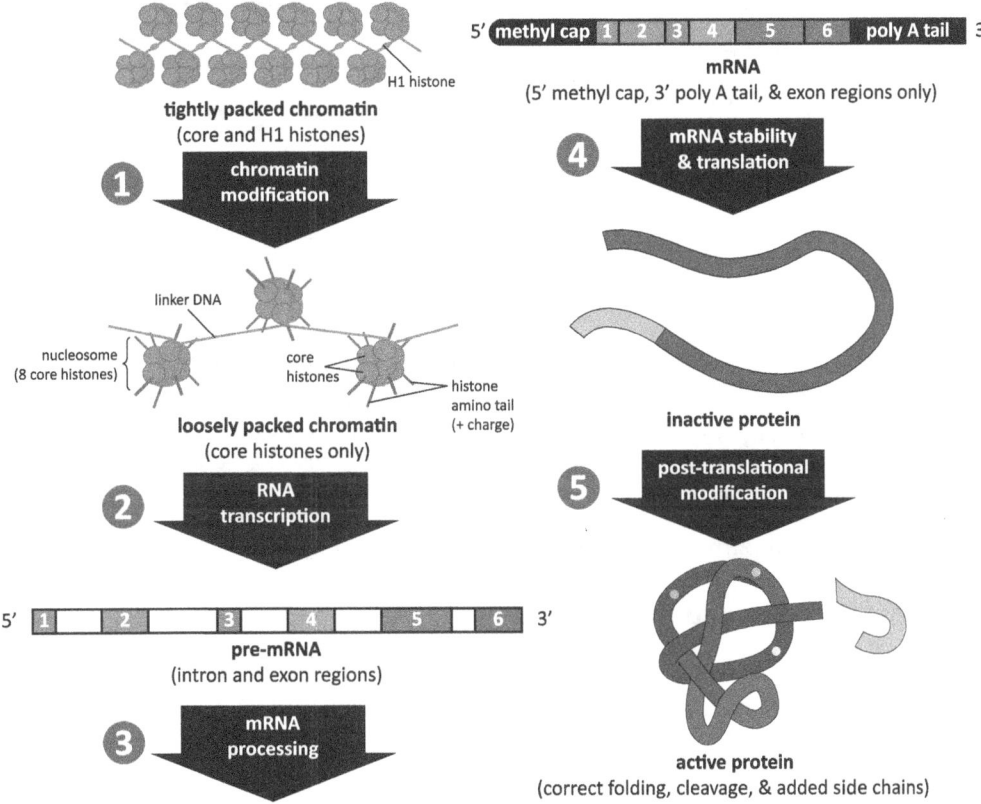

tightly packed chromatin
(core and H1 histones)

1 chromatin modification

linker DNA

nucleosome (8 core histones)

core histones

histone amino tail (+ charge)

loosely packed chromatin
(core histones only)

2 RNA transcription

5' | 1 | 2 | 3 | 4 | 5 | 6 | 3'

pre-mRNA
(intron and exon regions)

3 mRNA processing

5' methyl cap | 1 | 2 | 3 | 4 | 5 | 6 | poly A tail | 3'

mRNA
(5' methyl cap, 3' poly A tail, & exon regions only)

4 mRNA stability & translation

inactive protein

5 post-translational modification

active protein
(correct folding, cleavage, & added side chains)

THE PROKARYOTIC OPERON

The mechanism of gene regulation in bacteria is better understood than that for Eukaryotic cells. The bacteria cell uses **operons** to regulate its genes. An operon is the part of the DNA that allows for the regulations of a specific gene or set of genes. A generalized structure of the operon is illustrated in the following figure.

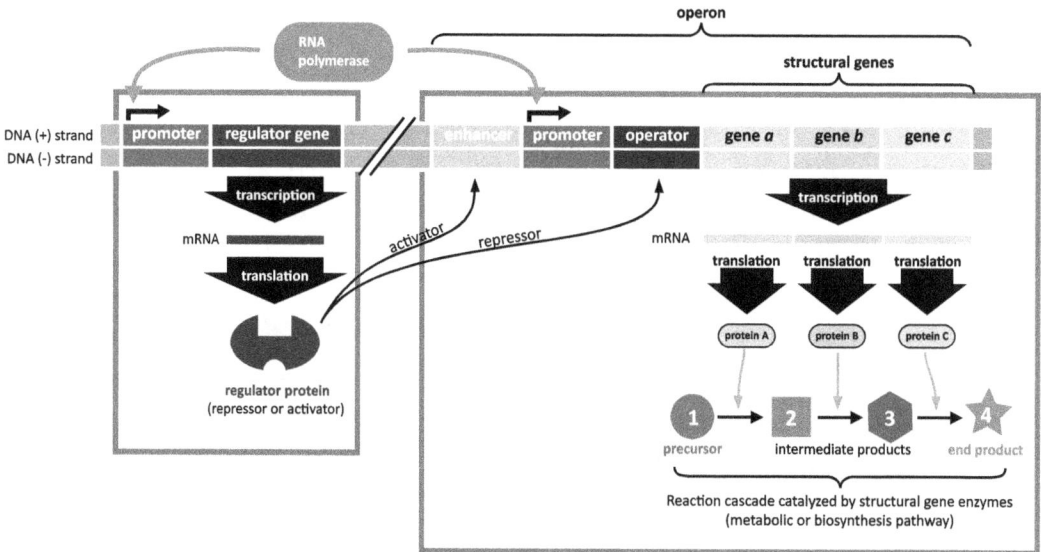

All operons include one or more **structural genes** (a, b, c), an **operator**, and a **promoter**. The **enhancer** is not part of the operon. It is not always involved in regulating the operon and, if so, is usually located far away from the operon. RNA polymerase binds to the promoter and transcribes the structural genes to mRNA. The mRNA is later translated by ribosomes to the structural proteins that are the workhorses for the specific biochemical pathway that is linked to the operon.

The regulatory genes and their proteins

Each operon is attached to a regulatory gene that may be located anywhere within the genome – regulator genes do not need to be near the operon that they regulate. The regulatory gene codes for a regulator protein that is either a repressor or an activator.

1. **Repressor** proteins bind to the operator and block RNA polymerase from transcribing the structural genes. Because they prevent transcription, repressors are negative regulators.
2. **Activator** proteins bind to the enhancer and increase RNA polymerase's affinity for the promoter. Activators positively regulate the structural genes by accelerating their transcription. In the absence of the activator, RNA polymerase can still bind the promoter, but with very low affinity, resulting in basal transcriptional activity.

The regulatory genes have their own promoters, but they lack operators and enhancers – they do not have their own regulators. The absence of an operator or an enhancer means that regulatory genes are continuously expressed (constitutive) at low levels.

Inducible operons

An inducible operon is OFF by default, but can be turned ON (induced) when the gene product is needed. There are 2 types of inducible operons:

- **Positive inducible operons** are regulated by activators that are by default inactive and are unable to bind the enhancer. A substance called the **inducer** allosterically interacts with the activator and cause it to undergo a conformational change (shape shift) to its active state. Inducers are usually substances that are substrates of the structural proteins. When in its active state, the activator binds the enhancer with a greater affinity; this interaction has a positive affect on transcription by increasing the binding of RNA polymerase to the promoter and acceleration the structural gene transcription.

- **Negative inducible operons** are regulated by repressor that are by default active and bind the operator to block the path of RNA polymerase, preventing structural gene transcription. The inducer allosterically causes conformational change to the inactive state so that the repressor loses its affinity for the operator, allowing RNA polymerase to access and transcribe the structure gene.

Repressible operons

A **repressible operon** is ON by default, but can be turned OFF (repressed) when the gene product is no longer needed. There are 2 types of repressible operons:

- **Positive repressible operons** are regulated by activators that are by default active and can bind the enhancers to accelerate transcription. Another substance called the coactivator (typically the product of a linked biochemical pathway) can allosterically deactivate the activator to turn OFF transcription. Note that the inactivation of the activator does not completely block transcription, but it does dramatically reduce RNA polymerases affinity for the promoter.

- **Negative repressible operons** are regulated by repressors that are by default inactive and unable to bind the operator to turn OFF transcription. Another substance called a **corepressor** (typically the product of a linked biochemical pathway) can allosterically activate the repressor so that it gains an affinity for the operator to repress transcription. When sufficient products are made, they tell the repressor to block their own production (negative feedback).

Operon	+ or -	ON or OFF	Native regulator	Allosteric actor
Inducible	positive	OFF	activator (inactive)	inducer (protein substrate)
	negative	OFF	repressor (active)	inducer (protein substrate)
Repressible	positive	ON	activator (active)	coactivator = product (blocks activator)
	negative	ON	repressor (inactive)	corepressor = product (activates repressor)

NOTES

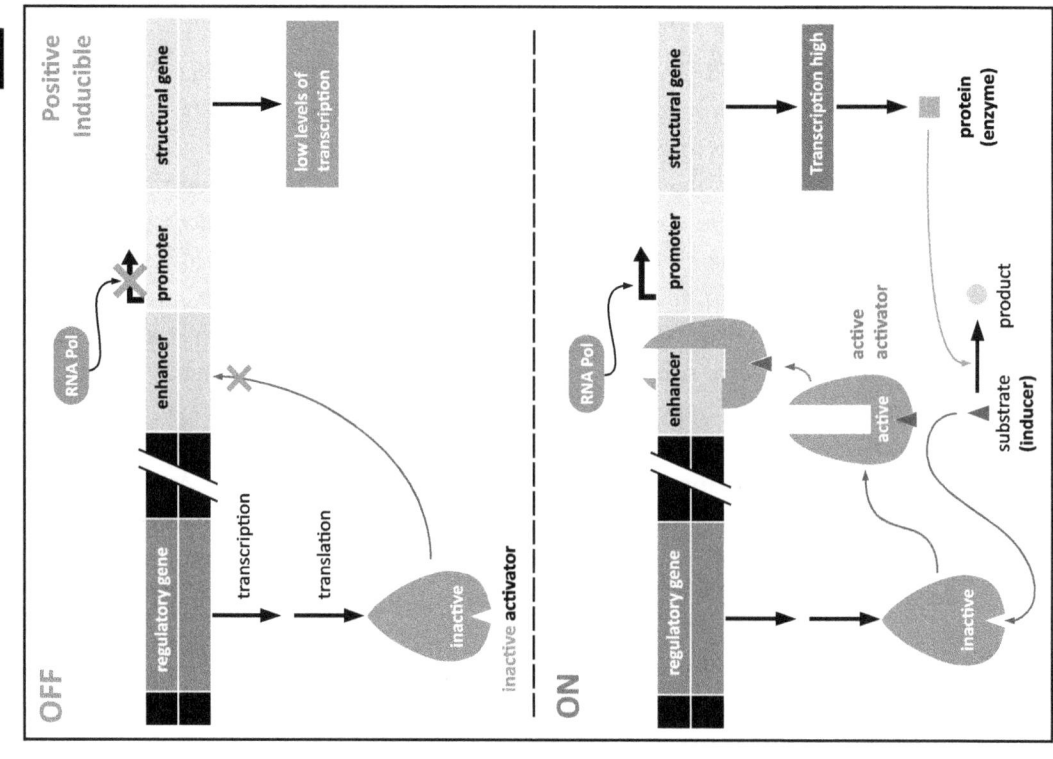

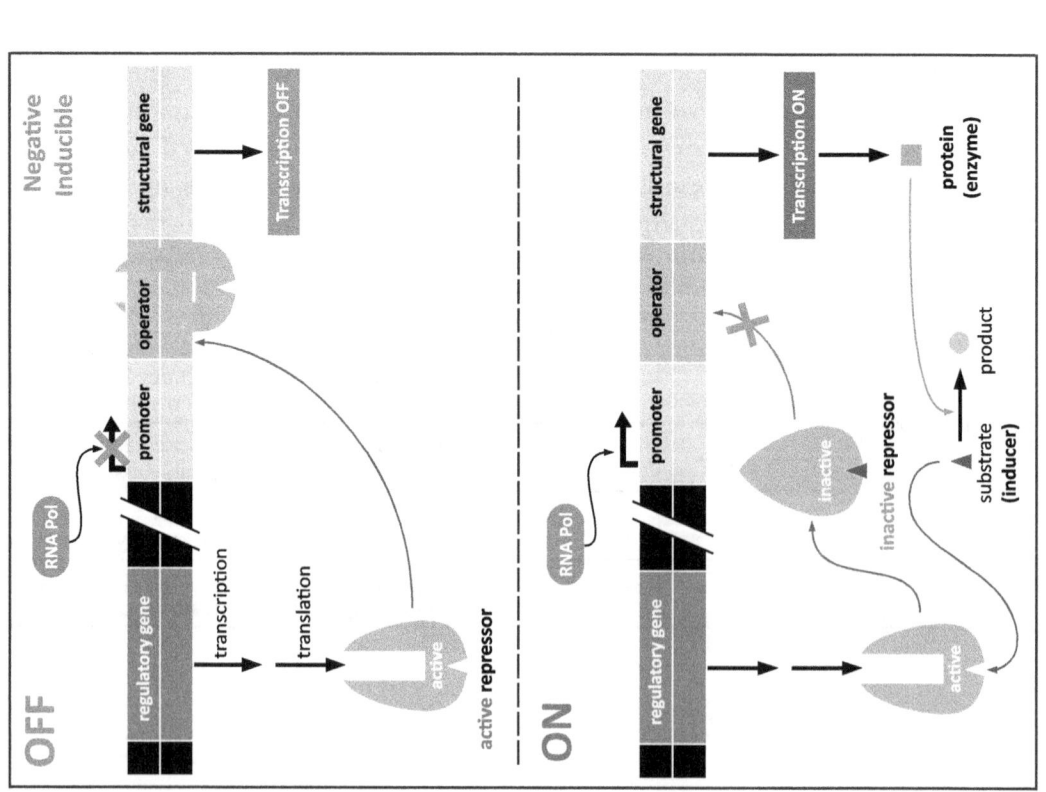

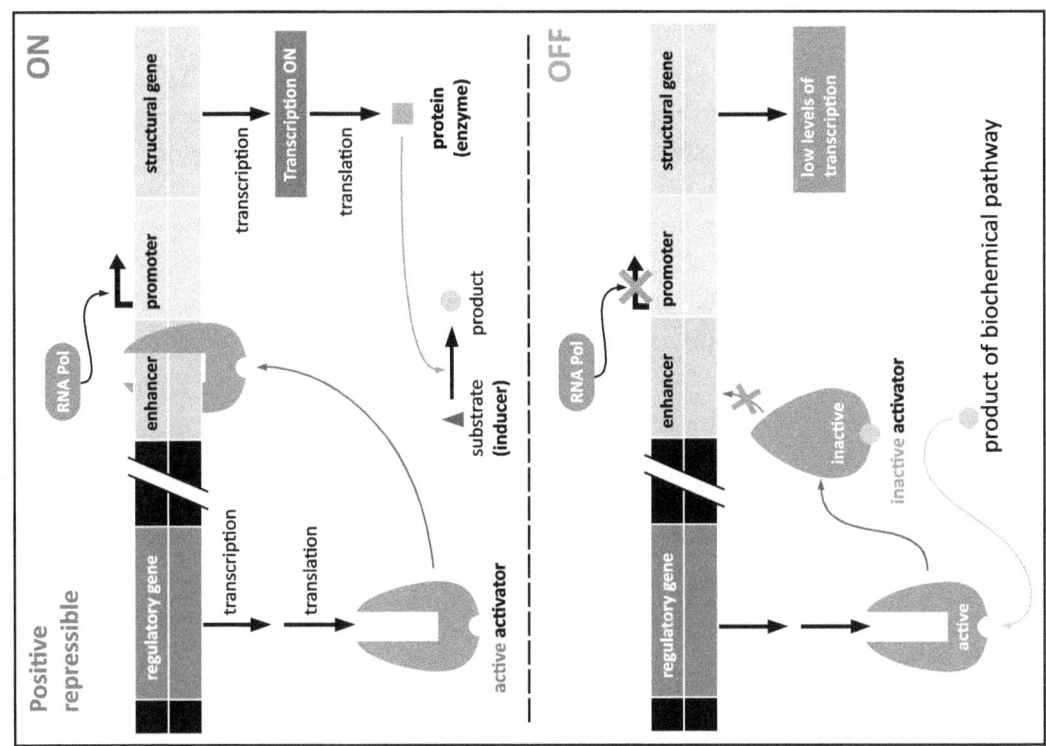

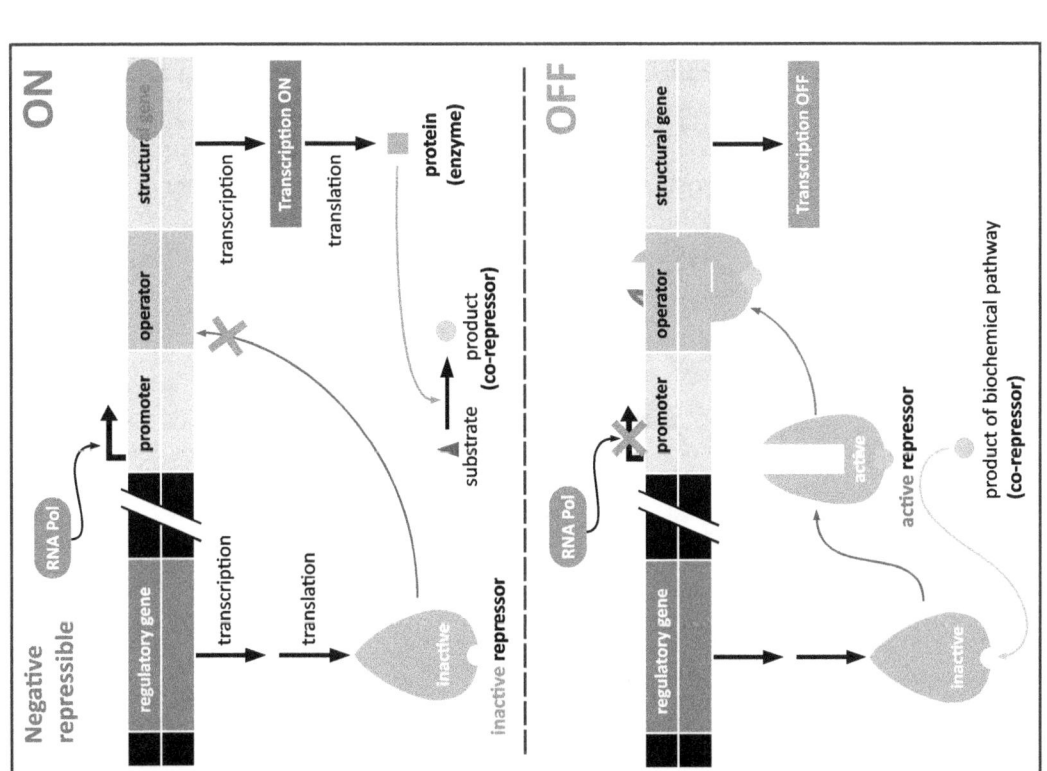

The lac operon (a negative inducible operon)

In bacteria, the lac operon has 3 structural genes (Lac Z, Lac Y, and Lac A) coding for the 3 proteins (**permease, β-galatosidase**, transacetylase) that are involved in the metabolism of the milk sugar lactose in bacteria. The role of transacetylase is not yet fully understood. Permease is a membrane channel protein that transports lactose into the cell (recall sugars do not simply diffuse across lipid bilayers). β-galatosidase is the enzyme that breaks down lactose once it has been transported into the cytoplasm. Important to note that it catalyzes 3 separate reactions: a) the conversion of lactose to the inducer called allolactose; b) conversion of lactose to galactose and glucose; and c) conversion of allolactose to galactose and glucose.

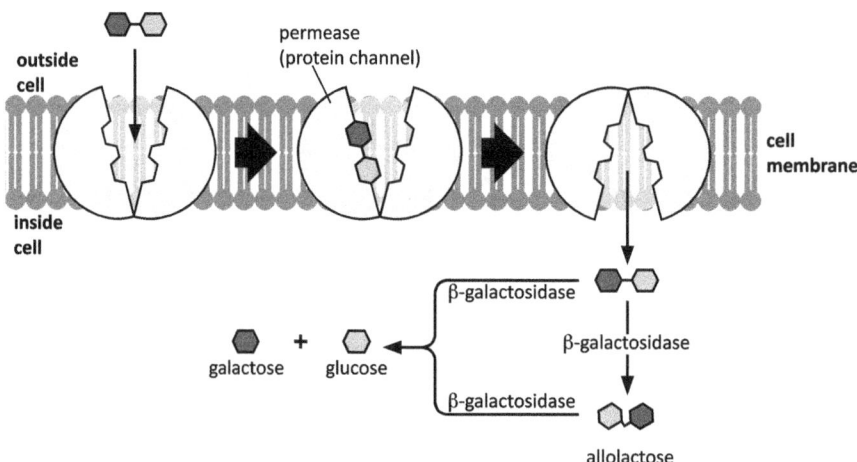

As illustrated in the figure, allolactose is the inducer that allosterically inactivates the repressor (Lac I), thereby inducing the lac operon. This only happens if lactose is present. Initially, the cell membrane has low density of permease to take in a few lactose molecules. Likewise, in the cytoplasm, there is low concentration of β-galatosidase to synthesize allolactose. The production of allolactose induces the production of more permease and β-galatosidase, thereby amplifying its own production (positive feedback).

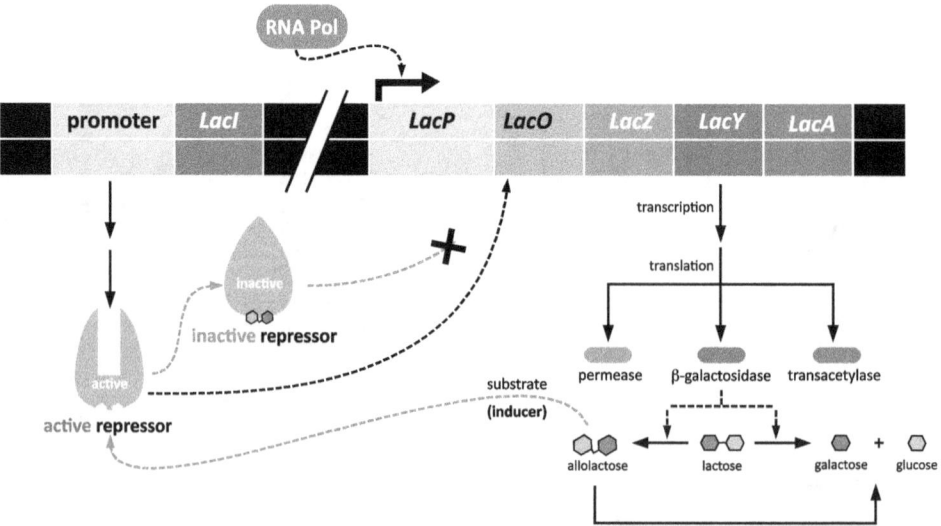

GENE REGULATION IN EUKARYOTES

Gene regulation in eukaryotes is more complex than in prokaryotes. The difference is due to the complexity of the eukaryotic genome that includes:

1. Multiple DNA strands, each containing a unique set of genes. Some combinations of cooperative genes are not linked on a single operon, and may even be located on different chromosomes. **Multiple gene coordination** require the simultaneous regulation of cooperative genes in response to the same signal.
2. The tight **chromosome-packing** of DNA strands by histones to form the chromatin structures can negatively regulate transcription by blocking the RNA polymerase and other transcription factors accessing the operon.
3. In multicellular organisms, all of the cells are genetic clones that undergo **cell specialization** to acquire their tissue-specific functions. This involves various mechanisms that lead to the activation of different combinations of genes.
4. The requirement of an orderly assembly of the organism from the zygote to embryo (**embryonic development**), and the maturation from the embryo to a reproductively viable adult means that embryonic development must have a special set of biochemical mechanisms and genes to manage the stages of embryonic development.

Multiple genes coordination

Unlike prokaryotes where coordinated genes are grouped together in a single operon, in eukaryotes, many genes coding for coordinated proteins are located on different chromosomes. To resolve this problem, eukaryotes use short consensus DNA sequences in the promoter region called **response elements** that work similarly to enhancers. A given response element can simultaneously bind activators and other transcription factors to a set of cooperative genes that share the same response elements, thereby inducing transcription of their mRNA at the same time. One gene may be regulated by several response elements to respond to several different stimuli, or multiple genes can be regulated by the same response element to respond to the same stimulus.

Chromosome-packing

The tight packing of DNA by histones to form chromatin structures can block transcription by preventing transcription factors, including RNA polymerase, from binding to the DNA. Histone methylation – the addition of methyl groups (CH_3-) to specific histone amino acids – reduces the affinity of histones for the DNA, loosening up the chromatin. While methylation positively regulates gene transcription, dimethylation acts to inhibit transcription. A similar affect occurs when histones are acetylated by the addition of acetyl group (CH_3CO-).

Cell specialization & development

Cell specialization and development relies on **epigenetics**, which is a long-term heritable DNA modification that occurs through methylation rather than changes in nucleotide sequence. Simply put, epigenetic modification of DNA can result in non-Mendelian gene expression such that if a dominant allele of a gene becomes methylated, it is **silenced**, resulting in the expression of the recessive allele. Specific epigenetic processes include:

1. **Paramutations**, where one allele causes methylation on the other alleles. This process can cause non-Mendelian expression where a paramutagenic recessive allele becomes expressed over the silenced mutagenic dominant allele.

2. **Genomic imprinting** is a species-specific process of gene silencing through DNA methylation. Imprinting within a given species is gamete specific, meaning that all egg cells share a methylation pattern, and likewise for sperm cells. For example, if gene A is imprinted in sperm cells, all sperm cells for that species should have the same gene silenced. Consequently, individuals for that species will only express the maternal allele for that gene. Similar to paramutation, this process can lead to non-mendalian expression of a gene – if the dominant allele is silenced, the recessive allele will be expressed.

3. **Imprint reprogramming**: Since each individual receives a normal and a silenced copy of the imprinted genes, the imprinting process needs to be reset during gamete production so that the gamete specific pattern is passed to progeny.

4. **Totipotent reprogramming**: With the exception of these imprinted genes, and a few other environmental factor induced methylated genes, after fertilization, the genome is reprogrammed with a methylation pattern that is necessary for totipotency (the ability of a cell to differentiate into all other cell types). Although not fully understood, most imprinted genes seems to be involved in embryonic growth and development, including the development of the mammalian placenta and gene associated with suckling.

5. **X-chromosome inactivation**: In females, only one X chromosome is activated - the other is permanently turned off. There is some evidence that X-chromosome inactivation is a result of genomic imprinting with the paternal X-chromosome being silenced by methylation.

Epigenetic is not yet fully understood, but it is believed that different cell types express different sets of genes due to differences in DNA methylation patterns that are passed down to daughter cells of the same type. Cell differentiation involves gene regulation for tissue specific proteins production. DNA methylation (different from histone methylation) is the major mechanism that is used silence genes for tissue specific gene regulation. This silencing is mitotically heritable and can result in the production cells that share similar functions. This is important for tissue and organ development, and plays a major role in defining the distinguishing cellular features of different organs and tissue types.

The maternal effect and environmental influence on phenotype

The maternal effect is the non-Mendalian expression of genes that is caused by the maternal environment and genotype. It occurs when the mother secretes mRNA and proteins into the egg, causing the organism to display a phenotype that is more similar to its mother rather than to it's inherited set of genes and alleles. The maternal effect may be adaptive in that it prepares the organism to more effectively deal with environmental stresses.

Apoptosis in embryonic development

The development of a multicellular organism begins with a single zygotic cell (fertilized egg), that gives rise to the diverse array of cell types of the whole organism. The first cell is epigenetically reprogrammed to be totipotent, meaning that it has the potential of differentiating into all of the other types of cells. The process of cell differentiation and organism development is highly organized in a temporally (time sensitive) and tissue coordinated manner to ensure that the correct cell type is produced at each stage of embryonic development. Some cell types are only needed for a brief period. Once they have outlived their use, apoptosis is initiated and the cell self-destructs.

Homeotic genes managed the stages of embryonic development

There are lots of factors that are taken into account during embryonic development. For example, there must be cellular processed that determine where in the body plan to develop the various organs, and how many fingers, legs, and eyes to produce, including the relative placement of these body parts to each other. All of the information needed to instruct the normal development process is coded in the DNA, specifically in a series of body plan genes called the **homeotic genes** that oversee the entire process. One set of homeotic genes includes the **hox genes** that are important for body segmentation. Think of the protein products of hox genes as factory bosses guide the assembly of the organism during embryonic development.

Structure and evolution of hox genes

All hox genes contain a 180-nucleotide consensus sequence called a homeobox that codes for a conserved homeodomain region of their specific protein. The homeodomain region is believed to help the protein bind to the DNA where it acts as a transcriptional regulator for developmental genes. As illustrated in the following figure, homeotic genes are clustered together on the chromosomes, with their order on the chromosome determining where along the body they act.

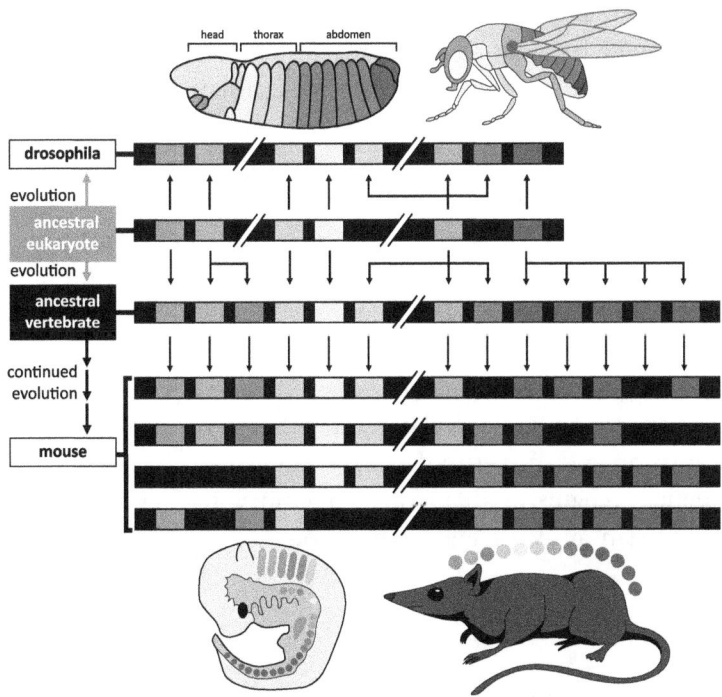

Hox genes have been highly conserved throughout evolution and may have originally evolved even before the divergence of the major eukaryotic lineages. Evidence for this comoes from genetic homology in the hox genes of insects, mammals, reptiles, birds, plants, and most other eukaryotic taxa. Hox genes are critical to the organism development because they regulate many other developmental genes. As such, small changes (mutations) in these genes can lead to drastic alterations in the physical structure of an organism. For example, mutation experiment in the hox genes of drosophila has produced flies with extra limbs, and wings in place of antennae.

These gene modification experiments suggest that hox genes were critical in a number of speciation events that lead to the divergence of the major eukaryotic lineages. Small modification, including hox gene duplication on different chromosomes, and the emergence of new hox genes via mutations in existing ones, may have altered the body plans of preexisting species to give rise to new species. As illustrated above, the evolution of mouse and drosophila has resulted in dramatic phenotypic divergence with only minor changes in the hox genes. Where in drosophila, the hox genes have been preserved on a single chromosome, in mouse these genes have been duplicated within and between chromosomes. In both species, new forms have also emerged by mutations in those likely present in the original common eukaryotic ancestor.

SUMMARY OF KEY CONCEPT

- Gene regulation is a process that cells use to control when and how a gene is expressed.
- Genes are regulated at several points of information flow from the DNA to phenotype.
- Gene regulation benefits the organisms by increasing energy efficiency in organisms through gene induction when the protein products are needed and gene repression when not needed, allowing cell specialization that leads to intercellular and interspecies variation, and allowing organisms to respond to their environment.
- Agents of gene regulation include regulatory DNA sequences (promoters, enhancers, consensus sequences, response elements), regulatory genes (repressor, inducer, activators, regulatory RNAi), and DNA & chromatin modification mechanisms (methylation or acetylation)
- Gene regulation in bacteria and viruses involve positive/negative inducible and repressible operons, and continuously expressed regulatory genes.
- Gene regulation in eukaryotes involves transcription factors, activators, repressors, response elements, and chromatin and DNA modification, and hox genes.
- Eukaryotes typically do not have operons but have evolved several mechanisms to ensure proper coordinated regulation of cooperative genes, and cell differentiation and organism development.
- Epigenetic mechanisms are involved in embryonic development and can cause non-Mendelian mode of gene expression.

CHECK YOU UNDERSTANDING

1. The expression of genes is regulated through

   (A) methylation and acetylation processes at the chromatin level.

   (B) RNA interference mechanisms.

   (C) alternative cleavage of the poly-A tail

   (D) All of the above.

2. Which of the following statements is incorrect?

   (A) Eukaryotic gene regulation typically uses operons to simultaneously activate a set of functionally related genes.

   (B) Gene A and gene B are located on separate chromosomes, but the same metabolic pathway requires their protein products. Gene A and B likely share the same promoter response element.

   (C) Epigenetic modification by methylation of the DNA is not heritable because it does not alter the DNA sequence.

   (D) Apoptosis is a mechanism that cells use strictly for preventing damaged cells from reproducing and possibly causing cancer.

3. A gene that is always expressed likely

   (A) lacks a promoter region.

   (B) has an enhancer that is permanently attached to an activator.

   (C) lacks a repressor.

   (D) codes for a regulator protein or RNAi.

4. The trp operon in bacteria contains 5 structural gene (trp A, B, C, D, and E) of the tryptophan synthase enzyme that is responsible for the biosynthesis of tryptophan. The trp operon's activity is inversely correlated with tryptophan concentration in the cell. Which of the following is likely true about this operon?

   (A) The trp operon is regulated by an activator.

   (B) Tryptophan is an inducer of the operon.

   (C) The trp operon contains an operator downstream from the promoter.

   (D) The trp operon is under a positive feedback control.

5. The regulator of the trp operon is a multi-subunit protein that normally oscillates between an active and inactive conformation. Which of the following can be assumed about this protein?

   (A) A corepressor allosterically alters it's activity.

   (B) High concentrations of tryptophan stabilized its active conformation.

   (C) Low concentration of tryptophan causes the regulator to bind the enhancer sequence.

   (D) All of the above is incorrect.

6. Several mutant bacteria strains have been identified that have alteration in DNA regions associated with the Lac operon. They include: Lac I⁻ mutant that makes a repressor protein that is unable to bind the operator; Lac Oᶜ mutant that contains an operator that is unable to bind the repressor; Lac Iˢ mutant that produces a repressor that remains permanently bound to the operator; and the Lac Z⁻ that produces a defective β-galactosidase enzyme that is unable to metabolize lactose. Which of the following scenario correctly predicts to expected result? Not that the normal operon is Lac I⁺ O⁺ Z⁺ and F' signifies the presence of a transforming plasmid in the bacteria cell.

   (A) In the absence of lactose, the lac operon in the Lac I⁻ O⁺ Z⁺ mutant is repressed.

   (B) In the presence of lactose, glucose is produced by the Lac Iˢ Oᶜ Z-, F' O⁺ Z⁺ transformed mutant.

   (C) In the presence of lactose, the Lac Iᶜ O⁺ Z⁺, F' I⁺ Z⁻ mutant constitutively expresses β-galactosidase, but produces no glucose.

   (D) All are correct predictions.

7. Which of the following statements about epigenetic is correct?

   (A) Genomic imprinting is species and sex specific.

   (B) Imprint reprogramming occurs prior to fertilization.

   (C) Totipotent programming occurs during embryonic induction.

   (D) All of the above are correct statements.

ANSWER AND SOLUTIONS

1. (D) is correct.

2. (B) is correct. Note that (D) is wrong because apoptosis is also used during morphogenesis in embryonic development.

3. (D) is correct. (A) is wrong because the promoter region is required for gene expression. Without a promoter, RNA polymerase cannot bind the DNA. (B) only happens rarely when the enhancer undergo a mutation that increases its affinity for the activator. (C) is wrong because operons that lack repressors can be positively controlled by activators.

4. (C) is correct. All of the other options are wrong because it is assumed that the operon is negatively controlled since as the product (tryptophan) levels rise, repression increases. Tryptophan is therefore likely the corepressor that activates the repressor when enough trptophan has been made, resulting in a negative feedback control of its own production.

5. (A) is correct.

6. (C) is correct because the plasmid Z⁻ is constitutively (or always) expresses since it lacks an operator for the bacteria Is to bind. However, the plasmid Z⁻ codes for a non-functional β-galactosidase that is unable to metabolize lactose. The bacterium's Z⁺ is permanently repressed because is has a O⁺ that is attached to the mutant Is repressor. (A) is incorrect because I- cannot bind the operator to repress. (B) is incorrect because Is is permanently attached to F'O⁺ to repress F'Z⁺. Meanwhile Oᶜ cannot bind Is, allowing Z⁻ to be constitutively expressed. However Z⁻ codes for a nonfunctional β-galactosidase that produces no glucose from lactose metabolism.

7. (D) is correct.

22 Cell communication

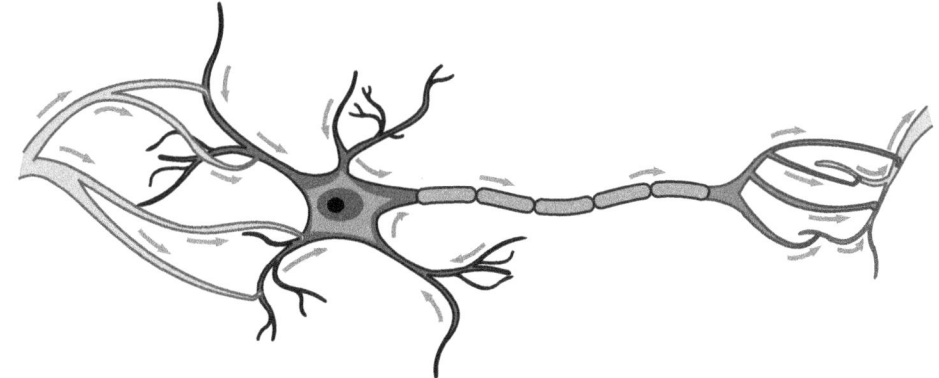

ALL STUDENTS MUST BE ABLE TO ANSWER THESE QUESTIONS

1. Why is cell communication important to prokaryotes? Eukaryotes
2. What are the three generalized steps of signal transduction?
3. How is conformational change in proteins related to their activity?
4. What are different mechanisms for activating and inhibiting multi-subunit proteins?
5. How are positive an negative feedback used to regulate enzyme activity?
6. What are six ways in which living systems use signal transduction to communicate? Reflect on the provided examples.
7. What is quorum sensing in bacteria, and how does it work? Reflect on autoinducers, membrane receptors, population density, and threshold concentration.
8. How do prokaryotes use chemotaxis to gather information and coordinate movement?
9. How is direct cell-to-cell contact and chemical messengers used in the immune response?
10. What is the difference between water and lipid soluble hormones?
11. How does the neurohormone and neuroendocrine pathways compare?
12. What is the difference between posterior (ADH, oxytocin) and the hypothalamic tropic hormones?
13. How do inhibitory and stimulatory hypothalamic tropic hormones act on the anterior pituitary?

ALL STUDENTS MUST BE ABLE TO COMPLETE THE FOLLOWING TASKS

1. Use examples to describe each of the 3 steps in the cell signaling pathways involving receptor tyrosine kinase and G protein-linked receptors, and explain the specific mechanism used to activate second messengers in each of these pathway, including the kinase phosphorylation cascade.
2. Describe and explain the 7 steps of the nerve impulse transduction, including the mechanism of depolarization, repolarization, hyperpolarization, and restoration of the resting potential.
3. Use the oxytocin response to explain the neurohormone pathway.
4. Explain the affects of ADH and oxytocin on their respective effector cells..

Big Idea 3B2, 3D, 3E2a-c

LIGAND RECEPTORS
Membrane-bound ligand receptors are transmembrane proteins. They typically have 2 domains: a) the first domain is the extracellular ligand-binding region that is exposed to the outside of the cell; and the second is cytoplasmic enzymatic region that is exposed to the inside. When the ligand binds the extracellular domain, it triggers a conformational change in the enzymatic domain from an inactive to an active state.

Biological systems rely on cells to coordinate activities in response to both internal and external conditions. Prokaryotes use simple signaling pathways to sense their external environment and respond to it in a manner that improves their fitness. To further enhance survival and reproductive success, prokaryote can reach higher levels of biological organization by sharing information with neighboring cells. In multicellular organisms, cell communication facilitates intercellular coordination of processes that are critical to embryonic development and organismal maturation. Without this ability to gather and share information, living organisms are unable to respond appropriately to conditions that threaten their internal order.

STAGES OF CELL COMMUNICATION

In both prokaryotes and multicellular eukaryotes, cell communication follows a basic 3-step process that first evolved in our common ancestor and remains highly conserved across biological domains. The 3 general stages are:

- Signal reception
- Signal transduction
- Cell response

Signal reception occurs when a mechanical or chemical signal is received by a membrane bound or cytoplasmic receptor protein. Once received, the signal is passed along through a process called **signal transduction**, which continues until the targeted **cell response** is initiated.

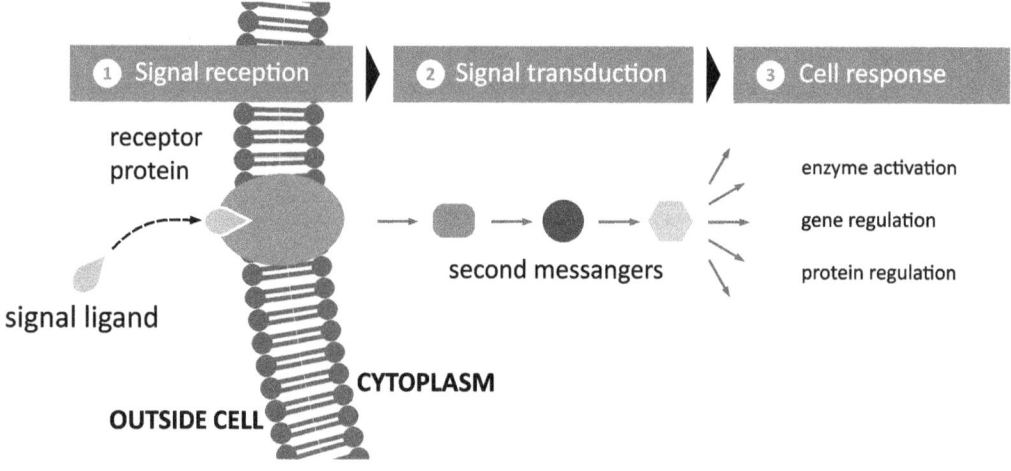

ROLE OF ENZYMES IN CELL COMMUNICATION

Proteins are the molecular workhorses of the cell, often functioning as cell messengers and effectors of cell response. In many instances, the functional activities of proteins can switch between active and inactive states when changes in their 3-D structures occur. Membrane bound signal receptors are usually such shape shifting proteins that undergo a conformation change from an inactive to active state once contact is made with the signaling ligand. In other instances, enzymatic proteins act as second messengers that are activated when they receive the signal. Once activated, enzymatic protein can induce the active states of other second messengers through a cascading series of reactions that eventually leads to the targeted cell response.

Mechanism of enzyme action

Enzymatic proteins, including the shape shifting membrane bound ligand receptor, all share a similar mechanism of action that can be summarized in the following three steps:

- A specific substrate(s) binds to the active site of the enzyme.
- This establishes the enzyme-substrate complex that lowers the activation energy of the enzyme-catalyzed reaction and allows the reaction to occur faster. The product is formed from the substrate(s) at the active site.
- Once formed, the product is has no affinity for the active site and is released

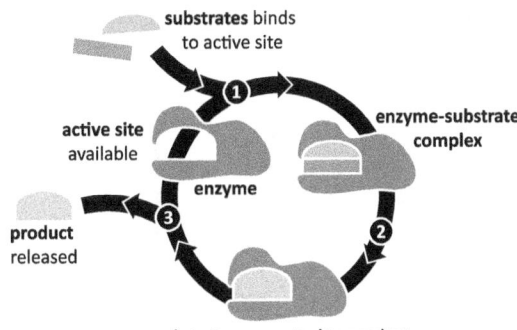

Substrate specificity

Enzymes display substrate specificity, meaning that they bind to a specific type or class of molecules. The active site contains the substrate binding site and is the part of the enzyme molecule where the biochemical reaction to convert the substrate to product occurs. The active site works to lower the activation energy of the reaction so that the reaction occurs quicker.

Activation energy

Activation energy is the energy need to start a reaction. The active site of enzymes lowers the activation energy to accelerate the reaction. This works for both energy absorbing endergonic and energy releasing exergonic reaction.

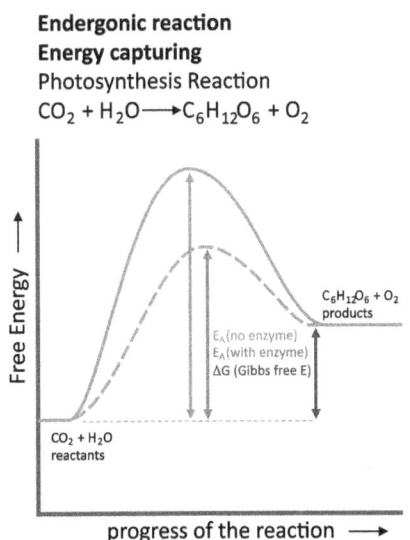

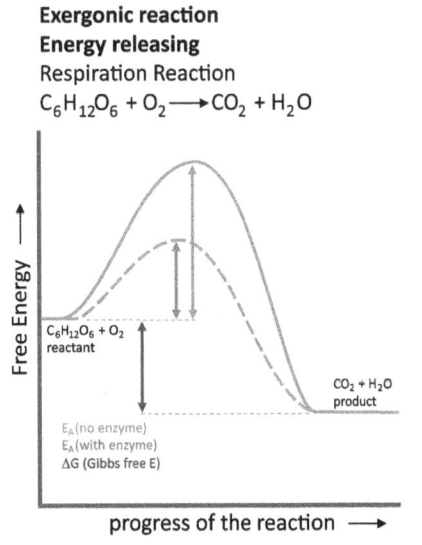

ENZYME ACTION
Note that the mechanism of enzyme action is also part of the key concepts covered in Big Idea 4. It is included in this chapter because a clear understanding of enzymes is necessary to fully appreciate cell communication.

COFACTORS & COENZYMES
Some enzyme require inorganic substance called cofactors in order to become active. For example, iron is a cofactor for hemoglobin - without iron, hemoglobin is unable to bind oxygen.

Organic cofactors are called coenzymes.

Positive and negative regulation of enzymes

Enzymes are regulated by positive and negative control mechanisms. Enzymes are positively regulated by inorganic substances called **cofactors** or by organic substance called **coenzymes**. They are **negatively regulated** by competitive inhibitors and non-competitive allosteric inhibitors.

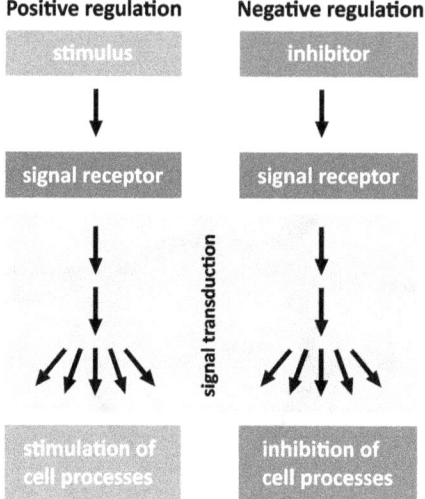

A **competitive inhibitor** has an affinity for the enzyme's active site and competes with the substrate for the enzyme. Interaction of the enzyme with competitive inhibitor does not result in any reaction, but it does prevent the substrate from binding. This enzyme-inhibitor interaction is reversible, depending on the relative concentrations of the substrate to the inhibitor and their relative affinities for the active site. The higher the substrate's relative concentration and affinity, the greater the chance of reversal.

A noncompetitive **allosteric inhibitor** binds to a site other than the active site to cause a conformation shift an inactive state of the enzyme. This is similar to the how the enzymatic region of a membrane-bound ligand receptors is allosterically activated when the ligand binds to the extracellular region.

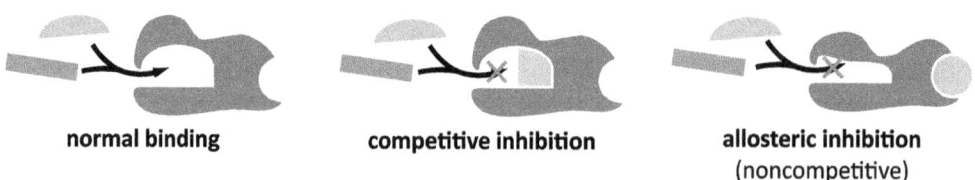

Regulation of multi-subunit enzymes

Some enzymes are made of multiple protein subunits, each folded from a single poly-peptide. The various subunits combine to form the multi-subunits enzyme that oscil-lates between an active and inactive conformation. Regulators (activators or inhibitors) can bind to grooves between the subunits to stabilize the active or inactive states. In some instances, the binding of a substrate to the active site on one subunit can act to stabilize the active state of the others without the need for an activator.

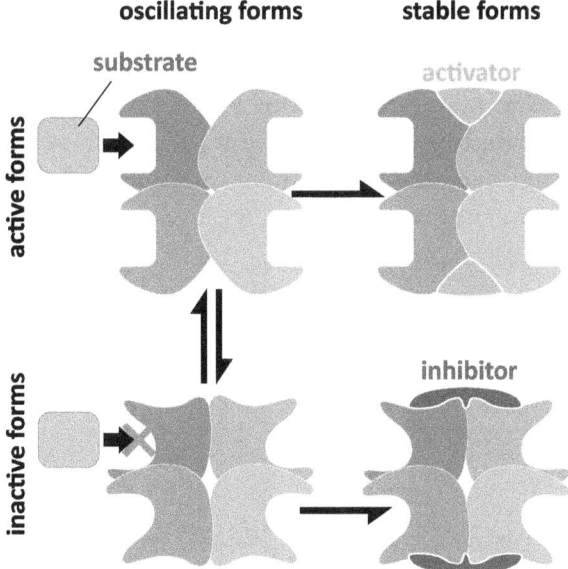

Feedback Regulation

Many biochemical pathways are self-regulated through a feedback mechanism where the end product of the pathway either inhibits or further induces enzymes of the pathway. In **positive feedback** the product returns to an earlier part of the pathway to activate enzymes thereby increasing its own production; and in **negative feedback** it inhibit enzymes to stop its production.

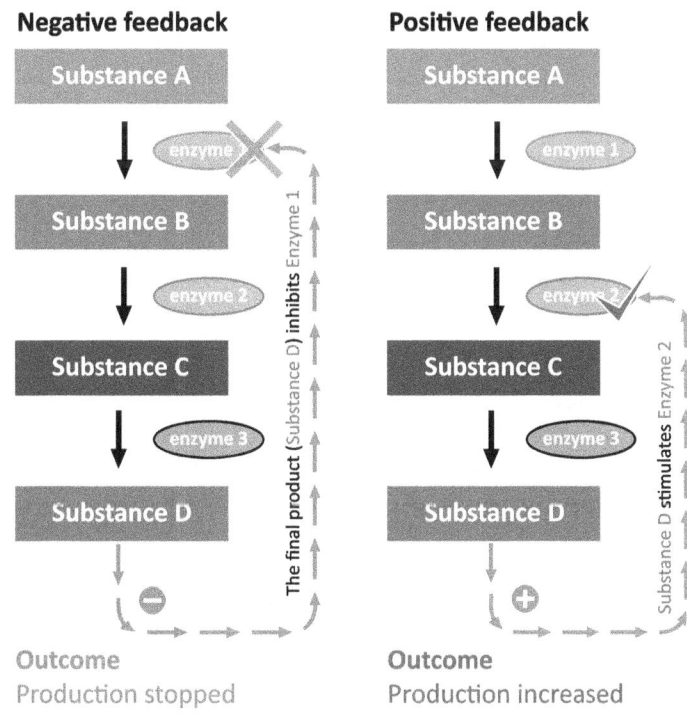

Temperature and pH can affect enzyme activity

Enzyme activity is affected by temperature and pH. Each enzyme has an optimal temperature and pH for maximum activity. Deviation away from the optimal value causes the enzyme activity to progressively decrease until all activity seizes. The graph below on the left shows the relative activity of human amylase as the temperature varies. The graph on the right shows the activity of the digestive enzyme, pepsin.

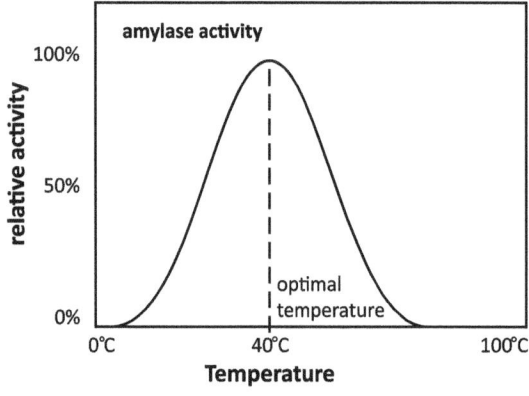
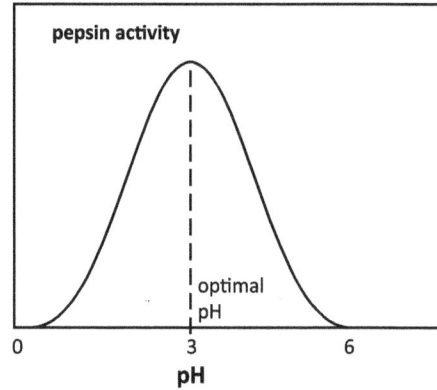

FACTORS AFFECTING REACTION RATES

The rate of an enzyme catalyzed reaction in affected by the concentration of enzyme and substrate. The higher the [enzyme] the more quickly the reaction can reach completion because there are more enzymes available to convert the substrate to product. If the enzyme concentration is kept steady while the substrate concentration increases, the reaction rate will display a sigmoid curve trend - a sharp increase in reaction rate that then levels off. This happens because at low [substrate], there are fewer enzyme-substrate complexes forming. As the substrate level increase, so does the number of enzyme substrate complexes. At high [substrate], all of the enzymes are in an enzyme-substrate complex and the maximum reaction rate is achieved. At this point, any additional substrate does not increase the reaction rate.

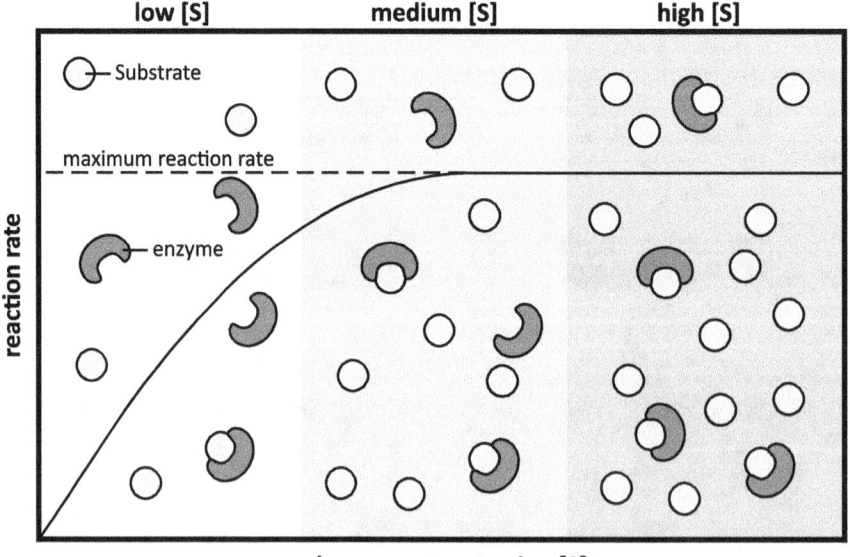

SIGNAL TRANSDUCTION

There are positive and negative regulatory mechanisms that ensure the step-by-step progression of each phase of the signal transduction pathway. As already discussed, positive regulation stimulates cell processes, while negative regulation inhibits these processes. The support mechanism include (see figures on next page):

1. **Reversible activation of second messenger proteins**
   Many enzymatic protein are activated through reversible biochemical reactions or by specific ligand-enzyme interaction.

2. **Ligand induce receptor activation**
   Ligand induced conformation change in receptor proteins can lead to the reversible activation of their cytoplasmic enzymatic domains.

3. **G protein-linked receptors, G proteins, and the kinase cascade**
   When a G protein-linked receptor is activated by growth hormones, they in turn activate adjacent G proteins. G proteins than target cytoplasmic kinase proteins, initiating the kinase phosphorylation cascade that leads to a cellular response. Conformation change between active and inactive forms of second messenger proteins can occur through the kinase phosphorylation cascade where a G protein activates (via phosphorylation) the cytoplasmic kinase protein A. Kinase A phosphorylates kinase B, which phosphorylates kinase C, and the cascade continues until the targeted cellular response is initiated.

4. **Protein phosphorylation** and **diphosphorylation**
   Phosphorylation, the addition of phosphate, leads to protein activation, and diphosphorylation deactivates proteins.

5. **Small molecules and ions as second messengers**
   Besides proteins, some small molecules and ions can act as second messengers. Some examples include:
   - Cyclic AMP (cAMP) is a common second messenger that formed by the adenylyl cyclase.
   - cGMP is a similar second messenger as cAMP
   - Inositol triphosphate (IP_3) stimulates IP_3-gated Ca^{2+} channels to release calcium ions (Ca^{2+}) into the sarcoplasmic reticulum of muscle cells, where they acts as second messengers for muscle contraction.
   - In nerve cells the voltage-gated Ca^{2+} channels open when a nerve impulses arrives, causing the membrane to depolarize. Ca^{2+} ions rush into the nerve cells and act as second messengers, instructing the synaptic vesicles to fuse with the cell membrane to release neurotransmitters into the synaptic cleft. Neurotransmitters send the signal on to the next neuron when they binds to their receptors on ligand-gated Na/K channels.

6. **Minimal stimulus threshold**
   A minimal stimulus threshold concentration (often achieved through signal amplification) is required for cellular response in many signal transduction pathways, most significant being the nerve impulse, where a minimum membrane potential of -50 mV (resting potential is -70 mV) is required before the voltage-gated Na/K channels will open.

The G protein-linked receptor (receptor tyrosine kinase)

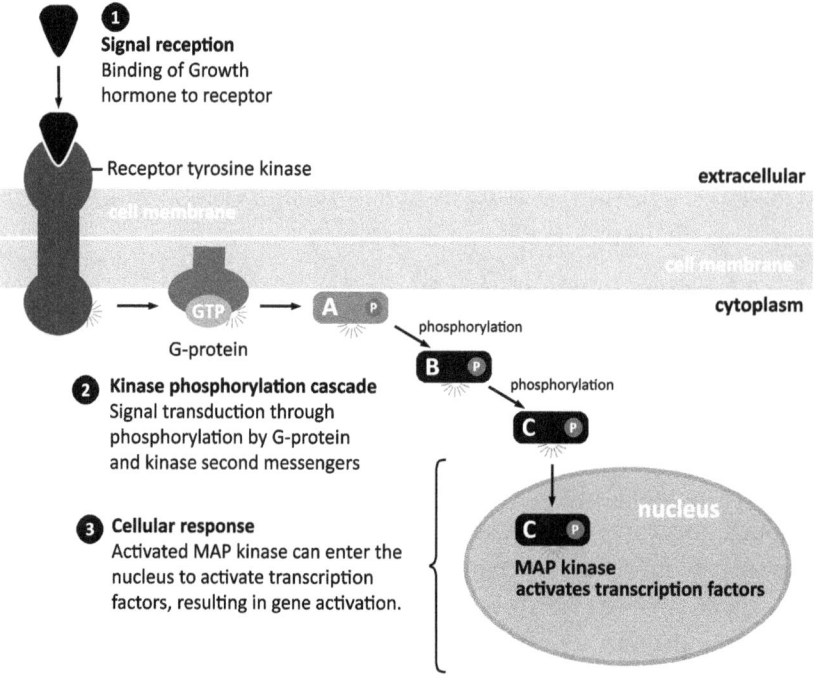

1 Signal reception
Binding of Growth
hormone to receptor

Receptor tyrosine kinase

extracellular

cell membrane

cell membrane

cytoplasm

G-protein

phosphorylation

phosphorylation

2 Kinase phosphorylation cascade
Signal transduction through
phosphorylation by G-protein
and kinase second messengers

3 Cellular response
Activated MAP kinase can enter the
nucleus to activate transcription
factors, resulting in gene activation.

nucleus

**MAP kinase
activates transcription factors**

Small second messengers molecules (IP$_3$)

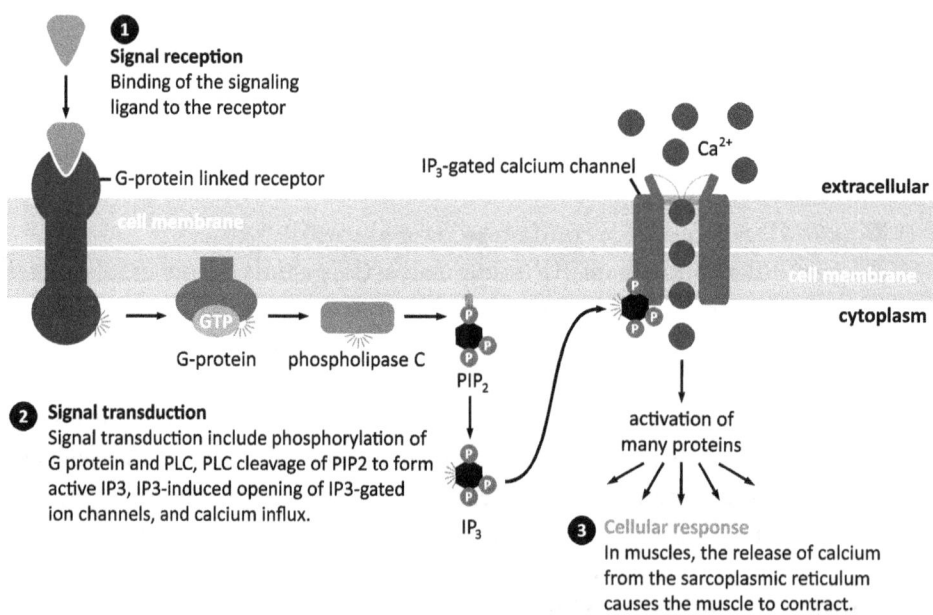

1 Signal reception
Binding of the signaling
ligand to the receptor

G-protein linked receptor

IP$_3$-gated calcium channel

Ca^{2+}

extracellular

cell membrane

cell membrane

cytoplasm

G-protein   phospholipase C

PIP$_2$

2 Signal transduction
Signal transduction include phosphorylation of
G protein and PLC, PLC cleavage of PIP2 to form
active IP3, IP3-induced opening of IP3-gated
ion channels, and calcium influx.

IP$_3$

activation of
many proteins

3 Cellular response
In muscles, the release of calcium
from the sarcoplasmic reticulum
causes the muscle to contract.

296

CELL COMMUNICATION IN PROKARYOTES

Prokaryotes rely on cell communication to launch a proper response to the external environment and to coordinate activities among cells of a growing bacterial colony.

Quorum sensing

Prokaryotic **quorum sensing** is a population-dependent mechanism for cell-to-cell communication that is highly conserved and shared across all prokaryotic taxa. Bacteria cells produce and secrete **autoinducer** molecules that are coded by a **constitutive gene**. As the bacteria population density increases, so does the concentration of autoinducers in the local environment. The autoinducers bind to membrane receptors located on the bacteria membrane. When a threshold concentration is detected, quorum-dependent genes are activated, resulting in the expression of group-behavior phenotype.

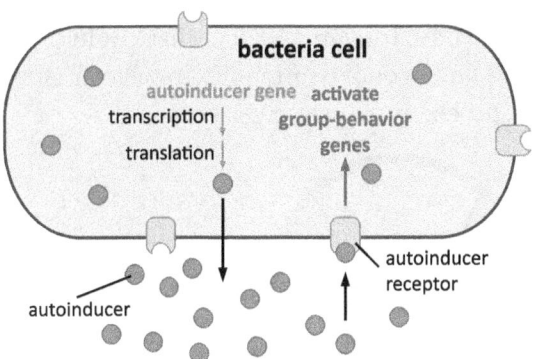
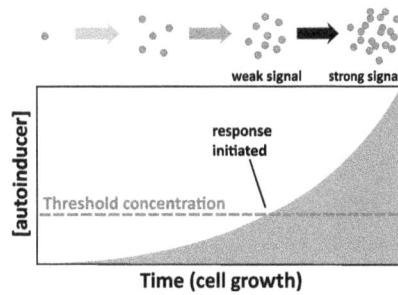

Other examples of prokaryotic communication

- Bacteria cells can detect pheromones that are released by neighboring cells. This leads to the activation of reproductive and developmental pathways, including the regulation of genes involved in these pathways. In most cases bacterial pheromones are a types of autoinducer that acts along the pathway described above.
- Certain types of food chemical, environmental toxins, and secreted chemicals from neighboring cell can influence bacteria cell movement. This type of movement in response to a chemical stimulus is called **chemotaxis**. Chemotaxis involves signal transduction using membrane receptors and cytoplasmic second messengers that influence the activity of the structural proteins involved in cell movement. One such protein is the whip-like flagellum which rotates to propel the bacteria away from a potentially harmful chemicals such as phenols or toward food molecules. Some types of bacteria movement within a solution can involve both a random tumbling phase and a directed straight swim phase based on the concentration gradient of the stimulating chemical. The absence of a gradient induces random tumbling, while the detection of a gradient induces the straight swim phase toward the favorable chemical concentration - movement toward the lower toxin concentration or movement toward a higher food molecule concentration.

NOTES

PHEROMONES
Pheromones are important in behavioral and developmental signaling in prokaryotes, including 'sexual reproduction' via conjugation.

QUORUM SENSING
Quorum sensing in prokaryote is an example of cell communication over short distances using local regulators that specifically target cells within the local

CELL COMMUNICATION IN MULTICELLULAR EUKARYOTES

Cell communication in multicellular organisms is needed to coordinate internal cellular activities and support the function of the whole organism. For example, in some amphibians and reptiles the external temperature can signal the developing embryo to form either into a male or female. This temperature-dependent sex determination is due to the enzyme **aromatase** that converts testosterone to estrogen at higher temperatures. Higher estrogen concentration causes undifferentiated gonads to form ovaries rather than testes, thus more females are born in early spring see figure on opposite page).

Cell communication in multicellular organism can take on 3 different forms: 1) direct cell-to-cell contact (T cells and APC), and 2) signal transduction over long-distances (neuroendocrine pathway) or 3) short-distances (neuron-to-neuron synapse).

Direct cell contact in mammalian immune response

Vertebrates use 3 defense lines to protect against foreign invaders that could potentially disrupt internal homeostasis. The third line, acquired immunity, involved immune cells that communicate via direct contact and chemical messengers.

NOTES

area.

INNATE IMMUNITY
Barrier defense (non-specific) is the First line of defense against. It includes the skin (physical barrier), mucous membranes, hair & cilia (traps and drains out), acidic gastric juices (kills), tears, sweat & saliva (antimicrobial), and symbiotic bacteria (out competes pathogens).

The Second line is non-specific innate immunity and involves white blood cells that launch the inflammatory response often associated with allergies and skin infections due to minor cuts.

INFLAMMATORY RESPONSE
In innate immunity Chemokines released by damaged blood vessel cells cause the chemotaxis recruitment of phagocytes (macrophage and neutrophils) that actively kill pathogenic non-MHC cells, and professional inflammatory cells (eosinophils, basophils, and mast cells) that secrete histamines.
Histamines cause the blood vessels to dilate for increased blood flow to injured site, so that more phagocytes can join the fight.

Other secreted chemicals include pyrogens that increase the body's temperature to help kill pathogens or inhibit their growth, complement that lyse foreign cells, interferon that stop virus-infected cells from spreading the infection to other cells.

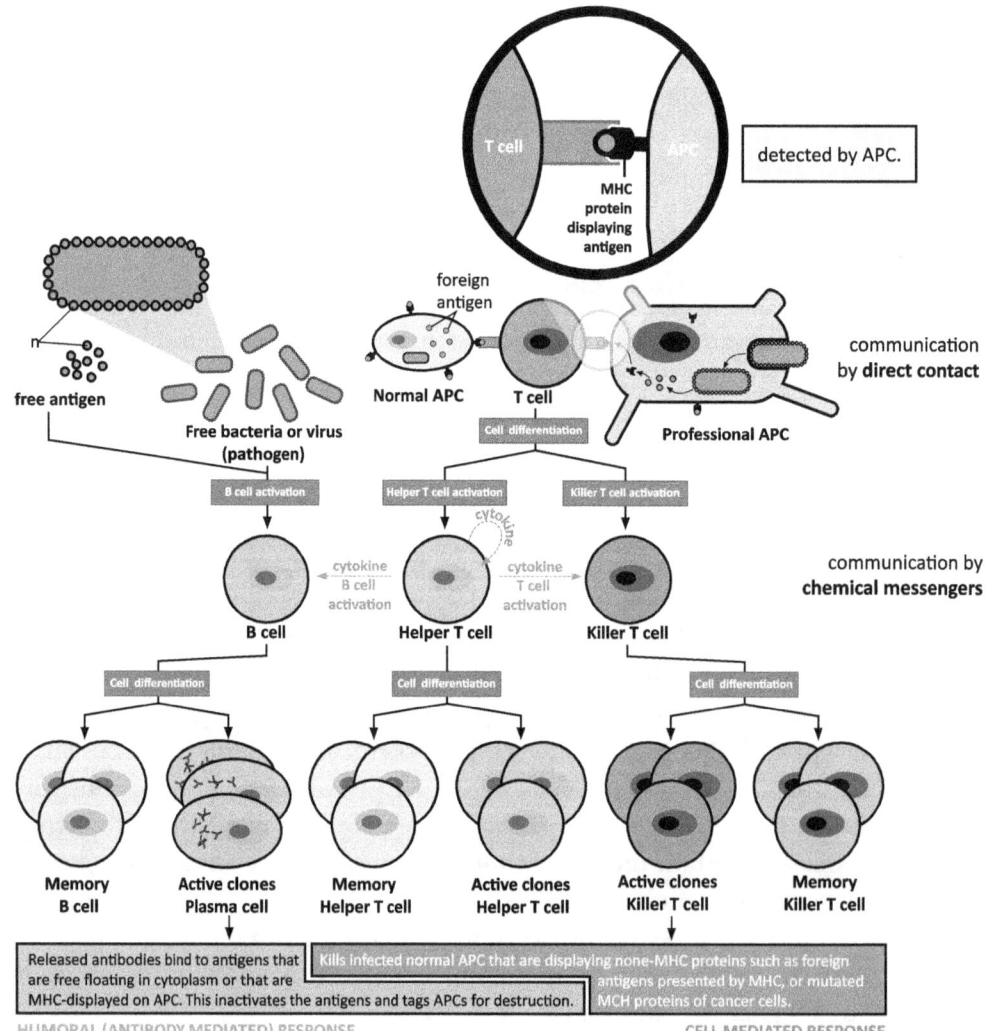

detected by APC.

communication by **direct contact**

communication by **chemical messengers**

HUMORAL (ANTIBODY MEDIATED) RESPONSE

CELL MEDIATED RESPONSE

298

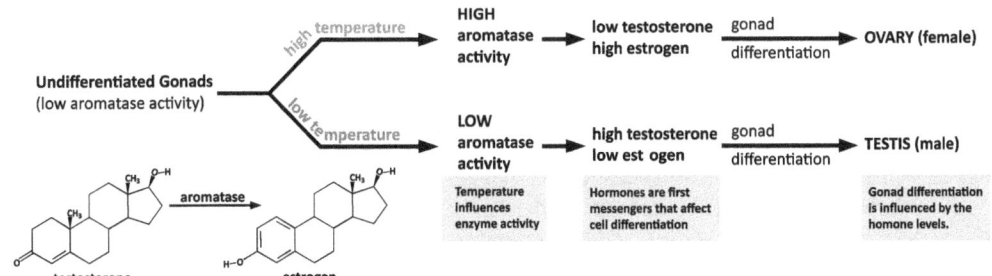

Long-Range communication

Hormones are important signaling molecules for long-distance communication. They are analogous to radio signals sent out into the blood and lymphatic systems by secretory cells. Once secreted, they communicate their regulatory messages throughout the body, specifically targeting cells that have complementary signal receptors for the hormone. There are 2 broad classes of hormones:

1. **Steroid hormones** like estrogen and testosterone are lipid soluble and hydrophobic. They can diffuse directly across the target cell membrane to reach their receptors in the cytoplasm or nucleus where they direct transcription.

2. **Peptide hormones** such as growth hormones, oxytocin, glucagon, and insulin are water-soluble. They can not diffuse across the target cell membrane and must instead bind their specific receptors on the extracellular side of the target cells. The signal is then passed on to cytoplasmic second messengers.

Neuroendocrine pathway

The neuroendocrine and neurohormone pathways are examples of long-distance communication in vertebrates. When sensory neurons are stimulated, they send a signal to the **hypothalamus** in the brain. **Neurosecretory cells** originating in the hypothalamus receive and relay the signal to the pituitary via hypothalamic hormones.

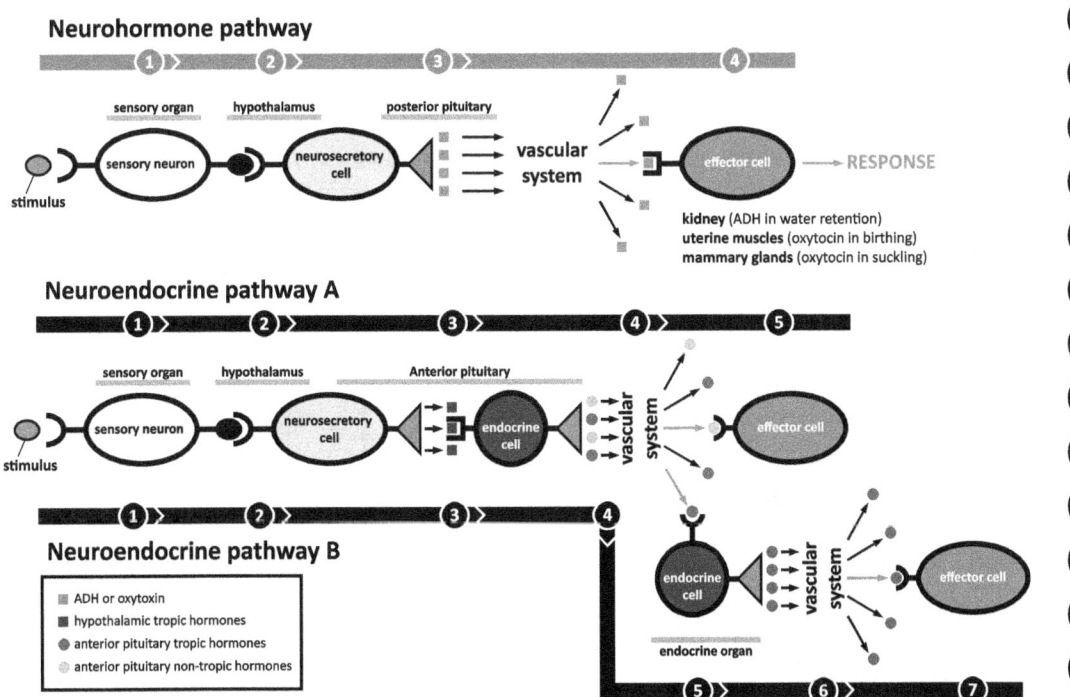

TROPIC HORMONES
Tropic hormones include all hypothalamic hormones that are released into the anterior pituitary and 5 anterior pituitary hormones, LH, FSH, TSH, ACTH, and GH (growth hormones have both tropic and non-tropic action).

ADH is also called vasopressin. It is secreted when pressure or osmoreceptors (sensory nerves) in blood detect a decreased blood volume, or increased osmotic pressure, respectively. ADH acts on kidney cells to increase water reabsorption thereby reducing urine volume. Meanwhile, the hypothalamus also triggers the sensation of thirst.

REGIONS OF BRAIN
Different regions of the vertebrate brain serve different functions (see figure to right). The hypothalamus and pituitary, and the brain stem are among the most primitive part of the brain.

Hypothalamic hormones are categorized into 2 groups:

1. **Posterior pituitary hormones (neurohormones)** are either **oxytocin** or antidiuretic hormone (**ADH**) that are produced by the hypothalamus and stored in the posterior pituitary until released into the blood. Once released, oxytocin travels throughout the body to reach their targeted effector cells.
2. **Hypothalamic tropic hormones** include releasing hormones and inhibiting hormones that act on the anterior pituitary to stimulate and inhibit the secretion of the endocrine hormones produced by the anterior pituitary.

The anterior pituitary secrete 6 different hormones, each made by different endocrine cell and secreted in response to a different hypothalamic hormone. Once secreted, pituitary hormones enter the blood stream to target either endocrine or non-endocrine organs, or both (growth hormones only).

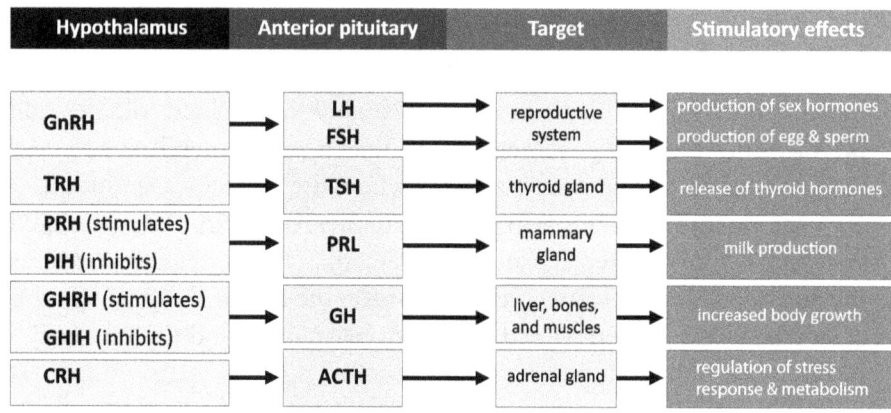

Hypothalamus	Anterior pituitary	Target	Stimulatory effects
GnRH	LH FSH	reproductive system	production of sex hormones production of egg & sperm
TRH	TSH	thyroid gland	release of thyroid hormones
PRH (stimulates) PIH (inhibits)	PRL	mammary gland	milk production
GHRH (stimulates) GHIH (inhibits)	GH	liver, bones, and muscles	increased body growth
CRH	ACTH	adrenal gland	regulation of stress response & metabolism

Neurohormone pathway (oxytocin and lactation)

Oxytocin and **antidiuretic hormones (ADH)** are the two neurohormones that are released from neurosecretory cells of the posterior pituitary. Once released these hormones enter the bloodstream and are sent throughout the body, coming into contact with all cell types. However, the hormones can only induce a response in targeted effector cells that have complementary signaling receptors for the particular hormone. Oxytocin release is stimulated by the onset of labor (induces uterine muscle contraction to facilitate childbirth), or the suckling of in fact on mother's breast (induces mammary glands to release milk). In both cases, it acts in a positive feedback loop, further stimulating its own release.

Neurohormal pathway

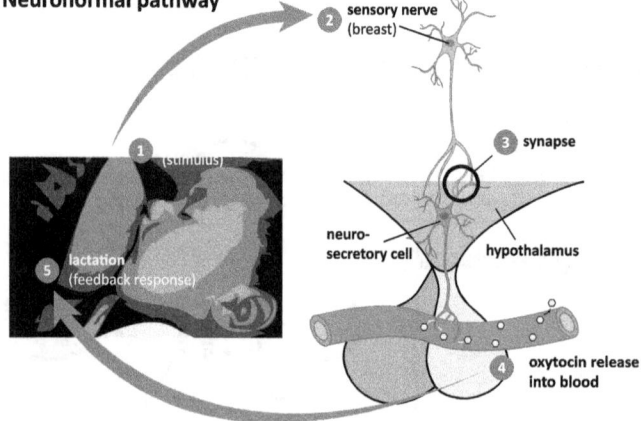

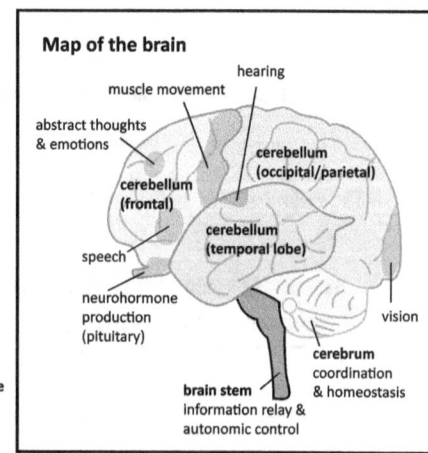

Map of the brain

Neuroendocrine pathway (thyroid regulation)

The thyroid hormones are critical to vertebrate development and maturation, exerting their influence through a negative feedback loop of the neuroendocrine signaling pathway.

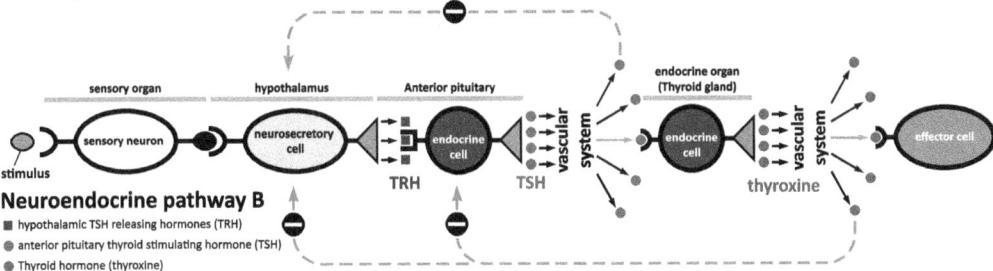

The thyroid signaling pathway occurs according to the following steps:

A sensory neuron is signaled by developmental, maturation or homeostatic stimulus.

1. The hypothalamus receives a signal from a sensory neuron.
2. TSH releasing hormone (**TRH**) from the hypothalamus stimulates the anterior pituitary to release the thyroid stimulating hormone (**TSH**).
3. Once secreted, TSH travels through the blood to target the thyroid gland.
4. TSH then signals the endocrine cells of the thyroid to produce and secrete their **thyroid hormones** into the blood stream.
5. The thyroid hormones then travel throughout the body via the blood.
6. Thyroid hormones act in two ways:
   - They signal their target effector cells to induce an appropriate developmental/maturation or homeostatic response.
   - They feedback inhibit their own production by negatively regulating the hypothalamus (inhibiting TSH release) and anterior pituitary (inhibiting TSH release).

Short-Range communication (neurotransmitters)

Nerve cells use neurotransmitters to communicate a nerve impulse (signal) between neurons. Nerve cell impulse (signal) is received by the dendrites. The cell body integrates the incoming signal and generates an outgoing signal that is sent along the axon. The axon transmits the outgoing signal to next neuron.

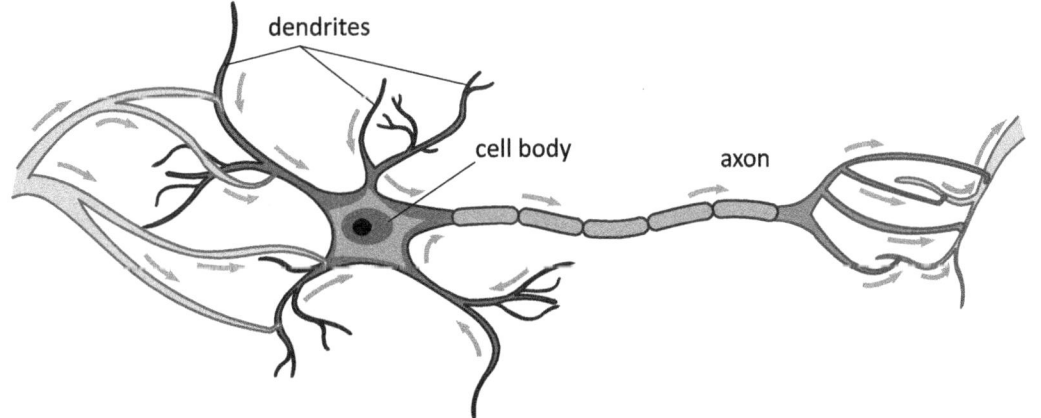

Steps leading to the propagation of nerve impulses:

1. **Nerve impulse arrives at the axon terminal of the presynaptic cell**
   Nerve signals arrive as action potential that is caused by diffusion of Na^+ into the neuron and K^+ out. This depolarizes the cell membrane by changing the membrane potential from -70 to 0 mV, and than to +30 mV. A minimum threshold membrane potential of -50 mV is needed to induce a transduction.

2. **Voltage-gated Ca^{2+} channels open**
   If the local depolarization increases the membrane potential to the minimum threshold level of -50 mV, voltage-gated calcium channels will open, allowing Ca^{2+} to move into the neuron.

3. **Synaptic vesicles fuse with presynaptic axon membrane**
   Ca^{2+} is an ion second messenger that binds to the **synaptic vesicles** and cause them to fused with the presynaptic axon membrane. This releases **neurotransmitters** from the synaptic vesicles into the **synaptic cleft**.

4. **Neurotransmitters bind ligand-gates Na/K channels**
   Neurotransmitters are released and diffuse across the synaptic cleft to bind their receptors on **ligand-gated Na/K channels**.

5. **Depolarization (Na^+ gates open/K^+ gates closed; -70 to +30 mV)**
   These Na/K channels have a Na^+ gate and a K^+ gate. The Na^+ gates are fast acting and the K^+ gates are slow acting. Fast acting Na^+ gates opens first, while K^+ gates remain closed. This causes the inside of the cell to receive lots of positively charged Na^+ ions cause and increase in the membrane potential from -70 to 0 (depolarized) and finally to about +30 mV.

6. **Repolarization (Na^+ gates closed/K^+ gates open; +30 mV to -90 mV)**
   At +30 mV, K^+ gates open and Na^+ gates close. The cell loses lots of positive charge as K^+ ions leave the cell. This repolarizes the cell so much so that it becomes hyperpolarized, undershooting the resting potential (-70 mV) to reach -90 mV

7. **Resting potential restored (Na^+ and K^+ gates closed; -90 mV to -70 mV**
   Resting potential is restored by Na/K pumps that maintain a high concentration of Na^+ outside and K^+ inside the neuron.

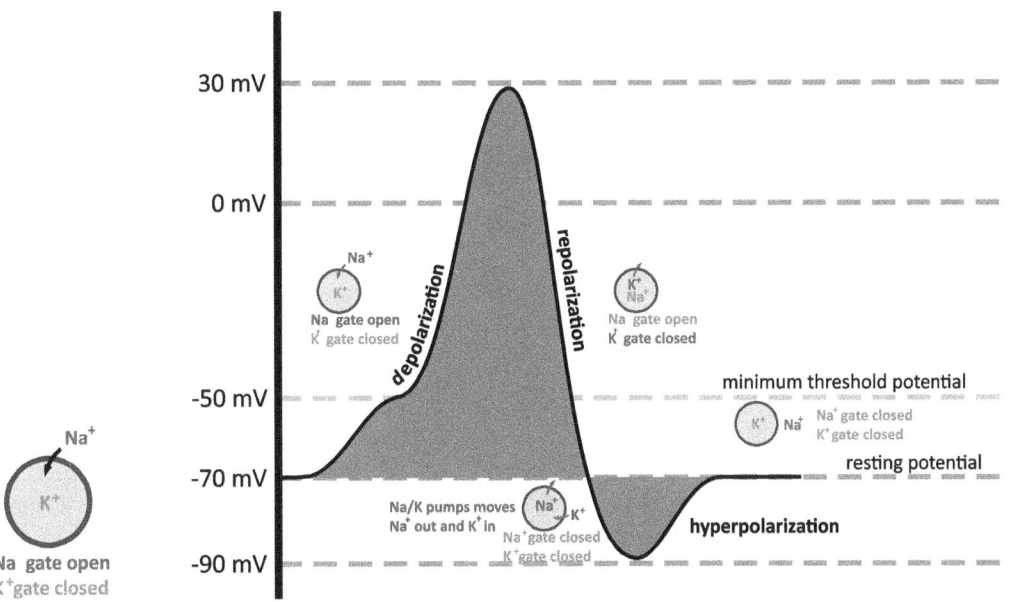

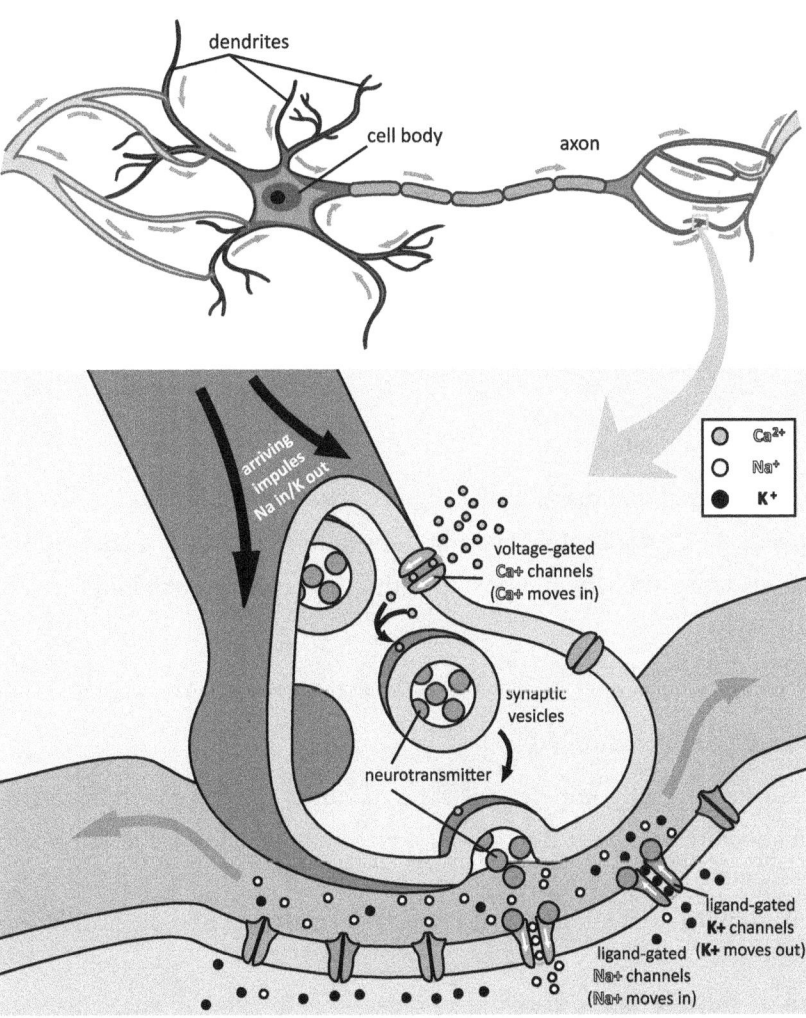

NOTES

NERVE IMPULSE

The negative charge of the cell is are due to negatively charged DNA and proteins that are present in the cell.

Do not confuse Na/K ion channels with the Na/K pumps. Ion channels facilitate diffusion from [high] to [low], while the pumps use energy from ATP to move particles from [low] to [high]. The k/Na pumps maintain the resting potential and are always active. They constantly move sodium out and potassium into the cell. However, when the channels are open (depolarization and repolarization), the pumps are overwhelmed and the resting potential is lost.

Remember the membrane proteins involved in the nerve impulse:
- ligand-gated Na/K channels on postsynaptic membrane
- voltage-gated Na/K channels on all nerve membranes
- Voltage-gated Ca channels on presynaptic axon membrane
- Na/K pumps on all nerve membranes

Neurotransmitters chart

Neurotransmitter	Mechanism of action	Effects
Acetylcholine (ACh)	Binds receptors on ligand gated channels	Activates muscles
Epinephrine (E) also called adrenalin	Binds G protein-linked receptors; involves G proteins and second messengers	Increases metabolism; regulates fight-or-flight response
Serotonin	Binds G protein-linked receptors; involves G proteins and second messengers	Regulates emotional states, mood, and body temperature.
Dopamine	Binds receptors on ligand gated channels	Controls movement, emotional response, and experiences of pleasure and pain.
Gamma-aminobu-tyric acid (GABA)	Binds to ligand-gated channels, or G protein-linked receptor.	Inhibits neuronal signal transduction

SUMMARY OF KEY CONCEPTS

- Gene regulation is mediated by signal transduction that involved intercellular and intracellular processes.
- Mechanisms of signal transduction that is shared across biological domains suggest a common ancestry, and is evidence of evolution.
- Intercellular communication occur via direct contact, and signal transduction over short and long ranges.
- The signal transduction pathway involve effector cell signal recognition (lingand binding, allosteric change in signal receptor), and signal propagation (signaling cascade, second messengers, phosphorylation cascade).

CHECK YOUR UNDERSTANDING

1. An enzyme is a protein that

    (A) catalyzes multiple reactions of a given metabolic pathway.

    (B) speeds up a reaction by lowering the reaction's activation energy.

    (C) has an active site which binds both the substrate and product with equally high affinity.

    (D) All of the above.

2. An enzyme can be inhibited by

    (A) a substance that competes with the substrate for the enzymes active site.

    (B) an allosteric inhibitor that binds to an allosteric site on the enzyme other then that active site.

    (C) an inhibitor that stabilized the inactive conformation of a multi-subunit enzyme.

    (D) All of the above.

NOTES

3. Cytokines are

    (A) substances secreted by foreign cells activate group behavior genes.
    (B) MHC molecules that are displayed on the surface the organism's cells and used for distinguishing self-cells from foreign cells.
    (C) substances secreted by T-cells to activate immune cells during an infection.
    (D) special classes of antibodies that target the cell membrane of prokaryotes.

4. Which of the following statements are true?

    (A) Lipid soluble hormones use simple diffusion to cross the cell membrane.
    (B) Steroid hormones can enter the cell to binds target receptor in the cytoplasm.
    (C) The hypothalamus is the part of the brain that regulates the neuroendocrine and neuroendocrine pathways.
    (D) All of the above statements are true.

5. When a predator approaches a zebra,

    (A) TRH is released from the hypothalamus.
    (B) ACTH binds its receptors in the adrenal gland.
    (C) Thyroid hormones are released, causing the body's metabolism to increase.
    (D) All of the above.

6. Hyperpolarization

    (A) is caused by the movement of sodium ions out of the neuron.
    (B) is driven by the pumping of sodium and potassium in opposite direction.
    (C) occurs when the membrane potential becomes positive.
    (D) None of the above statements is correct.

ANSWERS AND EXPLANATIONS

1. (B) is correct.

2. (D) is correct.

3. (C) is correct.

4. (D) is correct.

5. (D) is correct.

6. (B) is correct.

BIG IDEA 4

Interactions

- MACROMOLECULES
- SUBCELLULAR COMPONENTS
- GENES AND HE ENVIRONMENT
- ECOLOGICAL COMMUNITIES

23 Biological Interactions

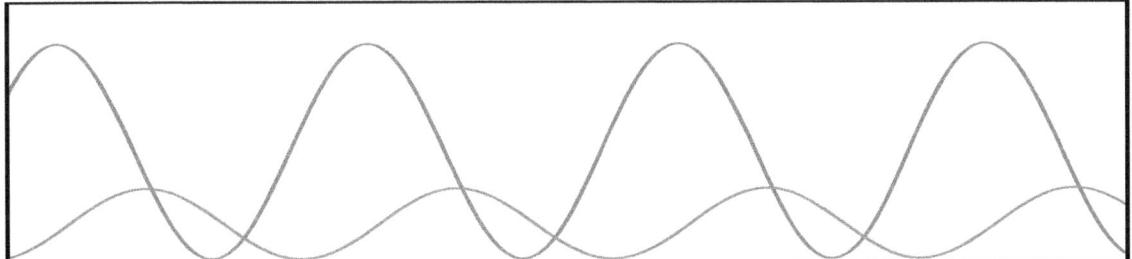

 Big Idea 4

All biological systems attain a degree of complexity that depends on the number of interacting component parts of the system. These interactions lead to the distinctive emerging property of the biological system that increase its ability to respond to changes in the local environment.

INTERACTIONS BETWEEN MACROMOLECULES

Macromolecules (nucleic acids, proteins lipids, and carbohydrates) are biological systems with distinct functions defined by the component parts that makes up their structure. For example, phospholipids have 2 fatty acids, a glycerol, and a phosphate-choline group, and the polymers (DNA, RNA, proteins, carbohydrates) have their distinct combination of monomers (nucleotides, amino acids, sugars). These various component parts are made of atoms that are linked together in specific relative positions to form the structure that supports a particular function (see L12) within the macromolecule.

Because polymers are made of repeating subunits (monomers), there is a specific structural pattern shared by all monomers of a particular type of polymer. For examples, all nucleic acids have a 3' and 5' end and amino acids have a amino (or N-terminal) and carboxyl (or C-terminal) end. The monomers are linked together end-to-end (5'-to-3' and C-to-N terminal), establishing directionality to the molecule. **Directionality** is a critically important part of all polymers. Without it, the enzymes involved in polymerizing DNA, RNA, and proteins will be unable to recognize where in the molecule to add another monomer.

Nucleic acids

DNA and RNA store genetic information based on the sequence and types of the nucleotides (component parts) in the molecule. All nucleic acids are linked in the 5' to 3' direction. Nucleotides are similar in that both are made of a 5-carbon sugar, a phosphate, and a nitrogenous base. They differ in the types of 5-carbon sugar (DNA = deoxyribose; RNA = ribose) and their combination of nitrogenous bases (DNA = GATC; RNA = GAUC). These differences give the 2 molecules their unique emerging properties related to storing and transmitting genetic information (see L16).

Proteins

Like nucleic acids, proteins are polymers that have specific sequences of amino acids (component parts). The polypeptide grows in the C-to-N terminal direction by forming peptide bonds between the amine and carboxyl groups of adjacent amino acids.

The 20 commonly occurring amino acids shared a general structure that includes a central carbon linked to an **amino group** (-NH₂), a hydrogen atom (-H), a **carboxyl group** (-COOH), and a variant side chain called the **R-group**. The side chain is what distinguishes one amino acids from another, giving each its unique chemical property. Side chains cause the amino acid to become hydrophilic by inducing polarization, or by having a positive

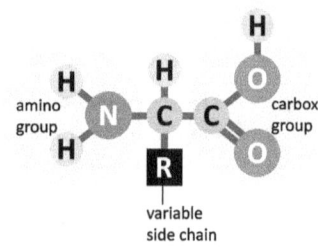

or negative charge. Non-polar and aromatic (contain 6-carbon rings with alternating double bonds) side chains induce the amino acid to become hydrophobic. Consequently, amino acids are classified as either hydrophilic (polar and charged side chains) or hydrophilic (non-polar and aromatic side chains).

As the polypeptide folds, the hydrophobic amino acids are typically buried inside the protein to hide it from water molecules that makes up most of the cell's cytoplasm. Charged and polar amino acids are hydrophilic and interact with water by forming hydrogen bonds in the α-helix and β-pleated sheets of the secondary structure. Thus, most amino acids have a hydrophobic inner core and a hydrophilic surface.

Carbohydrates

Carbohydrates are polymers of sugar monomers. There are 5 major classes of carbohydrate polymers: starches (plants storage), glycogens (animal storage), cellulose (plant cell wall), chitin (animal exoskeleton), and peptidoglycan (bacteria cell wall). The sugar monomers are linked together in the growing polymer via either an α- or β-glycosidic links, which determines the molecules directionality, structure, and function.

As shown in the following figure, storage carbohydrates (starches and glycogens) are formed exclusively with α-glycosidic linkage, which causes the molecule to take on a helical shape.

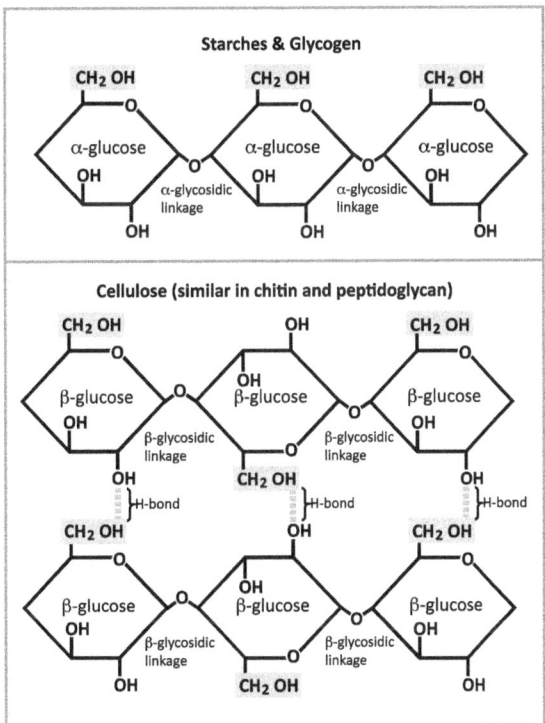

Structural polymers (cellulose, chitin, and peptidoglycan) are formed by β-glycosidic linkage that produces in a linear molecule. The strength of these structural carbohydrates comes from binding several polymer strands together. In chitin and cellulose this occurs via H-bonds between adjacent strands, while in peptidoglycan it occurs by peptide bonding between the amino side chains of adjacent strands.

NOTES

INTERACTIONS BETWEEN ORGANS AND ORGAN SYSTEMS
Other examples of of interactions between organisms component parts include (1) stomach and small intestines in digestion, (2) kidney and urinary bladder in electrolyte and water balance, (3) respiratory and circulatory system in oxygen and carbon dioxide transport, and (4) nervous and muscular systems in motor movement.

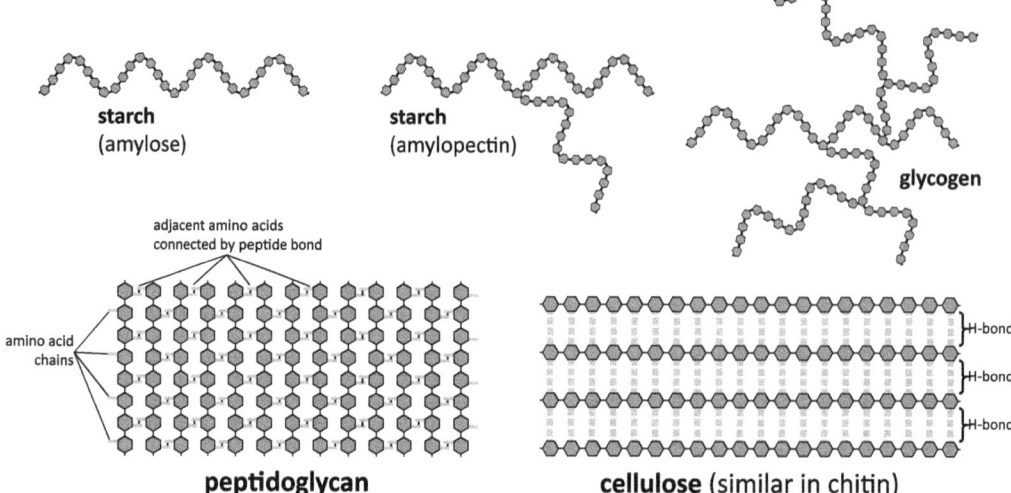

Storage carbohydrates form helical structures that are highly branched in animals (glycogen), and unbranched (amylose) or partially branched (amylopectin) in plants. The branching has an adaptive function by allowing animals to rapidly released energy from a single molecule. With an unbranched starch, the organism can at most release one glucose molecule at a time for cellular respiration. But, with a branched glycogen, animals can release as many glucose molecules as there are branches. In the event of a fight-or-flight response, this property allows for a burst of energy.

INTERACTION BETWEEN SUBCELLULAR COMPONENTS

The structure and function of subcellular components including the endomembrane system (nucleus, rER, sER, Golgi complex, vesicles, lysosomes, vacuole), other organelles, and cytoplasmic and extracellular matrix proteins enhance the ability of cells to internally organize and respond external conditions (see L13, 14,16). Furthermore, interactions between these various subcellular components give way to emerging cellular properties such as the ability to self-replicate through the highly orchestrated process of mitosis and meiosis.

INTERACTION OF GENES AND EXTERNAL STIMULI IN DEVELOPMENT

In multicellular organisms, cells collectively form tissues, which form organs, which form organ systems, and all of the organ systems together form the whole organism. Developmental factors and genes, including cytoplasmic determinants, embryonic inducers, and hox genes, determine how each of these biological systems develop (see L21). When a higher level of biological organization is established (i.e., cells to tissue, or organ system to organism), a new set of emerging properties become evident.

Plants organs (roots, stem, leaves)

The plant tissues of the roots, stem, and leaves each have a unique set of emerging properties. The roots grow down, absorb water and minerals from the soil, and typically store glucose produced by photosynthesis in the leaves; the stem grows up toward the light, provides structural support; and the leaves allows evapotranspiration via its stomata, performs photosynthesis, and exchanges gases with the atmosphere. Much of what the leaves do rely on a network of water transporting tubes called phloem and xylem that together form the water transporting vascular system. The Xylem transports water and minerals toward the leaves for photosynthesis and temperature regulation, while the phloem transports water and starch toward the roots for storage.

INTERACTIONS WITHIN ECOLOGICAL COMMUNITIES

A biological community is a collection of interacting populations of at least 2 species that occupy the same habitat at the same time. The structure and stability of ecological communities are measured and defined by its size (number of individuals) and biodiversity (number of species), with the more stable and resilient communities being larger and with greater biodiversity.

Mathematical modeling of populations

Population growth rate (dN/dt) is the change in population (dN) over a change in time (dt). In a closed population, the growth rate depends only on the number of births and deaths over time, or the birth rate (B) and death rate (D). Therefore,

$$dN/dt = B - D$$

The population growth rate can also be expressed using the per capita growth rate (r), which is a measure of the instantaneous birth and death rate (b and d).

$$dN/dt = rN = (b-d)N \qquad \text{\{N is the current population size\}}$$

Growth in an unlimited environment (exponential growth)

In an unlimited environment with excess resources, the population will achieve maximum instantaneous growth rate (r_{max}), and displays an exponential growth curve.

$$dN/dt = r_{max}N$$

Growth in a limited environment (sigmoid growth)

In an environment with limited space and food the population is restricted to a maximum size called the carrying capacity (K). As the population density increases, fertility and longevity decreases, causing the instantaneous growth rate to decrease toward zero, and the population to approach carrying capacity. Populations experiencing these conditions display sigmoid (or s-shape) growth curve.

$$dN/dt = r_{max} \, \frac{N(K-N)}{K}$$

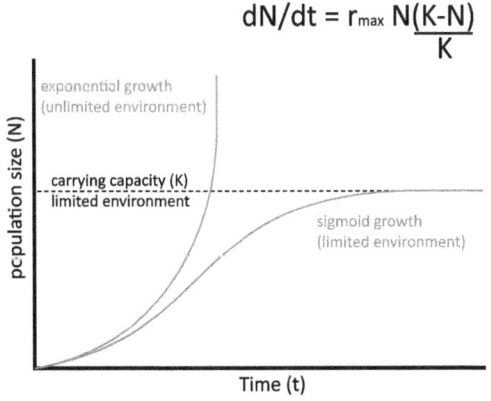

HUMAN IMPACT
Human activity is driving many species to extinction. The release of greenhouse gases from burning of fossil fuel is causing global warming, and lowering the pH of aquatic ecosystem. Organism cannot adapt rapidly enough to these changes, and are threatened with extinction.

Other human activity that is driving this trend include (1) loss of habitat due to deforestation from logging, slash/burn agriculture, urban expansion, infrastructure development (roads, dams, transmission lines), (2) introduction of invasive species that lack a natural competitor and/or predator, (3) loss of keystone species, and (4) introduction of exotic diseases (Dutch Elm, potato blight, small pox).

GEOLOGIC AND METEOROLOGIC IMPACT
For information on how the following natural event effect biodiversity see the associated lecture: (1) El Nino (see L07), (2) continental drift (see L03), and (3) meteor extinction of dinosaurs (see L01).

Logistic growth

Organisms do not respond instantly to changes in local resources (food and space). Prior to reaching carrying capacity, couples are still mating as though resources are plentiful, and newborns are on their way. Consequently, when carrying capacity is reached, the population may continue to grow as newborns arrive. This time lag in response to environmental changes will cause the population to shoot pass the carrying capacity (K). Once the K is exceeded, competition for the scarce resources will intensify and the death rate to spike, causing the population to shrink. The downward pressure on population size will reduce competition for resources as the population drops below K. However, there will also be a time lag in the response to changes in the availability in local resources and the population will continue to fall below carrying capacity. Eventually the population responds with and increased birth rate and decreased death rate. The population continues fluctuating around the carrying capacity in a pattern called logistic growth, always exceeding and falling below K, but never resting exactly on it.

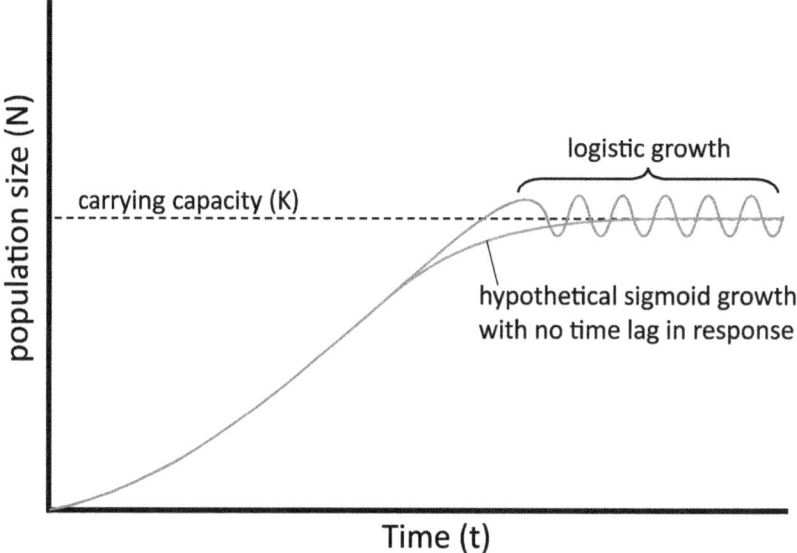

Population growth factors

The growth rate of the population is influenced by both density-dependent and density-independent factors. Density-dependent factors vary with the population, and include individual competition, territoriality, disease and predation, waste accumulation, and migration. Density-independent factors include catastrophic events and abiotic resources such as fires, floods, storms, and drought that are not influenced by population size. Additionally, human activities can impact the population size of other species through habitat displacement and destruction.

Modeling population growth rates

Different populations that occupy separate niches within the same local habitat tend to experience different growth rates and carrying capacities. As shows in the following figures, if the r is increased without changing K, the population will enter logistic growth more quickly. However, any increase in K has a positive impact on the growth rate without affecting how much time needed to enter logistic growth.

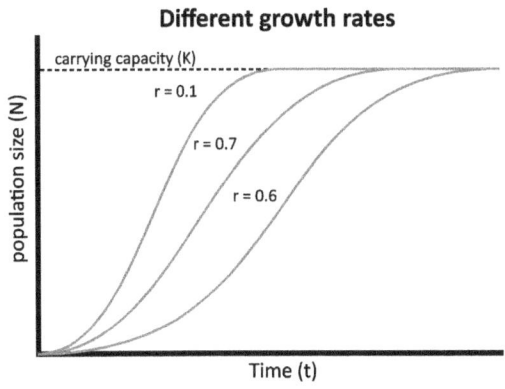

Different growth rates

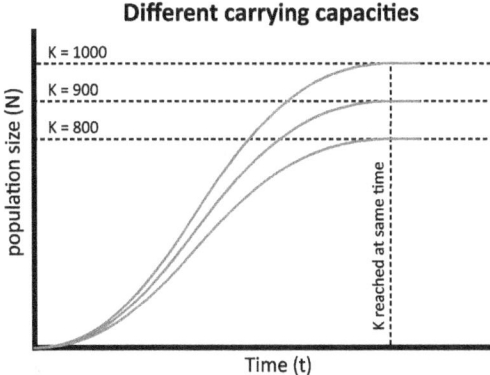

Different carrying capacities

NOTES

SYMBIOTIC RELATIONSHIPS
The predator-prey relationship is one of several types of symbiotic relationships, all of which are regulated by feedback control mechanisms. The 5 symbiotic relationships are (1) parasitism (one species benefits by slowly killing the other, (2) predation (one benefits by instantly killing the other), (3) competition, (4) mutualism (both benefit), (5) commensalism (one benefits without harm to the other). Note that each has either a positive, negative or neutral effect on the organisms involved.

Modeling community interactions (predator-prey relationship)

The predator is an organism that feeds on another organism (prey). Predation is most commonly used to refer to animal-animal interaction such as a wolf that preys on caribou. However, the term can extend to animal-plant interaction involving herbivores such as rabbits that feed on carrots.

Ecological communities have emerging properties that rely on interactions between whole organisms, and between populations of different species. The predator-prey relationship is a good model for studying these interactions. Since each organism is a part of the others environment, they have an impact on each other's population size. However, because energy flows through the ecosystem from the lower trophic level (prey) to the higher trophic level (predator), the total biomass of the prey is typically much higher than that of the predator. For predators and preys of comparable body sizes, such as a wolf and a moose, the same trend is observed for population sizes, with the prey population being considerable larger than the predator population.

Since the prey is a limited resource of the predator (not vice versa), the prey's population size determines the predator's carrying capacity ($K_{predator}$) such that as the number of predators increase, predation intensifies and the predator's food source (prey) is steadily reduced. This causes $K_{predator}$ to drop below the predator's existing population size. Once this happens, negative feedback will cause the predator to experience negative growth ($r < 0$), and its population will begin to shrink as the preys population rebounds. As the graph shows, the predator trails the prey population, rising and falling as $K_{predator}$ fluctuates.

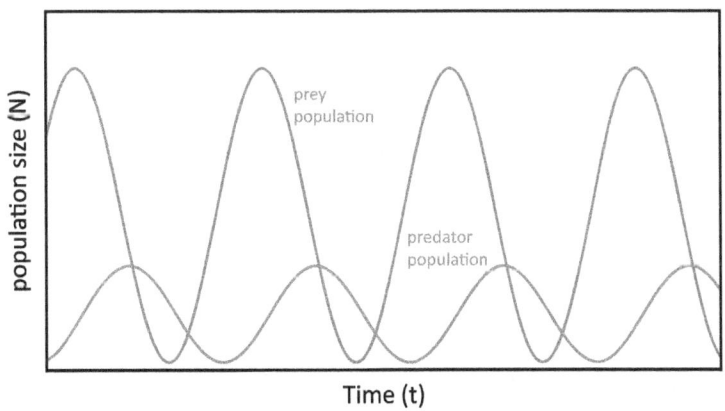

Matter and energy

Recall that energy flows linearly through the ecosystem from the lowest to the highest trophic, with roughly 10% transfer between levels (see L10). With access to sunlight, the limiting factor on the amount of energy that can enter the ecosystem is the amount of matter (water, carbon dioxide, and other essential minerals such as nitrogen and phosphorus) that is needed for photosynthesis. As discussed in L11, matter is a finite resource that must be constantly recycled in the ecosystem. Scavengers (e.g., vultures, hyenas, raccoons), detritivores (e.g., arthropods, slugs, earthworms), and decomposers (bacteria, protists, fungi) are the primary organisms responsible for recycling the matter that is stored in the tissue of dead organisms. The following food web shows the feeding relationship between organisms of a soil ecosystem.

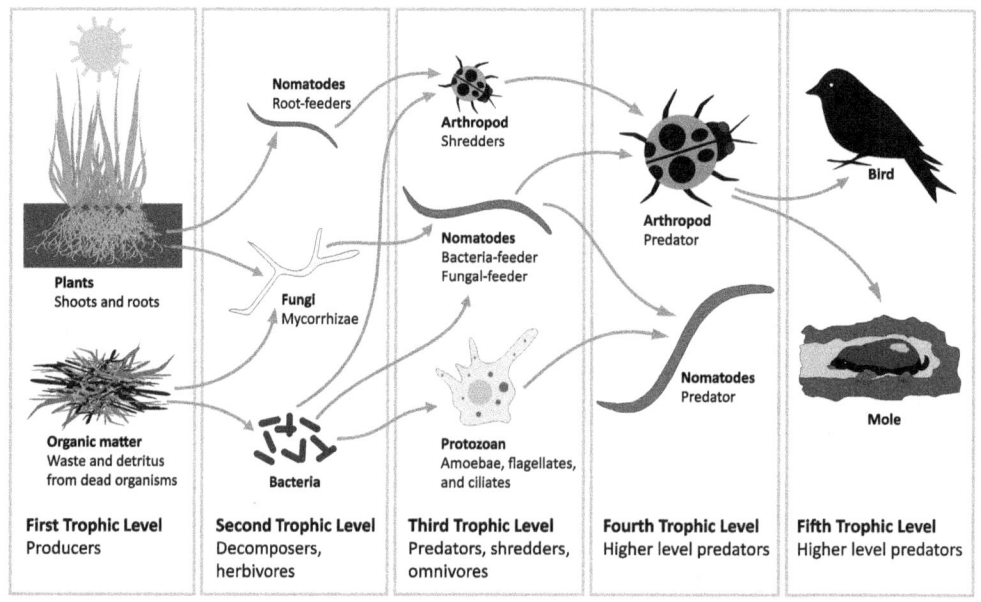

The food web contains several food chains that intersect as energy and matter is transferred between trophic levels. The following are examples of food chains within this ecosystem:

- Plants → root feeder nematode → shredder → predatory arthropod → bird
- Organic matter → bacteria → amoebae → predatory nematode → mole
- Plants → mycorrhizae → fungal feeder → predatory arthropod → mole

The complexity of the feeding relationship determines the resiliency and stability of the ecosystem. Because all energy enters the ecosystem through plants, the entire food web depends on the primary productivity from the first trophic level. If there is a catastrophic event such as a disease that reduces the producers population, the entire ecosystem suffers. More resilient ecosystems will therefore have significant biodiversity throughout the trophic levels, and particularly at the first trophic level.

Dominant and keystone species

Ecosystems tend to have a dominant species and a keystone species. The Dominant species is the one that outnumbers its competitor, or that contributes the most to the ecosystem's biomass. In many forest ecosystems, a specific type of tree is the dominant species. The keystone species is the species that has the greatest impact on the ecosystem relative to its population size or biomass. Although the keystone species make up a small part of their ecosystem, their removal can cause the ecosystem to undergo dramatic changes, or even collapse and go extinct. This is because they are positioned at a critical point in the food chain, and are needed for the effective and efficient flow of energy and matter between the other organisms of the system, and for feedback regulation of the size of each trophic level.

NATURAL VARIATION IMPACTS INTERACTIONS

Natural variation within biological system allows for as broader range of functions, while increasing the systems resiliency. At the cellular level, variation in the types of lipids (phospholipids, fats, steroids) allow the cell to acquire properties needed to form membrane barriers (phospholipids), communicate information (steroid hormones), and store energy (fats). A similar trend is observed with different types of antibodies that are used for defense against pathogens, with the diversity of MHC proteins that are used for cell recognition, and with the variation in chlorophyll pigments that allow the chloroplasts to absorb a broader range on the light spectrum.

Natural variation in alleles also impact phenotypes and natural selection. For example, genes with multiple alleles might lead to heterozygous advantage, or gene duplication might result in the emergence of a new gene with unique functions. A good example of this phenomena is the fish antifreeze gene that evolve after a digestive protein gene was copied and naturally selected to serve its new role in protecting arctic fish from freezing. Now the fish has two genes from one – the original digestive enzyme, and the antifreeze gene.

In additional to genes, natural variation can also be influenced by environmental factors that cause differential expression in the organisms genes, such as:

- Human height and weight can be affected by diet
- Flower color can be influenced by soil pH
- Foxes display seasonal changes in coat color
- Sex determination in some reptiles is temperature-dependent
- The density of plant hairs called trichomes can increase in response to herbivory
- Gene regulation in pGLO-transformed bacteria is influenced by exposure to the arabinose inducer
- Increase in melanin production is correlated with UV exposure
- Pheromones production and secretion in yeast and fungi occurs in response to the presence of a compatible mate

SUMMARY OF KEY CONCEPTS

- Interactions between macromolecules and organelles allow cells to increase their range of functions.
- Environmental factors can directly or indirectly effect gene expression, leading to variation in cell, tissue, and organ differentiation and development.
- Interaction between organs and organ system can lead to greater complexity in organisms, and can increase their range of function.
- The more species diversity within an ecosystem, and more community interaction that exists between species, the greater is the ecosystems resiliency.
- The predator-prey relation is a good model for study population dynamics.
- Population growth is regulated by density-dependent and density-independent factors, including human activity, and geological and meteorological events.
- Food webs and food chains are used to illustrate the complexity of energy and matter flow through the ecosystem.
- Competition and cooperation between component parts exist at all levels of biological organization, from the molecular level to the biome.
- Symbiotic relationships are interpopulation interactions that are under negative feedback control, and that contribute to the complexity of biological communities.
- Natural variation at all levels of biological systems influence ecosystem dynamics.

CHECK YOUR UNDERSTANDING

1. Once way that storage carbohydrates differ from structural carbohydrate is that

    (A) storage carbs have glucose monomers connected by a β-glycosidic linkage.
    (B) structural carbs have are made of α-glucose monomers only.
    (C) strand of structural carbs are linked together in parallel by H-bonds or peptide-bonds.
    (D) storage carbs are made of alternating α- and β-glucose monomers.

2. What is the growth rate of a population of 750 individual with a carrying capacity of 1000? Assume that the r_{max} is 2.

    (A) -150
    (B) 375
    (C) 1500
    (D) None of the above

3. What is the growth rate when the population reaches 1100?

    (A) -150
    (B) 375
    (C) 1500
    (D) None of the above

4. The growth rate of a population is usually never zero at carrying capacity because

    (A) Negative feedback control prevents a population from naturally reaching K.
    (B) Migration causes the population size to fluctuate.
    (C) Sudden changes in primary productivity at the first trophic level tends to alter K.
    (D) Population do not respond instantly to changes in the environment and tend to shoot past K.

5. A dominant species of an ecosystem is

    (A) the same as the keystone species.
    (B) the most abundant species.
    (C) the species that is typically few in number but with the greatest impact on the ecosystem stability.
    (D) always the species that has the largest biomass per individual than other species.

ANSWERS AND EXPLANATIONS

(A) (C) is correct.

(B) (B) is correct.

(C) $dN/dt = 2[750(1000-750)/1000] = 375$

(D) (A) is correct.

(E) $dN/dt = 2[750(1000-1100)/1000] = -150$

(F) (D) is correct.

(G) (D) is correct.

PRACTICE TEST

How to score the diagnostic test?

Score the two sections separately to get the raw score. For Section 1, divide the number of correct answers by 69 and multiply by 50. For example, if you received 35 correct answers on the multiple choice and grid-in questions of Section 1, the calculation is as follows:

$35/69 \times 50.4 = 25.4$

Scoring of Section 2 is more difficult since each question is graded using its own rubrics. However, our method should provide a reasonable assessment of your performance on this section. The long FRQ questions are weighted equal to the short FRQ questions, at 25 raw points for each part. Use you own judgement, guided by the *Answer and Explanation* section, as a guide to determine whether your response is excellent (A), good (B), pass (C) or fail (F). Once you have assigned these letter values to each question refer to the chart below to convert the letter grade to points and calculate your total raw score for section 2:

Letter grade	Long FRQ	Short FRQ
A	4.2	12.5
B	3.4	10
C	2.5	7.5
F	1.7	5
No response	0	0

For example, if on the long FRQ questions you received a letter score of B and C, and the short FRQ you received B, B, F, A, C, and no response for question 8, your total score for section 2 is calculated as follows:

$3.4 + 2.5 + 10 + 10 + 5 + 12.5 + 7.5 + 0 = 43.4$.

You will then add your scores from the two section to get your combined raw score and use the following table to determine your scaled score:

Raw score	AP score	Translation
65 - 100	5	Excellent
55 - 64	4	Good
45 - 54	3	Passed
40 - 43	2	Possibly passed
0 - 39	1	Failed

For example, if your raw score for Section 1 was 25.4 and Section 2 was 43.4, your combined raw score is 25.4 + 43.4 = 68.9, and your AP score is 5.

Section I

Time: 90 minutes
Part A: 63 multiple-choice questions
Part B: 6 grid-in questions

Directions (Part A - multiple choice): For each of the following 63 questions or statements, chose the best of the possible responses. When you complete these questions, continue on to Part B.

1. In the Miller-Urey experiment, an electric spark was applied to a chamber containing water vapor, ammonia, methane, and hydrogen gases. As the water vapor condensed, it dissolved some of the chemicals from the chamber and collected at the based of the experimental apparatus in a collecting trap. After some time, chemical analysis detected the presence of amino acids, nucleic acids, and other organic compounds. Which of the following statements best describes the hypotheses that the Urey experiment evaluated?
   (A) NH_4, CH_4, and H_2 gases are required for the formation of organic compounds.
   (B) The early Earth provided the right conditions for the formation of organic compounds from inorganic substances.
   (C) Excess free energy in the early Earth caused any inorganic molecules to become structurally unstable, thus preventing the formation of life-supporting organic compounds.
   (D) The origin of life according to the organic soup model required the preexistence of organic compounds.

2. In today's world, DNA is the most reliable genetic material. However, despite its ability to store genetic information with relatively high reliability, it is unlikely to have been the original heritable molecule because it
   (A) cannot be replicated.
   (B) lacks biochemical functionality.
   (C) is structurally unstable.
   (D) All of the above.

3. Deep sea hydrothermal vents are ideal candidates for where life may have originated because hydrothermal vents
   (A) have access to free energy from the Earth geothermal activity.
   (B) produce and release lots of precursor inorganic compounds that can form organic molecules.
   (C) limit diffusion to the surrounding and, therefore, allow local concentrations of precursor molecules to build-up.
   (D) All of the above.

GO ON TO THE NEXT PAGE

4. During his travels, Darwin observed a general pattern of over-reproduction across species in various geographic areas. However, despite a large number of newborns, the population size appeared to remain stable over time. He suggested that this trend was due to
   (A) high infant mortality due to disease and parasitism.
   (B) the immigration of predators that kept the population in check.
   (C) the availability of limited resources and intense competition for those resources.
   (D) intense sex selection, including intersex agonistic behavior that significantly reduces the male population.

5. Ecological performance is a measure of how well an organism is adapted to the local environment. Because it has a direct impact on the organism's fitness, it also influences natural selection. Which of the following factors determines ecological performance?
   (A) physiology and morphology
   (B) innate behavior
   (C) predators and competitors
   (D) all of the above

6. Which of the following statements refers to artificial selection, but not natural selection.
   (A) It requires a diverse starting population.
   (B) It relies on mutations.
   (C) The selective pressure comes from the breeder.
   (D) All of the above.

7. There is a population of 20 red and 30 white flowers of the same species that are in Hardy-Weinberg equilibrium. Assuming that the white allele is recessive. How many flowers are carriers (heterozygous) of the recessive allele?
   (A) 70
   (B) 47
   (C) 25
   (D) 17

8. Pangaea was a supercontinent that formed about 300 million years ago, but has since drifted apart to form the major landmasses of today. The break-up of Pangaea resulted in
   (A) a decreased in global biodiversity because the sudden movement of tectonic plates caused catastrophic climate change that drove many species to extinction.
   (B) the evolution of marine mammals that were able to cross large bodies of water between separated land masses.
   (C) an increase in global biodiversity as separated population gradually adapted to the changing climate created by drifting tectonic plates.
   (D) All of the above

GO ON TO THE NEXT PAGE

9. The grey wolf of North America and the now extinct Tasmanian wolf of Australia appear to share many morphological features despite their respective habitats being geographically far apart. The similarities between the two groups is due to

(A) shared derived characters that were present in a common ancestor before the isolation of Australia from the rest of the large land masses.

(B) evolutionary convergence when their respective ancestors adapted to similar niches in geographically isolated areas.

(C) the introduction of domesticated dogs to Australia by English settlers, some of which escaped to the wild and underwent further natural selection to become the Tasmanian wolf.

(D) the fact that the Tasmanian wolf is a close relative of the Eurasian grey wolf that reached Australia several thousand years ago by island hopping from southeast Asia through the island chains of Philippines, Malaysia, Indonesia and New Guinea, to finally reach Australia.

10. Based on the following figure, which fossil is the best candidate for an index fossil?

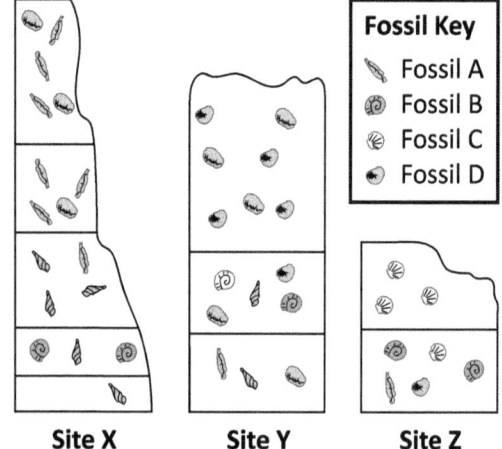

Site X   **Site Y**   **Site Z**

(A) Fossil A
(B) Fossil B
(C) Fossil C
(D) Fossil D

11. Scientist hypothesize that the evolution of the four chambered heart in birds and mammals is linked to the expression of the transcription factor, Tbx5, in the cells of the ventricular wall during embryonic development. In amphibians, Tbx5 is uniformly expressed throughout the ventricle; however, in birds and mammals, its expression is restricted to the cells of the left ventricular wall. This restriction in Tbx5 expression is believed to induce the separation of the amphibian-like ventricle into the two ventricles (right and left) of birds and mammals. If this hypothesis is correct, which of the following statements predicts the expected pattern of Tbx5 expression in reptiles?

(A) Homogenous expression of Tbx5 cells of the left and right ventricular wall.

(B) Expression of Tbx5 only in the cells of the right ventricular wall.

(C) A gradient of expression of Tbx5 in the ventricular wall that dissipates from left to right.

(D) A gradient of expression of Tbx5 in the ventricular wall that increases from left to right.

12. The gene for Rh factor is inherited independently from that of the ABO blood antigens. The products of both genes are displayed on the surface of the red blood cells and are involved in self-cell recognition. Rh⁺ individuals express the dominant *Rh* allele and display the Rh antigen, while Rh⁻ individuals express the recessive *rh* allele and produce a nonfunctional Rh antigen that is not displayed. If out of 1000 individuals, 225 are Rh⁺, what is the frequency of the dominant allele? Assume that the population is in Hardy-Weinberg equilibrium.

(A) 0.12
(B) 0.47
(C) 0.78
(D) 0.88

GO ON TO THE NEXT PAGE

13. Which of the following can affect the gene pool by introducing new alleles into the population?
    (A) sex selection
    (B) genetic drift
    (C) forest fire
    (D) none of the above

14. Prior to the industrial revolution, the white-bodied peppered moths were most common in the forest of the English countryside. By the end of the 19th century, black-bodied peppered moth had replaced the white-bodied moths as the most common type. The increase in black bodied moths were due to
    (A) Industrial soot-induced mutation of the white pigmentation allele to give rise to a new black allele.
    (B) Natural selection against white-bodied moths.
    (C) Darkening of white-bodied moths by soot from the English industrial plants.
    (D) Genetic drift of white pigmentation alleles as a result of chemical contaminants from surrounding industrial plants.

15. Of the following statements, which is an example of genetic drift?
    (A) The Pennsylvanian Amish population displays a 7% frequency for polydactyly relative to the 0.1% frequency for the general population of Pennsylvania. It is estimated that at the time when the ancestral Amish population immigrated to Pennsylvania in the 1700s, the general population in Europe also had a frequency of 0.1%, however this small immigrant population had at least one individual who carried the allele for polydactyly.
    (B) Although Native American are believed to have cross the Behring Straits land bridge from Asia to reach North American, the occurrence of Blood type B among this population is relatively rare compared to the closest Asian populations.
    (C) Excessive hunting in the 1890s reduced the Northern elephant seals population to as few as 20 individuals. Although the population has since rebounded, when compared to their southern relative, the gene pool of northern seals display much less genetic variation.
    (D) None of the above statements describes genetic drift.

16. The trp operon in bacteria contains 5 structural gene (trp A, B, C, D, and E) of the tryptophan synthase enzyme, which is involved in the synthesis of tryptophan. The trp operon's activity is inversely correlated with tryptophan concentration in the cell. Which of the following is likely true about this operon?
    (A) The trp operon is regulated by an activator.
    (B) Tryptophan is an inducer of the operon.
    (C) The trp operon contains an operator downstream from the promoter.
    (D) The trp operon is under a positive feedback control.

GO ON TO THE NEXT PAGE

17. Which of the following statements best describes the probable affect of global warming on the symbiotic relationship between pollinators and flowering plants.
   (A) The symbiotic relationship will shift from mutualism to commensalism as the growing period for flowering plants is extended by the warming climate and they become less dependent on pollinators.
   (B) The mutualistic relationship may collapse as the synchronization of the arrival of pollinators and the appearance of flower buds in spring becomes disrupted due to changes in local seasonal temperatures.
   (C) The relationship may collapse as pollinators go extinct due to their high sensitivity to small changes in ambient temperature.
   (D) There will be no changes in the relationship because the environment is changing slowly enough for both pollinators and flowering plants to effectively adapt

18. All of the following processes can act to preserve genetic variation in the population except,
   (A) Genes with multiple alleles that display a complete dominance pattern of expression.
   (B) Genes with multiple alleles where hybrid individuals display the highest fitness.
   (C) Genes with two alleles that display a complete dominance pattern of expression, and where the recessive phenotype confers a significantly greater fitness than the dominant phenotype.
   (D) Alleles of a gene that is responsible of coat pigmentation in a prey population confer improved fitness at the lower phenotype frequency, and reduced fitness at higher frequency.

19. The scarlet king snake, a nonvenomous species of snake that is commonly found in the United States, shares a similar stripe pattern to the deadly coral snake. This mimicry of the coral snake has been shown to protect the king snake from attacks by predatory birds. The divergence of scarlet king snakes from other king snake species likely occurred by
   (A) allopatric speciation with the coral snake.
   (B) sympatric speciation with the coral snake.
   (C) peripatric speciation with the coral snake.
   (D) parapatric speciation with the coral snake.

20. Which of the following will lead to the most dramatic adaptive evolution in a given species?
   (A) Over a period of 10 million years, a local temperate deciduous forest transforms to a tropical rain forest.
   (B) Over a period of 100 million years, a local temperate deciduous forest transforms to a tropical rain forest.
   (C) Over a period of 10 million years, a local temperate deciduous forest transforms to a tropical rain forest that experiences irregular catastrophic forest fires.
   (D) Over a period of 100 million years, a local temperate deciduous forest transforms to a tropical rain forest that experiences irregular catastrophic forest fires.

21. A new species is formed by chromosomal doubling of the parent species (2n = 6), followed by hybridization with a wild relative (2n = 10). What is the diploid number for this new species?
   (A) 11
   (B) 12
   (C) 22
   (D) 44

GO ON TO THE NEXT PAGE

22. The figure below is based on data from Peter and Rosemary Grant's research on the evolution of Galapagos finches. The data supports which of the following selection patterns?

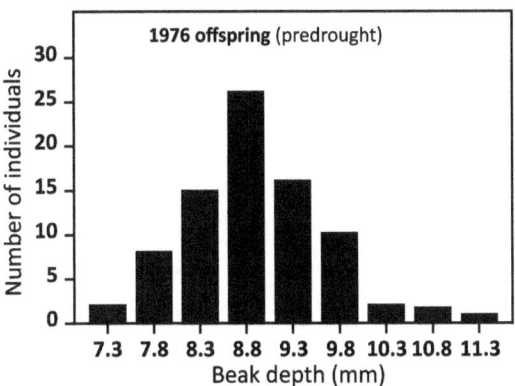

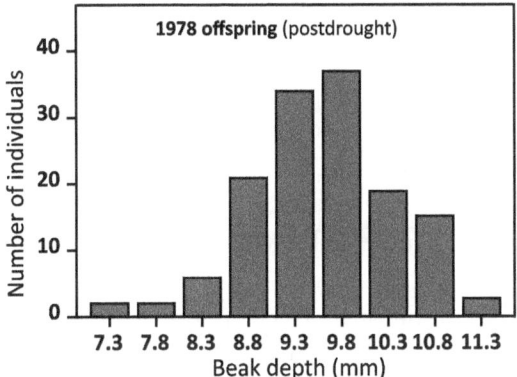

(A) Directional selection
(B) Stabilizing election
(C) Disruptive selection
(D) Sympatric selection

23. Which of the following statements is a correct description of the endosymbiotic theory?
(A) It explains how the organic soup gave rise to eukaryotic cells.
(B) It describes how macrophages are able to phagocytize amoeba.
(C) It describes how the chloroplast and mitochondria may have evolved in eukaryotic cells.
(D) It explains the consequences of prokaryotic phagocytoses of eukaryotes.

24. Which of the following provides the likely evolutionary sequence (first to last) for the following eukaryotic organelles?
(A) chloroplast > mitochondria > nucleus
(B) golgi > chloroplast > mitochondria
(C) rough ER > mitochondria > chloroplast
(D) mitochondria > vacuole > chloroplast

25. In an experiment, 10 caterpillars consumed 100 grams of collard greens. If the greens store 7 kilocalories per gram of total mass and caterpillar store 9 kcal/g, what is the average mass of each of the caterpillars? Assume that only 10% of the energy that enters a trophic level is passed on to the next trophic level.
(A) 128 g
(B) 78 g
(C) 0.8 g
(D) None of the above is correct.

26. 10 mL of solution A (conc. = 0 g/ml sucrose) is placed in a dialysis tube that is permeable to water, but not sucrose. The tube is than sealed and immersed into a beaker containing 10 mL of Solution B (conc. = 10 g/ml sucrose). What would you expect to observe at equilibrium?
(A) [solution A] > 10 g/ml
(B) [solution B] > 10 g/ml
(C) volume of solution A > 10 mL
(D) volume of solution B > 10 mL

27. Organisms use the excess energy than they consume for
(A) building energy storage molecules such as fats and carbohydrates.
(B) synthesizing structural and functional molecules that are involved in internal organization.
(C) producing heat.
(D) All of the above.

GO ON TO THE NEXT PAGE

28. Ectotherms are organisms that
    (A) have heat producing processes to maintain thermal homeostasis within a range of ambient temperature.
    (B) heat their bodies primarily through behavioral adjustments.
    (C) have the lowest energy efficiency.
    (D) consume a large amount of food relative to their body size.

29. Which of the following organisms would you expect to displays a Type I survivorship?
    (A) frogs
    (B) oysters
    (C) pine tree
    (D) elephants

30. Which of the following statements is true about secondary ecological succession of a temperate deciduous forest?
    (A) R-strategist will appear before K-strategist
    (B) K-strategist will appear before R-strategist
    (C) The dominant species following secondary succession is an r-selected species.
    (D) Secondary succession begins with species that are needed for soil deposition.

31. Which of the following is incorrect?
    (A) Annuals have a one-year life cycle and are typically the first plants to bud in the spring.
    (B) Vernalization prevents seed germination.
    (C) Dormancy stops or slows down growth.
    (D) Both germination and dormancy continues until the passing of winter.

32. When comparing an elephant and a deer, the
    (A) elephant's whole-body metabolic rate is lower than the deer's.
    (B) elephant is more energy efficient.
    (C) deer's mass-specific metabolic rate is lower than the elephant's.
    (D) deer and elephant have similar metabolic rates because they are both terrestrial mammals..

33. The start of the industrial revolution coincided with a rapid rise in human population. This was directly due to
    (A) increased fecundity coupled and infant mortality.
    (B) the availability of additional free energy to human population as organisms in higher trophic levels declined or went extinct, resulting in less competition for energy resources dropped.
    (C) a decrease in the number of armed conflicts, coupled with improved human longevity.
    (D) additional free energy from improvements in agricultural productivity.

GO ON TO THE NEXT PAGE

34. Which of the following statements about the carbon cycle is incorrect?
    (A) The majority of carbon enters the ecosystem as carbon dioxide for use during photosynthetic glucose production. Glucose is than used for different biochemical processes, including oxidative phosphorylation and macromolecule production.
    (B) The burning of fossil fuel releases lots of carbon gases into the atmosphere where, through the greenhouse effect, they increase the atmospheric temperature. This effect is moderated by the oceans, as they absorb a large portion of carbon dioxide, thereby helping maintain atmospheric thermal equilibrium at the expense of lower aquatic pH.
    (C) Most of the carbon in living systems is used for energy storage in nucleic acids.
    (D) Plants release carbon as carbon dioxide.

35. Over a 24-hour period, which of the following habitats should you expect to have the greatest fluctuation in ambient temperatures?
    (A) Temperate rainforest
    (B) Coastal marshland
    (C) Tropical rainforest
    (D) Tropical desert

36. The unique density property of water is critically important for the maintenance of the North Atlantic Gulf Stream. This movement of warm water and air northward from Florida and the Caribbean has the effect of moderating the climate of Western Europe and Britain by making the winters much less cold than is otherwise expected at that latitude. It is predicted that global warm will have an adverse effect on the Gulf Stream. Which of the following statements describe the most likely change to the climate of Western Europe and Britain?
    (A) The depth- and temperature-dependent circulation of the Atlantic water will stop or slow down, disrupting the Gulf Stream and causing Western Europe and Britain to experience ice age conditions.
    (B) Higher temperatures in the waters of Florida and the Caribbean will increase the flow rate of the Gulf Stream, thereby raising the northwestern temperature and causing Northern Europe and Britain to experience tropical conditions.
    (C) The increase in the water temperature of the North Atlantic will alter the relative temperature distribution along the path from The Caribbean to Scandinavia and cause the Gulf Current to reverse course. This will cause moderate increase in the temperature of Western Europe and Britain, while decreasing the temperature of the tropical Atlantic regions, including Florida and the Caribbean.
    (D) Global warming will bring minor increases in the temperature of Northern Europe on the scale of a couple of degrees. However, there will be no significant changes in the climate since the Gulf Stream operates in negative feedback loop that maintains conditions at the current set point.

GO ON TO THE NEXT PAGE

37. The ability of plants to move water and dissolved substances through the xylem from their root to their leaves relies on the
(A) cohesive and adhesive property of water molecules.
(B) capillary action of water molecules.
(C) surface tension of water.
(D) All of the above.

38. Soaps work well for washing away substances that are both hydrophilic and hydrophobic because
(A) soaps are amphiphilic.
(B) soaps interact well with water and hydrophilic substances.
(C) soaps interact well with hydrophilic substances.
(D) All of the above.

39. Which of the following statements about the cell membrane is true?
(A) Membranes are made of a double layer of phospholipids with the hydrophilic head connected on the interior.
(B) The membrane becomes more fluid as the number of saturated phospholipids increase.
(C) Membranes that contain a relatively large number of unsaturated phospholipids require cholesterol for stability.
(D) Saturated phospholipids have a bent shape and are less rigid.

40. When red blood cells are placed in distilled water, the cell
(A) shrinks in volume.
(B) lyses and dies.
(C) remains the same volume.
(D) expands to a maximum volume, but is protected from lyses by its cell wall.

41. Which of the following membranes will have the slowest rate of particle transport?
(A) a membrane that lacks aquaporin proteins, but has lots of other transport proteins.
(B) a membrane that lacks aquaporin proteins, but has a few of the other transport proteins.
(C) a membrane that has aquaporin proteins, but lacks other transport proteins.
(D) a membrane that neither aquaporin nor other transport proteins.

42. Which of the substances can enter the cell by simple diffusion?
(A) calcium ion
(B) peptide hormones
(C) glucose
(D) None of the above.

43. When white blood cells phagocytize pathogens, cytoplasmic primary lysosomes fuse with the phagocytic vesicles to form secondary lysosomes for the purpose of
(A) tagging the secondary lysosome with signaling proteins that instruct cellular processes to deport the pathogen from the body.
(B) quarantining the pathogen to prevent further infection.
(C) compartmentalization of the activity of hydrolytic enzymes that digest the pathogen
(D) determining the biochemical make-up of the pathogen and ensuring that the suspected pathogen is not a "self" cell.

GO ON TO THE NEXT PAGE

44. Natural selection have restricted cells from becoming infinitely small because cells that are too small
    (A) have a low rate of membrane import.
    (B) are limited in their ability to internally compartmentalize due to limited internal space.
    (C) form strong H-bonds with substances around them and become stuck.
    (D) All of the above.

45. When exposed to an experimental drug, otherwise healthy amoeba cells become more fluid and lose their ability to develop pseudopods that they use for "crawling" movement. This suggests that the drug acts on the amoeba's
    (A) ribosomes.
    (B) actin filaments.
    (C) endoplasmic reticulum.
    (D) Golgi apparatus.

46. Which of the following organelles is relatively abundant in cells of the thyroid gland?
    (A) smooth ER
    (B) Golgi apparatus
    (C) lysosomes
    (D) vacuole

47. During the ETC of photosynthesis,
    (A) water is the electron donor and NADP+ is the final electron acceptor.
    (B) NADPH is the electron donor and water is the final electron acceptor.
    (C) water is the electron donor and oxygen is the final electron acceptor.
    (D) oxygen is the electron donor and NADP+ is the final electron acceptor.

48. During the light reaction of photosynthesis,
    (A) each light absorbing pigment is able to pass its captured energy directly to the ETC.
    (B) the different types of antenna pigments evolved for the purpose of concentrating light absorbing molecules in each photosystem so that more green light is absorbed.
    (C) the energy collected in PSII and PSI are funneled from the antenna pigments to chlorophyll a, which then passes the energy on to the ETC.
    (D) when an electron from chlorophyll is excited and are passed to the ETC, NADPH donates one of its electron back to chlorophyll a.

49. Which of the following statements describes the direct impact of a drug that blocks the action of Fd-NADP+ reductase?
    (A) It blocks the light absorbing capabilities of the photosystems.
    (B) It prevents the transfer of electrons from photosystem I.
    (C) It blocks the donation of electrons from NADPH to photosystem II.
    (D) It inhibits ETC-coupled oxidative phosphorylation.

GO ON TO THE NEXT PAGE

The following figure shows the stem growth pattern in seedlings that have been manipulated by the various conditions shown. Note that opaque objects are non-transparent, and that gelatin is permeable to auxin, while mica is not. Refer to the figure when responding to questions 50-53.

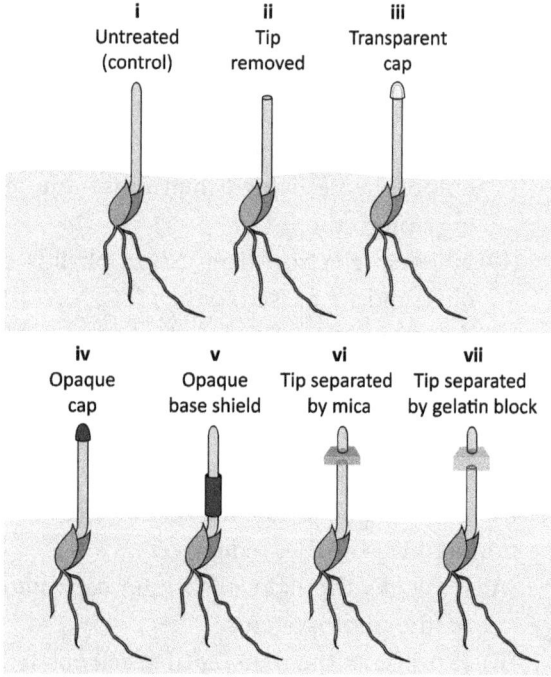

50. When the seedling detects directional light
(A) auxin molecules are moved from the shady to the sunny side.
(B) auxin molecules are moved from the sunny to the shady side.
(C) auxin molecules are sent toward the roots.
(D) more auxin molecules are produced on the shady side relative to the sunny side.

51. In which of the experiments should the seedling bend toward the light
(A) i, iii, v, and vii only
(B) ii, iv, vi only
(C) i, ii, iii, and iv only
(D) i, v, and vii only

52. The seedling from which experiment should have measurable concentrations of auxin?
(A) ii (tip removed)
(B) iv (opaque cap)
(C) vi (mica)
(D) vii (gelatin block)

53. Which of the following statements best summarizes the expected findings of the auxin experiment?
(A) Auxin is produced in the base stem of the growing seedling, but diffuses toward and concentrate on the sunny side where it stimulates cell growth.
(B) Auxin is produced in all stem cells of the growing seedling, but diffuses toward and concentrates at the tip where it stimulates cells division and the bending effect.
(C) Auxin is produced in the cells of the stem's tip, but diffuses towards and concentrates at the shady side of the stem where it stimulates cell elongation and the bending effect.
(D) The light stimulates cells to produce auxin and a concentration gradient is established from the sunny side (high) to the shady side (low). Because auxin negatively regulates mitosis and cell elongation, the growth rate on the sunny side is relatively lower and the stem becomes shorter on the sunny side, causing it to bend to the light.

GO ON TO THE NEXT PAGE

54. After the formation of the zygote, the process of cell division by cleavage leads to
(A) embryonic induction where tissue specific cells locally secrete signaling molecules called inducers that act regulate cell differentiation, and tissue and organ development among the locally determined cells.
(B) the activation of homeotic genes.
(C) cytoplasmic determination as certain maternally derived signaling molecules become unequally distributed among the daughter cells.
(D) apoptosis in cells that have outlived their utility during embryonic development.

55. The HER2 gene is a proto-oncogene that codes for a transmembrane tyrosine kinase receptor called human epidermal growth factor receptor 2. The HER2 protein functions as a mitotic-promoting factor in breast cells and supports the normal cell growth, division, and proliferation. p53 is a tumor suppression gene that is also involved in regulating cell division in breast tissue. What is the likely outcome of a chromosomal translocation of the HER2 gene to a chromosomal region with heavy histone methylation?
(A) The tumor suppression activity of p53 activity will decrease, leading to an increased risk of breast cancer.
(B) HER2 activity will remain unchanged because no genetic mutation has occurred. Since the gene remains fully intact, its regulation by transcription factors will not be affected.
(C) HER2 activity will decrease as a result of lower mRNA transcription, affecting the ability of the breast tissue to repair itself by replacing damaged cells through cell division.
(D) HER2 activity will increase because of increased mRNA transcription. This will increase the rates of cell growth, division, and proliferation and may lead to breast cancer.

56. A gene for coat color in cats is X-linked and codes for the dominant orange allele (X^B) and the recessive black allele (X^b). Which of the following genotypes is not possible for an orange cat?
(A) X^BY
(B) X^BX^B
(C) X^BX^b
(D) All of the above genotypes are possible for an orange cat.

57. Which of the following provides the correct sequence of events leading to the production of a mature mRNA?
(A) Transcription > splicing > addition of 5' cap > addition of poly A tail > export to cytoplasm
(B) Transcription > addition of 5' cap > addition of poly A tail > export to cytoplasm > splicing
(C) Transcription > addition of poly A tail > addition of 5' cap > export to cytoplasm > splicing
(D) Transcription > addition of poly A tail > addition of 5' cap > splicing > export to cytoplasm

GO ON TO THE NEXT PAGE

58. Which statement is correctly based on information from the following figure?

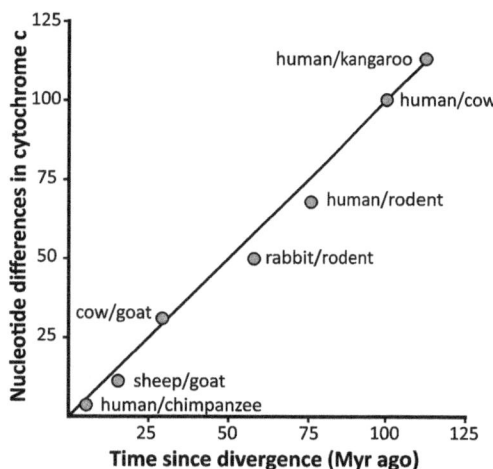

(A) Rabbit and rodent diverged more recently than the sheep and goat.
(B) The human and kangaroo are relatively closely related.
(C) The cytochrome C gene in human and chimpanzee has relatively few nucleotide differences between them.
(D) Human and kangaroo diverged less than 100 Myr ago.

59. The regulator of the trp operon is a multi-subunit protein that normally oscillates between an active and inactive conformation. Which of the following can be assumed about this protein?
(A) A corepressor allosterically alters the protein's activity.
(B) High concentrations of tryptophan stabilizes the enzyme's active conformation.
(C) The activity of the protein is inversely correlated with tryptophan levels.
(D) All of the above is incorrect.

60. The carrying capacity (K) of a population is the maximum population size supported by the local ecosystem. However, a population that expands to its carrying capacity will not abruptly acquire a growth rate, r = 0. Instead, it will continue expanding and will exceed its K value before entering logistic growth. Which of the following statements is true about logistic growth?
(A) During logistic growth, the population size oscillates around its carrying capacity with an ever decreasing frequencies until it acquires a growth rate, r = 0 at K.
(B) During logistic growth, the populations tend to shoot pass their carry capacity because organisms do not respond instantly to changes in the environment.
(C) Competition for resources always increases as the population approaches carry capacity.
(D) The carry capacity of a population is not affected by density independent factors such as floods or fires.

61. What is the growth rate of a population of 750 individual with a carrying capacity of 1000? Assume that the r_{max} is 2.
(A) -150
(B) 375
(C) 1500
(D) None of the above

62. Depolarization
(A) is caused by the movement of sodium ions out of the neuron.
(B) is driven by the pumping of sodium and potassium in opposite direction.
(C) occurs when the membrane potential becomes positive.
(D) None of the above statements is correct.

GO ON TO THE NEXT PAGE

7. In order for plants to grow tall, they needed to undergo adaptive evolution in favor of specialized tissue and processes for transporting water, minerals, and nutrients through the plant. The plant's xylem is one such tissue, and it transports water and minerals from the roots to the leaves.

(A) Describe the two emerging properties of water that the xylem uses to transport water, and explain how water transport from the roots to the leaves work. You may

GO ON TO THE NEXT PAGE

63. During the fight-or-flight response, the thyroid gland secretes its hormones that increases the organism's metabolism of carbohydrates for the purpose of ATP production. Which of the following carbohydrates is best for rapid energy release during the fight-or-flight response?
(A) chitin
(B) amylopectin
(C) glycogen
(D) cellulose

Directions (Part B - Grid-in): For each of the following 6 question, make the necessary calculations and record your answer as a numeric value on the provided grid.

64. The following data was generated from the mitosis lab where the effects of auxin and lectin on onion root growth was evaluated. Calculate the chi square (round to the nearest tenth) .

	Interphase	Mitosis
Control	194	32
Auxin treated	200	106

65. During the Hardy-Weinberg simulation lab, a student evaluated the effect of heterozygous advantage over 10 generations, with a starting p of 0.80 and ending p of 0.49. If the students keep the population constant at 26 individuals, how many homozygous individuals were present at generation 10? Report your answer as a whole number.

66. The table below show the trichome density (trichomes/cm) for each plant in the first generation of the artificial selection lab. Calculate the mean trichome density to the nearest tenth.

GEN 1 trichome density (trichome/cm)												
0	0	2	3	5	5	6	7	7	9	10	13	15
0	0	2	3	5	5	6	7	7	9	11	15	21
0	0	3	5	5	5	6	7	9	10	11	15	39

67. What is the probability (round to hundredth) that a cross between *AA BB Cc* and *aa Bb Cc* will produce an offspring that is heterozygous for all three genes?

68. In fruit flies, the white-eyed phenotype is due to a recessive mutant allele (X^W) located on the X chromosome. A white-eyed female and a normal male are crossed to produce the F1 offspring. From the F1 generation, a female and a male are crossed to produce the F2 generation of 224 individuals. How many white-eyed males are there in the F2 generation?

69. Catalase is the enzyme that catalyzes the metabolism of peroxide to water and oxygen. During an experiment, a student measures the volume (mL) of oxygen produced when the reaction mixture is kept on ice or at room temperature. The following data was collected at room temperature by measuring the volume of oxygen every 2 minutes from 0 to 10 minutes: 0.0 mL, 1.1 mL, 3.0 mL, 6.2 mL, 9.6 mL, 11.2 mL. What is the rate of the reaction (ml/min) from 4 to 8 minutes?

GO ON TO THE NEXT PAGE

Section II

Time: 90 minutes
Part A: 2 Long free-response questions
Part B: Short free-response questions

Directions (Part A - Long FRQs): Use complete sentences to answer each of the following questions. You may use diagrams and illustrations as supporting evidence of your mastery of the relevant concepts, but these alone are not sufficient. Be a detailed as possible without including information that is irrelevant to the question.

1. The non-venomous scarlet king snake shares noticeable superficial resemblance to the highly venomous coral snake. The similarity is primarily due to the colored ring-patterns displayed on their skin, which in coral snakes is red-yellow-black-yellow-red and in scarlet king snake it is red-black-yellow-black-red. It has been hypothesized that the scarlet king snake underwent adaptive selection to acquire its ring pattern, and that this mimicry help deter predatory bird attack.

   A study was conducted to investigate whether predatory birds have an innate general avoidance of all ring patterns that superficially resembled the coral snake's patterns, or whether they innately discriminate strictly against the R-Y-B-Y-R pattern of coral snakes. The study was conducted in an area native to venomous coral snakes, but that lacks native snake species with a superficially similar bicolor red-black ring pattern. Plasticine snake replicas were constructed with three different color patterns: the coral snake tricolor pattern (R-Y-B-Y-R), a non-native bicolor red-black pattern, and a monochrome brown pattern. 120 of each replica were placed randomly throughout the local forest and left for 48 hours before being surveyed for marks and scored as attacked or non-visually disturbed by non-avian attacks or other chance events. The following table shows the number of replicas that displayed signs of predatory avian attacks and non-visual disturbances.

Plasticine replicas	Predatory avian attacks	Non-visual disturbances
Brown	8	12
Bicolor	0	9
Tricolor	0	8

   (A) Why were brown and tricolor replicas used in this investigation?
   (B) Explain why the investigators used replicas with a non-native bicolor red-black ring pattern.
   (C) Using the provided data, construct the appropriate graph(s) for representing and comparing the data.
   (D) Does the data provide evidence that the predatory birds used generalized, specific, or random avoidance behavior toward ring patterns? Justify your response.
   (E) Does the data support an innate or a learned avoidance behavior? Justify your response.

GO ON TO THE NEXT PAGE

2. The figure below represents a sardaria spread that was produced by an F1 hybrid sporophyte.

(A) **Complete** the data table based on the sardaria spread.

	Phenotype	Tally	Total
Non-crossover asci	4:4 dark:light light:dark		
Crossover asci	2:2:2:2 dark:light:dark light light:dark:light:dark		
	2:4:2 dark:light:dark light:dark:light		

(B) **Calculate** the map units for the pigmentation gene, and **explain** what this value represents.

(C) **Sketch** a comparative bar graph of the relative percent of crossover to non-crossover asci.

Directions (Part B - Short FRQs): Use complete sentences to answer each of the following questions. You may use diagrams and illustrations as supporting evidence of your mastery of the relevant concepts, but these alone are not sufficient. Be a detailed as possible without including information that is irrelevant to the question.

3. A cross of a hybrid individual (Aa/Bb/Cc) with a recessive true-breeder (aa/bb/cc) produces the following offspring (letters indicate the observed phenotype): 190 ABC, 10 ABc, 1400 AbC, 100 Abc, 100 aBC, 1400 aBc, 10 abC, and 190 abc.

(A) Distinguish between the parental, double-crossover, and single crossover offspring.

(B) What is the order of genes A, B, and C on the chromosome? **Justify!**

(C) **Calculate** the map units between the three genes.

(D) **Diagram** the chromosome map and show the relative map units between the three genes.

GO ON TO THE NEXT PAGE

4. Temperature-dependent sex determination in some species of reptiles is regulated by the enzyme aromatase, which catalyzes the conversion of testosterone to estrogen.
    (A) What is the effect of temperature on aromase activity?
    (B) Explain how temperature influences impacts the amount of estrogen and testosterone present in developing embryos?
    (C) Explain how these sex hormones influence gonad development?

5. The following graph shows the variation in the body temperature of a specific organism as the ambient temperature changes from 2 to 40° C.
    (A) Base on the information provided in the graph, describe the most likely mechanism that this organism uses for maintaining internal thermal homeostasis. Justify your response!
    (B) If the organism's metabolic rate was measured over the same ambient temperature range, what do you predict will be the relationship between its metabolic rate and the ambient temperature.
    (C) What is one method that could be used to measure the organism's metabolic rate?.

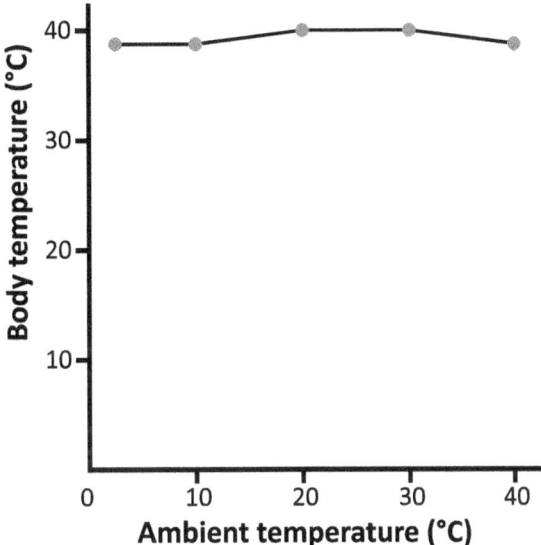

6. Epigenetic mechanisms are long-term heritable DNA modification that occurs through methylation rather than changes in nucleotide sequence. Epigenetic modification of DNA can result in non-Mendelian gene expression such that if a dominant allele of a gene becomes methylated, it is silenced, resulting in the expression of the recessive allele. Specific epigenetic processes include genomic imprinting, totipotent reprogramming, and X-chromosome inactivation.
    (A) Describe one of the above epigenetic processes.
    (B) Explain the developmental significance or the effects of the selected epigenetic process on the organisms.
    (C) Explain why is epigenetic modification not considered a mutation, although it can alter the organism's phenotype by effecting gene expression.

GO ON TO THE NEXT PAGE

chose to use a diagram as a supplement to your response.

(B) Describe the cohesive and adhesive properties of water, and explain how these properties allow plants to transport substances in addition to water.

8. Genetic variation in the gene pool can be preserved by diploidy, heterozygous advantage, frequency-dependent selection, and neutral variation.

(A) Describe one of the mechanisms listed above.

(B) Using a specific example of choice, explain how your selected mechanism has help preserve genetic variation is a particular gene pool.

DIAGNOSTIC TEST ANSWER FORM

NAME:_____

Section 1: 63 Multiple choice & 6 Grid-in questions

1 Ⓐ Ⓑ Ⓒ Ⓓ   31 Ⓐ Ⓑ Ⓒ Ⓓ   61 Ⓐ Ⓑ Ⓒ Ⓓ
2 Ⓐ Ⓑ Ⓒ Ⓓ   32 Ⓐ Ⓑ Ⓒ Ⓓ   62 Ⓐ Ⓑ Ⓒ Ⓓ
3 Ⓐ Ⓑ Ⓒ Ⓓ   33 Ⓐ Ⓑ Ⓒ Ⓓ   63 Ⓐ Ⓑ Ⓒ Ⓓ
4 Ⓐ Ⓑ Ⓒ Ⓓ   34 Ⓐ Ⓑ Ⓒ Ⓓ
5 Ⓐ Ⓑ Ⓒ Ⓓ   35 Ⓐ Ⓑ Ⓒ Ⓓ
6 Ⓐ Ⓑ Ⓒ Ⓓ   36 Ⓐ Ⓑ Ⓒ Ⓓ
7 Ⓐ Ⓑ Ⓒ Ⓓ   37 Ⓐ Ⓑ Ⓒ Ⓓ
8 Ⓐ Ⓑ Ⓒ Ⓓ   38 Ⓐ Ⓑ Ⓒ Ⓓ
9 Ⓐ Ⓑ Ⓒ Ⓓ   39 Ⓐ Ⓑ Ⓒ Ⓓ
10 Ⓐ Ⓑ Ⓒ Ⓓ   40 Ⓐ Ⓑ Ⓒ Ⓓ
11 Ⓐ Ⓑ Ⓒ Ⓓ   41 Ⓐ Ⓑ Ⓒ Ⓓ
12 Ⓐ Ⓑ Ⓒ Ⓓ   42 Ⓐ Ⓑ Ⓒ Ⓓ
13 Ⓐ Ⓑ Ⓒ Ⓓ   43 Ⓐ Ⓑ Ⓒ Ⓓ
14 Ⓐ Ⓑ Ⓒ Ⓓ   44 Ⓐ Ⓑ Ⓒ Ⓓ
15 Ⓐ Ⓑ Ⓒ Ⓓ   45 Ⓐ Ⓑ Ⓒ Ⓓ
16 Ⓐ Ⓑ Ⓒ Ⓓ   46 Ⓐ Ⓑ Ⓒ Ⓓ
17 Ⓐ Ⓑ Ⓒ Ⓓ   47 Ⓐ Ⓑ Ⓒ Ⓓ
18 Ⓐ Ⓑ Ⓒ Ⓓ   48 Ⓐ Ⓑ Ⓒ Ⓓ
19 Ⓐ Ⓑ Ⓒ Ⓓ   49 Ⓐ Ⓑ Ⓒ Ⓓ
20 Ⓐ Ⓑ Ⓒ Ⓓ   50 Ⓐ Ⓑ Ⓒ Ⓓ
21 Ⓐ Ⓑ Ⓒ Ⓓ   51 Ⓐ Ⓑ Ⓒ Ⓓ
22 Ⓐ Ⓑ Ⓒ Ⓓ   52 Ⓐ Ⓑ Ⓒ Ⓓ
23 Ⓐ Ⓑ Ⓒ Ⓓ   53 Ⓐ Ⓑ Ⓒ Ⓓ
24 Ⓐ Ⓑ Ⓒ Ⓓ   54 Ⓐ Ⓑ Ⓒ Ⓓ
25 Ⓐ Ⓑ Ⓒ Ⓓ   55 Ⓐ Ⓑ Ⓒ Ⓓ
26 Ⓐ Ⓑ Ⓒ Ⓓ   56 Ⓐ Ⓑ Ⓒ Ⓓ
27 Ⓐ Ⓑ Ⓒ Ⓓ   57 Ⓐ Ⓑ Ⓒ Ⓓ
28 Ⓐ Ⓑ Ⓒ Ⓓ   58 Ⓐ Ⓑ Ⓒ Ⓓ
29 Ⓐ Ⓑ Ⓒ Ⓓ   59 Ⓐ Ⓑ Ⓒ Ⓓ
30 Ⓐ Ⓑ Ⓒ Ⓓ   60 Ⓐ Ⓑ Ⓒ Ⓓ

64 [grid-in] 67 [grid-in]
65 [grid-in] 68 [grid-in]
66 [grid-in] 69 [grid-in]

ANSWERS AND EXPLANATIONS

Section I

1. (B) is correct. Stanley Miller experiment tested Alexander Oparin's hypothesis that the conditions of early Earth favored the formation of more complex organic compounds from simple inorganic molecules (i.e., hydrogen, ammonia, water, methane), and that living cells are not absolutely necessary to produce them.

2. (B) is correct. Although DNA has critical information storage functionality, it is relatively biochemically inert – meaning that DNA does not catalyze life supporting biochemical reactions or metabolic pathways. Option (A) is wrong because, although DNA does not self-replicate, the enzyme DNA polymerase can replicate it, which is a protein. Option (C) is wrong because, compared to RNA and protein, DNA is relatively structurally stable.

3. (D) is correct.

4. (C) is correct. Although high infant mortality (Option A) is a reasonable explanation for why populations remain stable, Darwin explained the survivorship of newborns as being typically due to competition. Option (B) is wrong because Darwin did not study the effects of immigration of predators on prey populations. Option (D) is wrong for a number of reasons: 1) Although Darwin later studies sex selection, he did not originally build his theory on the observation of sex selection, and 2) the growth rate of a population is not dependent on the number of available males, rather it is the number of females that are the limiting factor in the population growth rate. One male can father a virtual infinite number of newborn, while a single female can only produce a finite number depending on the species.

5. (D) is correct.

6. (C) is correct.

7. (D) is correct.
    $q^2 = 30/50 = 0.60$
    $q = (0.60)^{0.5} = 0.77$
    $p = 1 - 0.77 = 0.23$
    $2pq = 2 (0.77)(0.23) = 0.35$
    $n2pq = 50(0.35) = 17$

8. (C) is correct. As the major tectonic plates drifted apart to different parts of the planet, they each became subject to different environmental conditions that was most influenced by the angle of sunlight reaching the plate and local precipitation conditions. As some plates drifted away from the equator, they experiences greater seasonal fluctuation in temperatures (i.e., winter, spring, summer, fall conditions). Other plate crashed into each other to create tectonic uplifts that form mountains that trapped clouds on the ocean facing side of the range (i.e., south side of Mount Everest), while creating dry desert condition on the other side. There are numerous other possible scenarios, each requiring the local populations to adapt to the changing conditions. Thus, different populations of the same ancestral species were pressured to evolve with the unique characteristics that allow them to live in their new environment.

9. (B) is correct.

10. (B) is correct. Fossil B is present at all of the site, suggesting that it was widely distributed, and it is restricted to a single strata layer at each site, suggesting that it was short lived.

11. (C) is correct. Birds and mammals are believed to be descendent from some primitive reptile, and reptiles are descendent from a primitive amphibian. This means that reptiles are transitional between amphibian and mammals/birds. This is supported by the fact that reptiles have a ventricle that is intermediate between that of amphibians (3 chambered heart/ 1 ventricle) and birds/mammals (4 chambered heart/ 2 ventricles) in that it is partially separated by a septum. The expression of Tbx5 in reptiles should therefor be intermediate between that of amphibian (homogenous expression) and birds/mammals (expression in left ventricle only). An intermediate expression would mean a gradient of expression from the left to right ventricle, with the highest expression in cells of the far left ventricular wall, the lowest or no expression on the cells of the far right wall, and the cells in between showing progressively lower expression moving toward the left wall.

12. (D) is correct.
$Rh^+ = 225$;
$Rh^- = 1000 - 225$
$q^2 = 775/1000 = 0.78$
$q = (0.78)^{0.5} = 0.88$
$p = 1 - q = 1 - 0.88 = 0.12$

13. (D) is correct. Sex selection (option A) reduces genetic variation, as does all forms of selection, by selection against unfavorable alleles. Similarly, genetic drift (option B) is a random form of selection that removes "unlucky" alleles from the population. Recall that catastrophic events such as forest fires (option C) can cause genetic drift. Only immigration, mutations, or lateral gene transfer from other species can introduce new alleles into a gene pool.

14. (B) is correct. The decline of light colored lichen on forest trees coincided with a decrease in the adaptive advantage (camouflage) that white bodies moths had over dark colored moth. As the tree darkened, white moths became easier targets for predatory birds, while dark bodies moths became better camouflaged.

15. (D) is correct. Options (A) and (B) are examples of genetic drifts caused by the founder effect when a small segments of the original population migrate and become the founders of a new population. Option (C) is an example of the bottleneck effect when a catastrophic event significantly reduces a population size. In this example, human hunting had a catastrophic affect on the northern elephant seal population. In all three cases, the surviving alleles were selected by change and not as a result of their conferring any adaptive advantage.

16. (C) is correct. Base on the provided information, you should assume that the Trp operon activity decreases as the concentration of trp increases, and vice versa. This means that the operon is under negative feedback control. At high concentrations of tryptophan, the operon should be repressed either by deactivation of an activator, or by activation of a repressor. Since it product (tryptophan) and not the substrate that causes the change in activity, it is reasonable to assume that the product (tryptophan) is a corepressor. Therefore, the operon is regulated by a repressor that binds to it operator located downstream from the promoter.

17. (B) is correct.

18. (C) is correct. When the recessive allele confers a significantly greater adaptive advantage, both the heterozygous and the homozygous dominant will be selected against because

they both express the dominant phenotype. Consequently, overtime, the dominant allele will be eliminated. In options (A) and (B) the recessive allele will be preserved in heterozygous individuals who will show the same fitness as the homozygous dominant individuals. In option (D), frequency dependent advantage will cause both alleles to be preserved.

19. (B) is correct. The scarlet king snake likely evolved from a common ancestor with other king snakes when a sub-population of the parent species under went natural selection in favor of the coral snake pattern. Although geographic isolation in an area populated by coral snakes can accelerate the speciation process by allopatric speciation (option A) or peripatric speciation (option C), it is more likely that the divergence of scarlet king snakes occurred by sympatric speciation (option A) since scarlet kind snakes tend to live in areas shared by other king snakes, suggesting that they were not geographically isolation from the larger group. Instead, sympatric speciation requires speciation event requires the sub-population to begin occupying a unique niche from the larger group. Part of the diverging subgroup's new niche may have included different mating patterns, resource sequestration, and predatory defense mechanisms. Allopatric speciation. Parapatric speciation

20. (C) is correct. In all four scenarios, the climate changes in the same way, however, option C will cause the most dramatic change because the climate changes over a shorter period. Coupled with the occurrence of irregular forest fires, there will be a significant loss of genetic variation in the population in favor of those individuals that are uniquely adapted to the change. Option D that also experiences the same climate change and irregular forest fires is less dramatic because the change is more gradual and will result in higher genetic

preservation (or less genetic drift).

21. (C) is correct.
    Step 1: chromosomal doubling
    $2n = 6 \Rightarrow 4n = 12$; or $2n = 12$
    Step 2: hybridization between $2n = 12$ ($n = 6$ gamete) and $2n = 10$ ($n = 5$ gamete)
    $(n = 6) \times (n = 5) \Rightarrow n = 11$
    Step 3: hybrid non-disjunction during mitosis to become $2n = 22$.
    Step 4: Further self-fertilization to establish a starter population.

22. (A) is correct because the median shifted from about 8.8 to 9.8 mm.

23. (C) is correct.

24. (C) is correct. The endomembrane system, which includes the endoplasmic reticulum, evolved first, followed by the mitochondria, and finally the chloroplast.

25. (C) is correct. At the first trophic level (collard green) there is $100 \text{ g} \times 7 \text{ kcal/g} = 700$ kcal of stored energy. 10% of this energy, or 70 kcal, is transferred to the 2nd trophic level (caterpillars). The total mass of all the caterpillars together is $70 \text{ kcal} \div 9 \text{ kcal/g} = 7.8 \text{ g}$. Average mass $= 7.8 \text{ g} \div 10 \text{ caterpillars} = 0.78$ or 0.8 g.

26. (D) is correct. Solution B is hypertonic to solution A, therefore water will move from solution A to solution B and will cause the volume of solution B to increase from 10 mL. Option A is incorrect because the concentration of Solution A will remain 0 g/mL sucrose since sucrose is restricted from crossing the semipermeable membrane to be imported. Option B is incorrect because as solution B gains water, its concentration of

sucrose will decrease below its starting concentration of 10g/mL.

27. (D) is correct

28. (B) is correct. Ectotherms are "cold-blooded" organisms such as reptiles and amphibians whose body temperature adjusts to the ambient temperature. As a result, they must change their behavior by moving to areas that have the most ideal ambient temperatures. For example, reptiles might lie in the sun to heat their bodies, or in water to cool down.

29. (D) is correct. Elephants have a relatively low reproductive rate and parents invest heavily in the care of offspring. Like humans and domesticated pets, they have relatively low infant mortality rate, and because they have no natural predators, they tend to live to close to their maximum life expectancy (or old age).

30. (A) is correct. Primary ecological succession occurs in newly created habitats such as new islands formed by volcanic eruption, lava flow, or glacial retreat. Primary succession occurs slowly because new soil has to be deposited by pioneer species such as fungi and lichen before plants and animals can move in. Secondary succession occurs in previously established ecosystems that have experienced a significant catastrophic event such as a forest fire that killed off most of the preexisting species. Unlike primary succession, there is already deposited soil so plants can begin moving in, followed by animals. The first species to move in are plants with high reproductive rates and short life expectancy such as grasses. The first types of animals will also be those with high reproductive rates and short life expectancy, and those that are herbivores that feed primarily on grasses, such as insects. In both cases, these secondary pioneers are r-strategists (higher reproductive rates, short life expectancy, low parental care, high infant mortality).

31. (B) is incorrect. Vernalization prevents flowering (not seed germination) before the winter period has passed.

32. (B) is correct. Although elephants have a higher total metabolic rate, their mass-specific metabolic rate is lower, meaning that they use less energy per gram of body tissue than the deer.

33. (D) is correct.

34. (C) is incorrect. Most of the carbons in living systems are stored in carbohydrates and fats, and used in the synthesis of proteins. The amount stored in nucleic acids is relatively miniscule.

35. (D) is correct. Tropical deserts, such as the Sahara, reach peak temperature at mid-day when the sun is directly above (at 90 degrees) and drop significantly to chilling temperatures at night. The fluctuation in temperature is due to low humidity. In tropical rain forest, the high moisture content in the air allows for good heat storage throughout the day, and the slow release of this stored heat through the night. As a result, in tropical rain forests, the temperature is moderated throughout the 24-hour period, while in tropical desert there is greater temperature swings due to an inability to efficiently store heat during the day.

36. (A) is correct.

37. (D) is correct.

38. (D) is correct.

39. (C) is correct.

40. (B) is correct. Since distilled water is hypotonic relative to the RBC, water will move into the RBC until is expands beyond its membrane capacity and burst (lyse), destroying the cell. Unlike plant cells, RBC have no cell wall to protect them against lysis.

41. (C) is correct. Aquaporins work to equilibrate the concentration across a membrane by moving water molecules, while transport protein move particles. When both aquaporins and transport proteins are present, the membrane will quickly equilibrate. With only transport proteins the rate of particle transport will be higher than if aquaporins are also present since the concentration difference across the membrane will be quickly reduced by water transport. With only aquaporins, restricted particles will not be transported by facilitated diffusion, resulting in the slowest possible rate of transport by simple diffusion.

42. (D) is correct. Calcium ions are charged and charged particles must pass through protein channels. Glucose and peptide hormones are large and hydrophilic and are therefore unable to cross the hydrophobic tail region of the membrane core. Instead the are imported into the cell by vesicle mediated endocytosis.

43. (C) is correct.

44. (B) is correct.

45. (B) is correct. Pseudopod based movement rely on the extension of actin filaments that make-up part of the cell cytoskeleton.

46. (B) is correct. Endocrine glands produce and secrete their specific protein products into the blood stream. Secretion occurs Golgi apparatus derived secretory vesicles that fuse to the cell membrane and release their products through exocytosis.

47. (A) is correct. Recall that at PSII, water is split into H^+ and O_2, and 2 electrons from water are donated to the ETC. The ETC than delivers its electrons to $NADP^+$, which becomes NADPH.

48. (C) is correct. Only chlorophyll a can pass energy to the ETC.

49. (B) is correct. Fd-$NADP^+$ reductase is the enzyme that delivers the ETC electrons to $NADP^+$ - it reduces $NADP^+$ to NADPH.

50. (B) is correct. The high concentration of auxin on the shady side of the stem stimulates the cells there to elongate, causing the stem to bend toward the light.

51. (A) is correct. Auxin is produced in cells at the tip of the growing stem. Auxin than diffuses to the shady side of the stem where it stimulate cell elongation. Without any treatment (control) the seedling should bend toward the light. If the tip is removed (experiment ii) or shaded from light (experiment iv), no auxin should be produced and the seedling should not bend toward the light. In experiments iii (transparent cap) and iv (opaque base), the tip should still receive light, therefore auxin should be produced and bending should be observed. In experiments vi the stem is cut and reattached using a block of mica which prevents the diffusion of auxin from the tip to the rest of the stem, therefore no bending should be observed. In experiments vii the stem is cut and reattached using a block of

gelatin, which allows the diffusion of auxin from the tip to the rest of the stem, therefore bending should be observed.

52. (D) is correct. See the above explanation.

53. (C) is correct. See the above explanation.

54. (C) is correct.

55. (D) is correct. The translocation of the HER2 gene to a highly methylated chromosomal region will result in increased activity because such a region is loosely packed since methylated histones have a low affinity for the DNA. Loosely packed chromatin contains genes with high transcriptional levels.

56. (C) is not a possible genotype for an orange cat. $X^B X^b$ is a female cat that has both the orange and black allele. Due to X inactivation, heterozygous female cats will display patches of both back and orange, a phenotype commonly referred to as tortoise-shell or calico.

57. (D) is correct.

58. (C) is correct.

59. (D) is correct.

60. (B) is correct. Option C is wrong because when the population is already above carrying capacity, competition will be intense and the population will experience negative growth. This causes the population size to progressively decrease toward K, and as this happens, competition will decrease. The reverse occurs when the population is already below K – is it approaches K, competition will increase.

61. (B) is correct.
$$dn/dt = r_{max} N (K - N)/K$$
$$dn/dt = 2(750)(1000 - 750)/1000 = 375$$

62. (D) is correct. Depolarization occurs when the membrane potential moves from -70 mv (to -50 mv at threshold potential) and finally to 0 mv as a result of the opening of the sodium gates, which causes positively charged sodium ions to move into the neuron.

63. (C) is correct. The fight-or-flight response is an animal adaptation, therefore only those carbohydrates that are produced and/or metabolized by animals are relevant. Specifically, glycogen is the energy storage carbohydrates in animals and it is highly branched to allow for the release of multiple glucose subunits simultaneously from the same glycogen molecule.

64. The acceptable whole-number range is between 104 and 109.
Step 1: the expected frequency for the interphase and mitotic cells based on the control values: $f_{interphase} = 194/226 = 0.86$; $f_{mitosis} = 1 - 0.86 = 0.14$
Step 2: calculate the expected values using the expected frequencies and the experimental (auxin treatment) sample size: Expected (interphase) = 0.84(306) = 263; Expected (mitosis) = 0.14(306) = 43.
Step 3: complete the chi square table and add-up the chi values for interphase and mitosis: $[(200-263)^2/263]+[(106-43)^2/43] = 15.0 + 90.7 = 105.7$

65. Only the exact value of 13 is acceptable.
Number of homozygous = $n(p^2+q^2)$
= $26(0.49^2 + 0.51^2) = 13$.

66. Acceptable range is 7.2 to 7.6. Solve by adding all of the values and dividing by the sam-

ple size of 39.

67. Only the exact value of 0.25 is acceptable.
AA $\times$ aa $\Rightarrow$ 1.0 Aa
BB $\times$ Bb $\Rightarrow$ 0.5 Bb and 0.5 BB
Cc $\times$ Cc $\Rightarrow$ 0.5 Cc, 0.25 CC, and 0.25 cc
AaBbCc probability $\Rightarrow$ (1.0)(0.5)(0.5) = 0.25

68. Only the exact value of 56 is acceptable.
P cross $X^wX^w \times X^+Y \Rightarrow$ all X^wX^+ (female) & X^wY (male)
F1 cross $X^wX^+ \times X^wY \Rightarrow$ 0.25 X^wX^w, 0.25 X^+X^w, 0.25 X^wY 0.25 X^+Y
White-eyed males (X^wY) = 0.25(224) = 56

69. Acceptable range is 1.6 to 1.7.
Step 1: Determine the volume of oxygen produced at time of 4 and 8 minutes.
= 3.0 and 9.6 mL.
Step 2: Calculate the reaction rate.
R. rate = (9.6 − 3.0)/(8 − 4) = 1.65 mL/min

Section II

1. (A) Brown replicas were used as the positive control group and the tricolor replicas were the negative control group. The brown replicas represent the normal nonvenomous snakes that predatory birds would normally seek out and attack, while the tricolor represented the normally avoided venomous coral snakes. The investigators should have expected that the Brown replicas would have the highest frequency of predatory avian attacks (positive attack effect), while the tricolor replicas should have the lowest (negative attack effect).

(B) Since the experiment was investigating whether avoidance behavior in predatory birds was due to generalized or specific ring pattern avoidance, the scientists needed to use replicas that shared some ring pattern similarities, but that were not exactly identical to the negative control. They chose to use the red and black ring color of the venomous coral snake, but not the yellow. They also specifically selected a ring color pattern that was not present in any native snake species so as to control for possible error due to previously learned behavior. By so doing, they could argue with greater confidence that any difference between the bicolor and the controls was due to innate behavior.

(C) Graphs

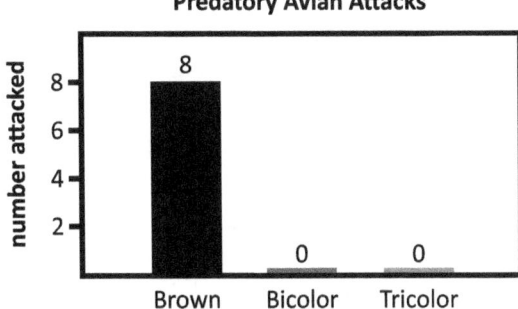

Predatory Avian Attacks

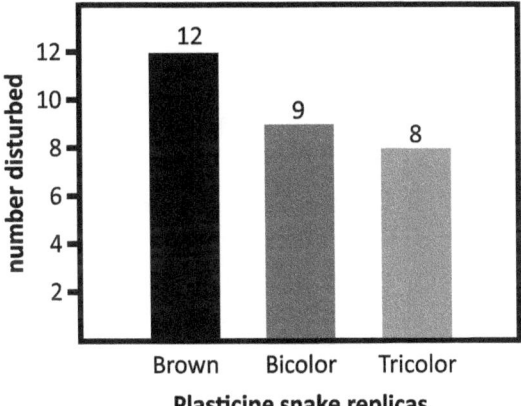

Non-visual disturbances

Plasticine snake replicas

(D) The data supports the conclusion that predatory birds were using innate generalized avoidance because they seemed to avoid the bicolor replicas at a similar rate as the negative control tricolor replica. It is unlikely that this avoidance behavior is learned for two reasons: first, birds that had a negative experience with a venomous snake would likely die and not have a second opportunity to avoid the snake's pattern via a learned experience, or to teach their offspring to avoid such patterns; and second, it is unlikely that the birds would have encountered such a pattern since the investigators specifically used one that was absent in all native snake species.

(E) Consequently, the predatory birds must be avoiding the bicolor ring pattern because they inherited an adaptive evolutionary trait that causes them to innately avoid all general ring patterns that share some similarities to those of venomous snakes such as the coral snake.

2. (A) In order to complete the data table, you must make a tally of the number of non-crossover and cross over asci in each of the individual. A non-crossover asci will display its 8 ascospores in a 4:4 ratio of gray (dark color) to tan (light color), while cross-over asci have a ratio of 2:4:2 or 2:2:2:2 with alternating coat color. Asci that have only one color are derived from homozygous parent cells and are not included in the tally. The first asci spread is show below. This asci has 16 non-crossover, 4 crossover, and 4 homozygous. The 4 homozygous are not included. The other 7 asci has the following tally count of non-crossover/crossover: 19/5, 16/6, 19/5, 20/4, 9/5, 19/5, 16/6. The total count of non-cross over from all 8 asci is 134, and crossover is 40 (there are 18 homozygous).

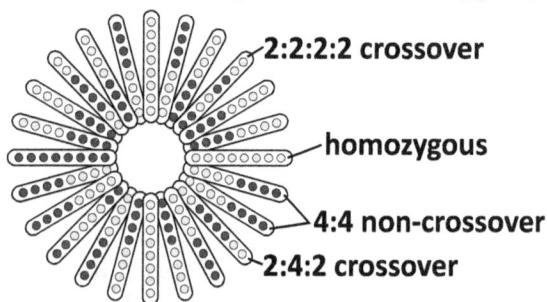

(B) Calculate the recombinant frequency. The map units is equal to half of the recombinant frequency. Therefore, there are 11.5 map units. The higher the map units, the further is the gene from the centromere region on the chromosome. This is because, the centromere seems to inhibit cross-

ing over so that the closer a gene is to the centromere, the lower the probability that a crossover event will occur between that gene and the centromere. Thus, the map units is a measure of the relative distance a gene is located to the centromere. It is also correlates to the crossover frequency of that gene.

$$\text{Recombinant frequency} = \frac{\text{\# recombinant asci}}{\text{total \# of asci}} \times 100$$
$$= 40(100)/(134 + 40)$$
$$= 4000/174 = 23$$

(C) Graph of relative percent of crossover versus non-crossover asci.

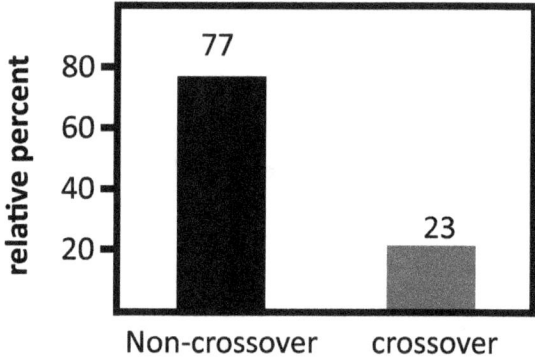

3. See chapter 18 for solution.

4. (A) As with all proteins, aromatase activity is influenced by temperature and pH. and is highest at its peak temperature. Any deviation (up or down) from that ideal temperature will result in lower activity. In temperature-dependent sex determination, aromatase activity decreases with the ambient temperature. (B) Since aromatase catalyzes the conversion of testosterone to estrogen, at peak temperature (assumed to be at the higher range of ambient temperature), more estrogen will be produced, while under cooler conditions, aromatase activity is lower and less estrogen is produced. Thus, at higher temperature the relative amount of estrogen to testos-

terone is higher in the developing embryo. The reverse is the case at lower temperature. (C) Testosterone stimulates the male sex characteristics, including testis development. Estrogen is responsible for female characteristics, including ovary development. Since at lower temperatures testosterone levels are higher, more males are produced, while at higher temperature more females are produced.

5. (A) The organism appears to maintain a constant internal temperature over a range of ambient temperature. This suggest that it is an endotherm that relies primarily on internal metabolic processes to produce heat. Some examples of specific heat producing processes include increased metabolism, shivering, increased heart rate, and dilation of capillaries. (B) Since at lower temperatures, the organism needs to produce additional heat in order to maintain internal thermal homeostasis, its metabolic rate will be higher at lower temperatures. (C) Metabolic rate can be measured via oxygen consumption, respiratory rate, heart rate, or pulse rate.

6. (A) See chapter 21.
    (B) See chapter 21.
    (C) Epigenetic modification is not considered a mutation because the nucleotide sequence of the gene is not altered, instead specific nucleotides are reversibly methylated, resulting in gene repression.

7. (A) The xylem uses capillary action and surface tension to move water. See chapter 11 for a complete explanation and diagrams. (B) See chapter 11 for a description and explanation of the cohesive and adhesive property. Together this two property makes water a universal solvent that can dissolve substances for transport through the xylem and phloem.

8. (A) See chapter 6 for description.
    (B) Answers will vary. See chapter 6 for examples.

Formulas & equations

STATISTICAL ANALYSIS AND PROBABILITY	
Standard Error	Mean
$$SE_{\bar{x}} = \dfrac{s}{\sqrt{n}}$$	$$\bar{x} = \dfrac{1}{n}\sum_{i=1}^{n} x_i$$
Standard Deviation	Chi-Square
$$s = \sqrt{\dfrac{\sum(x_i - \bar{x})^2}{n-1}}$$	$$\chi^2 = \sum \dfrac{(o-e)^2}{e}$$

s = sample standard deviation (i.e., the sample based estimate of the standard deviation of the population)
x = mean
n = population size
o = observed individuals with observed genotype
e = expected individual with observed genotype

Degrees of freedom equals the number of distinct possible outcomes minus one.

CHI-SQUARE TABLE

Degrees of freedom

p	1	2	3	4	5	6	7	8
0.05	3.84	5.99	7.82	9.49	11.07	12.59	14.07	15.51
0.01	6.64	9.32	11.34	13.28	15.09	16.81	18.48	20.09

LAWS OF PROBABILITY

If A and B are mutually exclusive, then P(A or B) = P(A) + P(B)
If A and B are independent, then P(A and B) = P(A) x P(B)

HARDY-WEINBERG EQUATIONS

$p^2 + 2pq + q^2 = 1$

$p + q = 1$

p = dominant allele frequency in the population
q = recessive allele frequency in the population

METRIC PREFIX

Factor	Prefix	Symbol
10^9	giga	G
10^6	mega	M
10^3	kilo	k
10^{-2}	centi	c
10^{-3}	milli	m
10^{-6}	micro	μ
10^{-9}	nano	n
10^{-12}	pico	p

Mode = value that occurs most frequently in a data set
Median = middle value that separates the greater and lesser halves of the data set
Mean = sum of all data points divided by number of data points
Range = value obtained by subtracting the smallest observation (sample minimum) from the greatest (sample maximum)

RATE AND GROWTH		Water Potential (Ψ)
Rate dY/dt **Population growth** $dN/dt = B - D$ **Exponential growth** $dN/dt = r_{max}N$ **Logistic growth** $dN/dt = r_{max}N[(k - N)/K]$	dY = amount of change t = time B = birth rate D = death rate N = population size K = carrying capacity r_{max} = maximum per capita growth rate of the population	$\Psi = \Psi_P + \Psi_S$ Ψ_P = pressure potential Ψ_S = solute potential The water potential will be equal to the solute potential of a solution in an open container because the pressure potential of the solution in an open container is zero.
Temperature Coefficient Q_{10} $Q_{10} = \left(\dfrac{k_2}{k_1}\right)^{\frac{10}{t_2-t_1}}$ **Primary Productivity Calculation** mg O_2/L x 0.698 mL O_2/L mg O_2/L x 0.536 mg Carbon fixed/L	t_2 = higher temperature t_1 = lower temperature k_2 = metabolic rate at t_2 k_1 = metabolic rate at t_1 Q_{10} = the factor by which the reaction rate increases when the temperature is raised by 10 degrees.	**The Solute Potential of a Solution** $\Psi_S = -iCRT$ i = ionization constant (this is 1.0 for sucrose because sucrose does not ionize in water) C = molar concentration R = pressure constant (R = 0.0831 liter bars/mole K) T = temperature in Kelvin (ºC + 273)
SURFACE AREA AND VOLUME		**Dilution used to create a dilute solution from a concentrated stock solution** $C_i V_i = C_f V_f$ i = initial (starting) C = concentration of solute f = final (desired) V = volume of solution
Surface Area and Volume Volume of a Sphere $V = 4/3 \, \pi r^3$ **Volume of a Rectangular Solid** $V = lwh$ **Volume of a Right Cylinder** $V = \pi r^2 h$ **Surface Area of a Sphere** $A = 4\pi r^2$ **Surface Area of a Cube** $A = 6a = 6s^2$ **Surface Area of a Rectangular Solid** $A = \Sigma$ (surface area of each side)	r = radius l = length h = height w = width a = surface area of one side of a cube s = length of one side of a cube A = surface area V = volume Σ = sum of all	**Gibbs Free Energy** $\Delta G = \Delta H - T\Delta S$ ΔG = change in Gibbs free energy ΔS = change in entropy ΔH = change in enthalpy T = absolute temperature (in Kelvin) **pH** = $- \log_{10} [H^+]$

GLOSSARY

Abiogenesis is a hypothetical, but empirically supported gradual evolutionary process through which life may have emerged from non-living matter.

Absolute dating is a radiometric method of determining the exact age of fossils.

Acquired immunity is the third line of defense that involves the development of specific antibodies against a specific foreign pathogen or antigen.

Activation energy is the energy need to start a reaction. The active site of enzymes lowers the activation energy to accelerate the reaction.

Activators are protein that stimulate transcription by interacting with the DNA enhancer regions to help RNA polymerase bind the promoter.

Active transport the movement of particles against their concentration gradient.

Adhesion occurs when two different substances are attracted to each other and form Van der Waals interactions.

Aerobic cells are those that can undergo aerobic respiration in the presence of oxygen to produce ATP.

Aerobic respiration, which includes the Krebs and Calvin cycles, is the oxygen dependent process of food metabolism to produce 36 ATPs. It will be discussed in more details later.

Algal bloom is a rapid increase in algae population in aquatic ecosystem that is often caused by agricultural run-off carrying fertilizer or animal waste to the waterways.

Alleles are different forms of a gene. When a gene first appears in the genome, there is only one form. Mutations produce additional alleles. Dominant alleles are typically symbolized with a capital letter and recessive alleles with a lowercase letter.

Allele frequency is a measure of how common one allele of a given gene is to all the other alleles of that gene. P represents that dominant allele frequency and q the recessive allele frequency. It is a value between 0 and 1

Allosomes are the sex chromosomes that are also know as the X and Y chromosome.

Allosteric inhibitors are noncompetitive inhibitors that bind to a site other then the active site. The negatively regulate enzymes by inducing a conformational change to the inactive state.

Amphiphilic substances have both hydrophilic and hydrophobic regions.

Anaerobic respiration, which includes glycolysis and fermentation, is the oxygen independent process of food metabolism to produce 2 ATPs. It will be discussed in more details later.

Analogous structures are comparative structures that have superficially similar appearance and serve similar functions, but that do not share an evolutionary origin.

Annuals are plants that complete their life cycle in one year. They mature, reproduce and die in a single year.

Antibiotic is a drug or chemical that specific targets and kills bacteria cells.

Antigen is an abbreviated form of 'antibody generator'. Antigens are substances (including proteins) that complement antibodies.

Antigen-presenting cells (APC) are body cells that display protein antigens that are not recognized by T cells as MHC self-protein on their cell surface. Non-professional APC that display foreign antigen either directly or in association with MHC glycoproteins are attacked by killer T cells. Professional APC (macrophages and dendritic cells) serve a special role of signaling T cells to differentiate and activate B and T cell that target the specific antigen - unlike non-professional APCs, they can induce a full immune response.

Antiporters carry two different particles - one at a time - in opposite directions.

Apoptosis is a self-induced mechanism that regulates the programmed destruction of a cell.

ATP is the primary fuel molecule that cells use to power most of their energy absorbing endergonic reactions. It is produced through the process of oxidative phosphorylation.

Autosomes are the non-sex chromosomes. In human, these are chromosomes 1 through 22.

Autotrophs (or producers) are the organisms of the first trophic level of an ecosystem that can harness energy directly from the sun or inorganic sources.

Auxin is a plant hormone that plays a role in cell and tissue growth.

Bacteria conjugation is the exchange of genetic material between two bacteria cell through direct cell-to-cell contact using a bridge-like connection call pilus.

Barr bodies are the inactivated X chromosome in female somatic cells.

Biennials are plants that complete their life cycle in two years. They reproduce in the second year and than die.

Biochemical functionality refers to the ability of organic molecules (RNA and proteins) to catalyze reactions that improve the information collection and self-replication properties of biological systems.

Biodiversity is a measure of the number and variety of organisms within a defined geographic area.

Biogeography is the study of how species ad ecosystems are distributed geographically and over geologic time.

Biomass is a measure of the amount of organic matter derived from a living organism.

Biostratigraphy is a relative dating method that correlates the relative age of fossils to their position in the rock strata.

Bottleneck effect is a kind of genetic drift caused by a catastrophic event that kills off a large portion of the population.

Catastrophism is an idea that evolution and extinction is heavily influenced by abrupt and brief periods of rapid and large scale environmental changes.

Centrioles produce the mitotic spindle fibers that separate the chromosomes or sister chromatids during cell division.

Chemosynthetic bacteria are chemoautotrophs that source their energy from inorganic chemicals such as hydrogen sulfide or hydrogen gas.

Chlorophyll-protein complex is a functional unit of the chloroplast photosystems of thylakoid membrane. They are responsible for harnessing light energy.

Chloroplasts are photosynthetic organelles present in plant cells. They repackage light energy in glucose molecules through a reaction that occurs in reverse to aerobic respiration.

Chordata is a phylum that includes all animals with a notochord - the shared derived character. In vertebrates, the notochord develops into the spinal cord.

Clade is a group that contains 2 or more taxa that share a common ancestor.

Cladogram is a branch of the phylogenetic tree that shows the relationship between a group of organisms.

Closed systems are systems that exchanged energy, but not matter with the surrounding.

Cohesion occurs when two identical substances are attracted to each other and form Van der Waals interactions.

Common ancestor is a group that split at a node, and diverged into two or more sister groups. Any shared characteristics between sister groups are absent in the outgroup because they emerge only with the common ancestor of the sister groups.

Competitive inhibitors compete with the substrate for the active site of the enzyme.

Constitutive genes are those that typically have no regulators of their own and are therefore always expressed. However, because they lack positive regulators (activators), they can only be expressed at low levels.

Constitutive is a terms used in when describe a gene that is always expressed.

Continental drift is the movement of the continents relative to each other.

Continental plates are the major tectonic plates that form the continental landmasses.

Convergent evolution is an evolutionary process in which two or more distantly related species begin occupying similar niches and evolve similar adaptations by modification of unrelated ancestral features.

Crossover is an event that occurs prior to the formation of gamete cells in which segments of homologous chromosomes are exchanged.

Cyanobacteria are photosynthetic prokaryotes that are believed to have been the ancestral cell of chloroplasts. They are also the source of the oldest discovered fossils that date back to 3.5 billion years.

Cytoplasmic determinants are substances originating from the maternal gamete (unfertilized egg) that determines the fate of cells in the developing embryo.

Denitrifying bacteria release nitrogen from the soil back into the atmosphere.

Derived character is a feature that is shared by related species and that originally appeared in their common ancestor.

Diffusion is the movement of particles from [high] to [low] particle concentration.

Diploid is a term that refers to cells that have 2 copes of each autosome. Somatic cells are diploid.

Diploidy is where each individual of the species inherits 2 copies of each gene.

Directional selection favors one extreme on the phenotype spectrum.

Disruptive selection favors both extremes on the phenotype spectrum.

Divergent evolution is an evolutionary process that begins with the splitting of a group into two or more subgroups that then continue evolving independently of each other. Over time the groups evolve with unique adaptations that cause their previously shared features to take on new forms with different functions.

DNA methylation is a method of silencing genes so that expression is blocked.

Domains are the first three branches of the phylogenetic tree, bacteria, Achaea, and eukaryote. All organisms are grouped into one of the domains.

Dominant allele is the allele that is always expressed in both the homozygous and the heterozygous individuals.

Ecological performance refers to how well the individual benefits from its phenotype by improved fitness within its local environment.

Ectotherms are organisms such as reptiles, amphibians, and fish that maintain optimal temperature by adjusting their behavior to take advantage of the external surrounding temperature.

Electron transport chain includes a series of compounds and proteins that transfer electrons from the electron donor (chlorophyll a in photosynthesis, or NADH and FADH$_2$ in respiration) to the electron acceptor (NADPH in photosynthesis, or oxygen in respiration).

Electronegativity is a measure of the attraction that an atom has for electrons.

Embryo is a diploid eukaryotic organism that is still going through the stages of development. In placental mammals, the embryo stage ends at birth.

Embryonic induction is the process involving cell inducer molecules that guides groups of cells along specific developmental pathways that determine the types of cells and tissue that their descendant cells will become.

Emigration is the movement of individuals out of the population.

Endergonic reactions are reactions that do not occur spontaneously because they require energy input from the surrounding. Many endergonic reactions take in energy by absorbing heat from the surround (endothermic).

Endocytosis is the exergonic import of large substances.

Endomembrane system includes many of the membrane bond organelles that are suspended in the cytoplasm. The major components are the nuclear membrane, the ER, Golgi, and transport vesicles.

Endonucleases are enzymes that cut within DNA (exonuclease cut at the ends of DNA).

Endosymbiotic theory suggests that the chloroplast and mitochondria are evolved from aerobic and photosynthetic prokaryotes, respectively, that were engulfed with a primitive eukaryote.

Endotherms are "warm-blooded" organisms such as mammals and birds that use internal metabolic processes for maintaining optimal temperatures.

Ethylene is the plant hormone that stimulates the ripening process.

Evolution is a change in the characteristics of a population or species over time.

Exergonic reactions are reactions that occur spontaneously by releasing energy to the surrounding, usually as heat.

Exocytosis is the exergonic export of large substances.

Exonucleases are enzymes that cut at the ends of DNA (endonucleases cut within DNA).

Extant species are species that are still surviving and can be observed today.

Extinct species are species that are no longer alive.

Facilitated diffusion is the diffusion of particles through a membrane protein channel.

Fats are lipids that contain 3 fatty acids bonded to 1 glycerol

Fecundity is the reproductive rate of an organism. Superfecundity is the tendency to produce more offspring than the environment can support. When there is superfecundity in a population, fewer offspring survive.

Fermentation is an anaerobic conversion of pyruvate into lactate or ethanol in order to replenish NAD+ for further glycolysis.

Fertility is the ability of a viable organism to reproduce.

Fitness is a measure of an individuals ability to survive, attract a mate, and reproduce.

Fluorine dating is a relative dating method based on the absorption of fluorine from the soil, with older fossils having higher fluorine content.

Fossil record includes all of the discovered and yet to be discovered fossils.

Fossils are the preserved remains of traces of dead organisms.

Founder effect is a kind of genetic drift caused by the migration of a small group out of the original population to found a new population. The new population will experience a sudden change in its allele frequency from the original population, and will reflect the allele frequency of the founder individuals.

Free energy is the energy that is neither stored in the bonds of molecules, nor being used.

Frequency-dependent selection is where the fitness of a phenotype declines as it becomes increasingly common in the population.

G proteins are peripheral membrane proteins that are activated by G protein-linked receptors and that act to induce the kinase phosphorylation cascade.

G protein-linked receptors are receptor proteins that activate G proteins.

Gamete cells are the haploid sex cells (egg and sperm), while somatic cells are all of the other diploid cells that make up the organism.

Gamete isolation is where morphologic difference between the gamete cells (sperm and egg) prevents fertilization from occurring.

Gene is a part of the DNA that codes for a specific protein. Genes are the basic unit of inheritance that makes up the coding regions of the DNA.

Gene pool refers to the net total of all genetic information or genes of a particular population.

Gene regulation is the process that cells use to trun on and off genes,

Gene shuffling is a term that is analogous to genetic recombination that occurs as a result of crossover between homologous chromosomes.

Gene silencing means the gene is present in the DNA in a functional state, but expression is blocked by some biochemical mechanism such as DNA methylation.

Genetic homology is the degree of similarity in the DNA or protein sequence between two species. Genetic homology increases with the relatedness of compared species.

Geographic distribution refers to the different regions and localities that organisms occupy.

Geology is the study of the physical nature of the Earth.

Gibbs free energy (G) is the amount of energy available for use by biological systems.

Glucagon is a pancreatic hormone that acts to raise blood glucose liver by stimulating cells in the liver and adipose tissue convert glycogen and fats to glucose. Glucagon's effect on blood glucose levels is opposite of insulin.

Glycogen is a carbohydrate polymer (or polysaccharide) that animals use for storing excess consumed glucose. Glycogen is produced from excess glucose and stored in the liver.

Glycolipids are lipids that have a carbohydrate attached to it. They are typically displayed on the ECM side of the cell membrane where they function in cell recognition.

Glycolysis is an anaerobic breakdown of glucose into 2 pyruvates, resulting in the net production of 2 ATP molecules.

Glycoproteins are proteins with carbohydrates attached. They have a host of different function roles in cells, including cell recognition.

Gradualism is the idea that a descendent species evolves through slow and steady divergence from an ancestral species.

Granum is a stack of thylakoid discs.

Habitat is an ecological area where an organisms lives.

Habitat isolation is where there is a physical geographic barrier that separates subgroups.

Half-life is the amount of time it takes for half of the radioisotope to decay. For example, since C-14 has a half-life of 5730 years, a sample of 100 g of C-14 will be reduced to 50 g in 5730 years, and 25 g in twice that time (11,460 years).

Haploid is a term that refers to cells that have only one copy of each chromosome. Gametes become haploid after meiosis when the homologous pairs are separated.

Heat (q) is a measure of the amount of thermal energy being transferred. It is kinetic energy that is not being used for work.

Hemizygous is a terms that refers to having only one copy of a gene. Males are hemizygous for all sex-linked genes because they have only one copy of each sex chromosome.

Hemophilia is a medical condition where patients fail to produce key blood clotting factors and are unable to form a stable blood clot.

Heterotrophs (or consumers) are organism that get their energy by feeding on others.

Heterozygote advantage occurs when heterozygous individuals have a higher fitness than homozygous individuals.

Heterozygous is when both inherited alleles of the gene are different.

Homeostasis is a self-regulating process used by biological systems to maintain a constant and stable internal environment.

Homeotic genes are also called hox genes. They regulate the orderly process of embryonic development. Hox genes are highly conserved throughout eukaryotic domains and may have been critical for the emergence of new species.

Homologous structures are comparative structures that have superficially dissimilar appearance and serve different functions, but that share a common evolutionary origin.

Homozygous is the state when both inherited alleles of the gene are identical.

Hox genes are the overseers of development, and are commonly referred to as the body plans genes because they instruct when during embryogenesis, and where in the body to assemble a specific body part.

Humus is decomposed organic matter.

Hybrid breakdown is where the hybrid is viable and fertile, but the offspring from the second or later generation has a developmental defect that causes the lineage to breakdown.

Hybridization occurs when two contrasting true-breeder are crossed. Their offspring are hybrids of the true-breeding parents. The hybrid offspring of the P generation form the F1 generation. The offspring of the F1 generation form the F2 generation.

Hydrogen bonds, or H-bonds, are the strongest type of Van der Waals interaction, and are formed when hydrogen of one molecule interacts with nitrogen, oxygen, or fluorine of another.

Hydrophobic substances are uncharged particles and non-polar substances that do not dissolve well in water.

Hyperthyroidism is a medical condition present in patients suffering from Graves' disease where the thyroid gland becomes over stimulated and secretes large amount of thyroid hormones into the blood. It is believed to be an autoimmune condition causes by the production of antibodies that bine the TRH receptor on cells of the thyroid gland, instructing these cells to produce and releases their T3 and T4 hormones.

Hypothalamus is the part of the brain links the endocrine system and its hormones to the nervous system.

Hypothalamus is the part of the brain responsible for maintaining internal homeostasis. It links the endocrine and nervous systems.

Immigration is the movement of individuals into the population.

Index fossil is a fossil used in relative dating that is of a known species that was short lived, but that had a wide geographic distribution.

Inducers are chemical signaling molecules that are produced and secreted by local embryonic cells in order for them to communicate and coordinate the growth and development into a structured tissue and organ.

Innate behavior is any behavior that is directly genetic and not derived from experience.

Insulation is the act of preventing the loss of heat or energy.

Insulin is a pancreatic hormone that acts to lower blood glucose liver by stimulating cells in the liver and adipose tissue to synthesis fats and glycogen from blood glucose. Insulin's effect on blood glucose levels is opposite of glucagon.

Intersexual selection occurs when one sex influences the reproductive success of individuals of the other sex.

Intrasexual selection occurs when individuals of the same sex influences the reproductive success of each other.

Isolated systems are those that exchanged neither energy nor matter with the surrounding.

Isolated systems exchange neither energy nor matter with their surrounding.

K-strategists are typically large organisms with low reproductive rates and high survivorship.

Kinases are common second messenger proteins that activate other proteins by phosphorylation. Several kinases can participate in a kinase phosphorylation cascade whereby the activation of one kinase by G proteins leads to a domino effect in the activation of the other kinases.

Legumes is a large plant family of bean-like plants, including peas, beans, lentils, and soybeans.

Ligand is the signaling molecule that can bind the membrane bound ligand-receptor.

Linked genes are those that are located on the same chromosome.

Lumen is the fluid of the thylakoid.

Lytic enzymes are a class of proteins that are typically found in lysosomes, and that breaks down macromolecules such as proteins, nucleic acids, and carbohydrates. Each specific enzyme is specialized for digesting specific types of molecules.

Macrophages are a type of white blood cell that can phagocytize waste and foreign particles, bacteria, and cancer cells.

Marsupial mammals are a subclass of mammals with a poorly developed placenta, and that gives birth to a premature embryo that remains attached to a nipple in a pouch located on the mother's lower belly where it continues developing.

Mass extinction refers to the sudden disappearance of 50% or more of the existing species. Mass extinction always coincides with rapid and severe environmental change.

Mass-specific metabolic rate is the whole-animal metabolic rate divided by body mass.

Mechanical isolation is where a morphological difference between the subgroups makes copulation impossible. Mechanical isolation is often due to large differences in body size, or sex organs.

Micelles are single phospholipid layers that form a spherical structure.

MicroRNA (miRNA) are regulatory RNA sequences that remove the poly A tails from mRNA after it has been translated in order to reduce the mRNA stability and tag it for endonuclease destruction.

Mitochondria are organelles present in all eukaryotic cells that produces ATP during aerobic respiration when glucose $(C_6H_{12}O_6)$ and oxygen (O_2) are taken in and carbon dioxide (CO_2) and water (H_2O) are released.

Morphology refers to the form and structure that confers a specific function in organisms.

Mutations are changes made to the sequence in the heritable molecule (DNA or RNA) that can occur as a result of replication error, or mutagens in the environment.

Mutualistic symbiosis is a relationship between two organisms where they both benefit.

Negative feedback control is a regulatory mechanism that reverses the effects of a pathway that causes a deviation from the homeostatic set point. It results in a return to the set point value.

Negative mutations reduce the organism's competitive advantage.

Negative regulation can occur by both competitive inhibition and allosteric inhibition.

Neurosecretory cell is the hormone-secreting cell of the hypothalamus. They release their hormones in either the posterior or anterior pituitary.

Neurotransmitters are ligands that are released by the presynaptic cells and that bind to their ligand receptors on the ligand-gated Na/K

Neutral variation is a genetic variation that appears to confer no selective advantage or disadvantage.

Niche is the specific combination of factors that describes how a particular organism is specially adapted to its particular habitat. A niche is defined by the organism's innate behavior, its physical adaptations, and other biotic and abiotic factors that the organism needs to be successful in its habitat. Each organism has a unique niche so that no two organisms can occupy the same niche.

Nitrogen-fixing bacteria are bacteria cells that can directly sequester nitrogen from the atmosphere and deposit it into the roots of legumes or the soil.

Null mutations are negative mutations that result in a complete loss of a functional protein product.

Open systems are systems that exchanged both energy and matter with the surrounding.

Organic compounds are carbon-containing molecules that form the basic building blocks of living organisms. The 4 major organic compounds are nucleic acids, amino acids, lipids, and sugars.

Organic soup model, or primordial soup model, and refers to a hypothetical concentrated mixture of precursor molecules that reacted to form the first organic compounds.

Osmoregulation is the process of regulating an organism's osmotic pressure to maintain homeostatic water content. In animals, the kidneys play a major role in removing excess water by diluting the urine, and reabsorbing water by concentrating it.

Osmosis is the diffusion of water across a semipermeable membrane from [low] to [high] particle concentration.

Outgoup is the taxon that shares the fewest number of characters with the other groups of a cladogram.

Oxidative phosphorylation is the metabolic process that used energy released from respiration to produce ATP.

Pancreas is an endocrine gland that produces and secretes the hormones called insulin and glucagon, which regulate blood glucose levels.

Pangaea was a supercontinent that formed roughly 300 million years ago, but has since broken up as a result of continental drift.

Parapatric speciation occurs when three or more populations occupy continuous geographic area with no specific geographic barriers between them, and adjacent groups interbreed but not distant groups.

Passive transport the movement of particles in the direction of their concentration gradient, or water (osmosis) in the direction of the water potential gradient.

Perennials are plants that continue grow and reproduce for many years.

Peripatric speciation occurs when a small subgroup becomes geographically isolated from the larger group and experiences an intense genetic drift relative to the larger group.

Peripheral membrane proteins are hydrophilic proteins that bind only to the surface of the cell membrane.

Phagocytosis (cell eating) is the endocytosis of large particulate matter such as whole bacteria cells and food. It is the process of membrane invagination that cells use to import large particular matter or whole cells.

Phenotype is the observable characteristics (physiology, innate behavior, and morphology) of an individual that is determined by its inherited genes. **Phenotype frequency** is a measure of how common one phenotype of a given trait is to all the other phenotypes of that trait. It is a value between 0 and 1.

Pheromones are chemicals that are produced and secreted into the environment by an organism; and they illicit a specific response from other individuals of the same species.

Phospholipid bilayer is a double layer sheet of phospholipids, where the adjacent layers are attracted to each other and held together by their hydrophobic tails.

Photoperiodism is the seasonal flowering of plants that depends on the number of hours of plant exposure to light.

Phototropism is the growth of plants toward a light source.

Phylogenetic tree is a diagram that illustrates the evolutionary lineages of biological species and the relationships of these lineages and species to each other.

Physiology deals with the way that the structures of living systems function to support life.

Pinocyctosis (cell drinking) is the endocytosis of fluid that contains restricted material such as large molecules.

Placental mammals are a subclass of mammals with a well-developed placenta for a longer gestation period, which allows for a more fully developed infant.

Polar molecules have distinct partially positive and partially negative poles. They are different from ions, which have either a full positive or full negative pole.

Polyadenylation is the addition of poly A tail during mRNA processing which protects the mRNA from cytoplasmic degradation by endonuclease enzymes.

Polymers are molecules that form by linking together smaller subunits.

Population is a group of individuals of the same species that lives in the same geographic area and at the same time.

Positive feedback control is a regulatory mechanism that amplifies the effects of a pathway that causes a deviation from the homeostatic set point. It results in a further deviation from the set point value. In positive feedback, the product of a biochemical pathway typically acts to further stimulate the pathway.

Positive mutations produce proteins that improve the organism's competitive advantage in the environment.

Post-zygotic barriers prevent individuals from different species from producing a viable and fertile hybrid offspring. These barriers do not necessarily prevent copulation, but due genetic incompatibility between the gametes the embryo either fail to fully develop, or the hybrid lineage is unsustainable.

Posterior pituitary hormones are the neurohomones (oxytocin and ADH) that are produced by the hypothalamus and released in the posterior pituitary.

Pre-zygotic barriers prevent individuals from different species from copulating. These barriers can potentially be overcome by artificial insemination or in vitro fertilization.

Primary consumers are organisms that feed exclusively on producers.

Protobiont (or protocell) is the hypothetical RNA based life-form that preceded DNA based life.

Punctuated equilibrium is the process of evolution that involves extended periods of little or no change in the population followed by short period of rapid change that is caused by changes in the environment.

R-strategists are typically small organisms with high reproductive rates and low survivorship.

Radiocarbon dating is an absolute dating method that used the radioactive decay of carbon-14 isotopes into nitrogen-14 with a half-life of 5730 years.

Recessive allele is only expressed in the homozygous recessive state where 2 copies are present.

Recessive alleles are the alleles that are only expressed in the homozygous state.

Reduced hybrid fertility is where the hybrid is viable but infertile and unable to continue the hybrid lineage.

Reduced hybrid viability is when the hybrid zygote fails to develop fully, and dies in the embryonic stage.

Relative dating is a method of dating fossil relative to the age of an index fossils and the fossils position in rock layer, or fluorine absorption. This method is not used to determine the exact age of the fossil.

Repressor proteins block transcription by binding to the operator and blocks RNA polymerase from reading the genes.

Response elements short DNA consensus sequences located within promoter regions that work like enhancers. They are shared by genes that work cooperatively in specific biochemical pathways. They bind transcription factors to positively regulate genes.

Resting potential is the membrane potential when the cell is not transducing a nerve impulse and relevant ion channels are closed. It is -70 mV with the cell be negative relative to the extracellular fluid.

Ribozymes are RNA molecules that can catalyze some biochemical reactions. They can also self-replicate.

Saturated lipids contain only single bonds in the hydrocarbon chains and form linear molecules that do not bend.

Secondary consumers are organisms that receive energy from the first and second trophic level. They include carnivores and omnivores.

Set point refers to the homeostatic value for internal conditions in biological systems. It may be a measure of species-specific body temperature, pH, or electrolyte concentration, or some other variable at the ecosystem level, or other variables that defines a normal healthy biological system.

Sex-linked genes are genes that are located on the sex chromosomes.

Sexual dimorphism is a term that is used to describe any observable differences between the sexes of a given species that is not directly related to primary sexual characteristic. Sexual dimorphic features do not typically occur until the onset of puberty.

Signal transduction is the process that cells use to communicate a specific signal along a pathway.

Sister groups are all of the taxa of a cladogram, with the exception of the outgroup.

Small interfering RNA (siRNA) are small regulatory RNA sequences that work in a similar manner to miRNA. However, siRNA tags foreign RNA (rather than native RNA) for destruction by exonucleases.

Solute is any substance in a mixture other than the solvent.

Solvent is the substance the dissolves something else. When mixing two substances, it is the one that is most abundant. The solvent dissolves the less abundant solute.

Species is a biological level above a population. It is the largest grouping of individuals of a specific type of organism. A species may comprise of multiple populations, whose members can interbreed naturally to produce a viable offspring.

Stabilizing selection favors the intermediate phenotypes on the phenotype spectrum.

Starch is a carbohydrate polymer (or polysaccharide) that plants use for storing excess glucose. Starch is formed when glucose subunits are linked together.

Strata is a layer of sedimentary rock that may have fossils of organisms trapped from the time when the strata was deposited.

Stromal lamellae are the bridge between granum that is used for exchange of nutrient and material.

Structural genes code for RNA or protein products other than a regulatory protein. They typically code for enzymes or structural proteins that confer morphological or functional characteristics of the organism.

Structural stability refers to the molecules resistance to degradation, mutation, or structural alteration of function 3-D shape (proteins only). Adaptive radiation is the segmentation (splitting) of a population into multiple subgroups, followed by the evolutionary divergence of the subgroups to form new species.

Substrate is the substance that binds to the active site of an enzyme. It is the reactant of the enzyme catalyzed reaction, and is used to form the product.

Superfecundity is the tendency to produce more offspring than the environment can support. When there is superfecundity in a population, fewer offspring survive than are born.

Surrounding is the locally defined part of the universe outside of the system.

Sympatric speciation occurs when a subgroup from the larger population adapts to an alternative niche and diverges without becoming geographically isolated from the original group.

Symporters are transport proteins that carry two different particles simultaneously in the same direction across the membrane.

Synaptic vesicles contain neurotransmitters and are concentrated at the axon terminal. In the presence of calcium, they fuse with the presynaptic membrane of the axon.

System is a defined part of the universe that is being studied.

Taxon (plural = taxa) is a group of organisms that have a set of shared derived characters - they share a common ancestor. A taxon can exist at any of the 7 levels of the taxonomic hierarchy below kingdom.

Taxonomy is the scientific discipline that focuses on grouping organisms based on there shared characteristics.

Tectonic uplift is the lifting of tectonic plates that causes a geographic area to rise above sea level to form mountains.

Temporal isolation is where the two subgroups have different mating times or seasons.

Tetrapods are vertebrate that have four appendages (legs, arms, snakes, birds).

Thermodynamics is the study of how heat ad energy flows within and between systems.

Thylakoid membrane is the membrane of an internal compartment of chloroplasts called the thylakoid, where light absorption takes place.

Thylakoids are the light absorbing parts of the chloroplasts, and where the light reactions occur.

Thyroid gland is an endocrine gland that secretes the thyroid hormones T3 and T4 that regulate cell metabolisms and the fight-or-flight response.

Totipotency is the ability of embryonic cells in the early stages of segmentation to differentiate into any of the fully differentiated cells of the organism

Transitional fossils are fossilized remains of transitional species.

Transitional species are extinct species that are inferred from the fossil record, and that have features that are intermediate of an ancestral species and an extant descendant species. Transitional species are used to infer the evolutionary progression from a distant ancestor to its extant descendant.

Transmembrane proteins, or integral membrane proteins, are proteins that transcend the full length of the cell membrane. They tend to have both hydrophobic regions that interact with the middle hydrophobic tail regions of the phospholipid bilayer, and hydrophilic cytoplasmic and extracellular regions.

Trophic level defines the groups of organisms in an ecosystem that receive their energy from a similar source.

Tropic hormones are those that target other endocrine glands.

True breeders are individuals who are homozygous for a particular trait or set of traits. They produce offspring that look like themselves. They are used to form the P generation.

Type I survivorship pattern occurs for organisms with the highest survivorship that typically die as a result of old age.

Type II survivorship pattern occurs for organisms with the medium survivorship that typically die as result of predation. Their survivorship is age independent.

Type III survivorship pattern occurs for organisms with the lowest survivorship that typically dies prior to reaching reproductive age.

Uniformitarianism is the idea that the same natural geologic forces that has shaped the Earth and its environment in the past, are still in operation today.

Uniporters are membrane transport proteins that can carry only one type of particle across the membrane.

Unsaturated lipids contain one or more double or triple bonds in their hydrocarbon chains and form molecules that bend at the double and triple bonds.

Van der Waals interaction is an attraction between different molecules or substances that cause them to stick to each other. Compare to intramolecular interaction (covalent bonds) the intermolecular Van der Waals interactions form weak bonds.

Vasoconstriction is the narrowing of blood vessels due to the contraction of the smooth muscle cells that surround them. When blood vessels constrict, they transport a smaller volume of blood.

Vasodilation is the widening of blood vessels due to the relaxation of the smooth muscle cells that surround them. When blood vessels dilate, they can transport a larger volume of blood.

Vernalization is a process through which a plant gains the ability to flower in the spring. It requires a prolong exposure to cold weather.

Vestigial organs are a type of homologous structure that has lost is functional utility.

Viability is the ability of a zygote to develop into an adult living organism.

Whole-animal metabolic rate is the amount of energy that the organism consumers per hour.

Work (w) is a process that uses kinetic energy to move matter.

X chromosome inactivation is the random process of inactivating one of the two X chromosomes in the female somatic cells.

Zygote is the earliest developmental stage of an embryo. It begins when the gamete cells fuse to form the fertilized egg.

www.ingramcontent.com/pod-product-compliance
Lightning Source LLC
Chambersburg PA
CBHW080633180526
45168CB00008B/3157